Syzygies and Homotopy Theory

T0211998

Algebra and Applications

Volume 17

Algebra and Applications aims to publish well written and carefully refereed monographs with up-to-date information about progress in all fields of algebra, its classical impact on commutative and noncommutative algebraic and differential geometry, K-theory and algebraic topology, as well as applications in related domains, such as number theory, homotopy and (co)homology theory, physics and discrete mathematics.

Particular emphasis will be put on state-of-the-art topics such as rings of differential operators, Lie algebras and super-algebras, group rings and algebras, C^*-algebras, Kac-Moody theory, arithmetic algebraic geometry, Hopf algebras and quantum groups, as well as their applications. In addition, Algebra and Applications will also publish monographs dedicated to computational aspects of these topics as well as algebraic and geometric methods in computer science.

F.E.A. Johnson

Syzygies
and Homotopy
Theory

 Springer

Prof. F.E.A. Johnson
Department of Mathematics
University College London
London, UK
feaj@math.ucl.ac.uk

ISSN 1572-5553   e-ISSN 2192-2950
Algebra and Applications
ISBN 978-1-4471-5812-7   ISBN 978-1-4471-2294-4 (eBook)
DOI 10.1007/978-1-4471-2294-4
Springer London Dordrecht Heidelberg New York

British Library Cataloguing in Publication Data
A catalogue record for this book is available from the British Library

Mathematics Subject Classification (2000): 16E05, 20C07, 55P15

Printed on acid-free paper

Springer is part of Springer Science+Business Media (www.springer.com)

or

Du côté de Chez Swan

*To the memory of
my parents*

Preface

The underlying motivation for this book is the study of the algebraic homotopy theory of nonsimply connected spaces; in the first instance, the algebraic classification of certain finite dimensional geometric complexes with nontrivial fundamental group G; more specifically, directed towards two basic problems, the $\mathcal{D}(2)$ and $\mathcal{R}(2)$ problems explained below.

The author's earlier book [52] demonstrated the equivalence of these two problems and developed algebraic techniques which were effective enough to solve them for some *finite* fundamental groups ([52], Chap. 12). However the theory developed there breaks down at a number of crucial points when the fundamental group G becomes infinite. In order to consider these problems for general finitely presented fundamental groups the foundations must first be re-built ab initio; in large part the aim of the present monograph is to do precisely that.

The $\mathcal{R}(2)$–$\mathcal{D}(2)$ Problem   Having specified the fundamental group, the types of complex we aim to study are, from the point of view of homotopy theory, the simplest finite dimensional complexes which can then be envisaged; namely n-dimensional complexes X with $n \geq 2$ which satisfy

$$\pi_r(\widetilde{X}) = 0 \quad \text{for } r < n, \qquad\qquad (*)$$

where $\widetilde{X}$ is the universal cover of X. These restrictions alone are not sufficient to specify the next homotopy group $\pi_n(\widetilde{X})$; nor, however, is the choice of $\pi_n(\widetilde{X})$ entirely arbitrary. We shall explain in detail throughout the book how to parametrize the possible choices for $\pi_n(\widetilde{X})$ as a module over the group ring $\mathbf{Z}[G]$ and the extent to which an admissible choice determines the homotopy type of X.

Given a complex X as above we can construct the cellular chain complex

$$C_n \xrightarrow{\partial_n} C_{n-1} \xrightarrow{\partial_{n-1}} \cdots \xrightarrow{\partial_2} C_1 \xrightarrow{\partial_2} C_0,$$

where $C_r = H_r(\widetilde{X}^r, \widetilde{X}^{r-1}; \mathbf{Z})$ is a free $\mathbf{Z}[G]$-module with basis the r-cells of X. By the Hurewicz theorem, the conditions (∗) above force

$$H_r(C_*) = \begin{cases} \mathbf{Z} & r = 0, \\ 0 & 1 \leq r < n, \\ \pi_n(X) & r = n, \end{cases}$$

so that we may extend the above chain complex to an exact sequence

$$C_*(X) = (0 \to \pi_n(\widetilde{X}) \to C_n \xrightarrow{\partial_n} C_{n-1} \xrightarrow{\partial_{n-1}} \cdots \xrightarrow{\partial_2} C_1 \xrightarrow{\partial_1} C_0 \to \mathbf{Z} \to 0).$$

By an *algebraic n-complex* over $\mathbf{Z}[G]$ we mean an exact sequence of $\mathbf{Z}[G]$-modules

$$A_* = (0 \to J \to A_n \xrightarrow{\partial_n} A_{n-1} \xrightarrow{\partial_{n-1}} \cdots \xrightarrow{\partial_2} A_1 \xrightarrow{\partial_1} A_0 \to \mathbf{Z} \to 0)$$

in which each A_r is finitely generated and free over $\mathbf{Z}[G]$. An algebraic n-complex A_* is said to be *geometrically realizable* when there exists a geometric n-complex X of type (∗) such that $C_*(X) \simeq A_*$. One may then ask the obvious question:

$\mathcal{R}(n)$: Is every algebraic n-complex geometrically realizable?

For $n \geq 3$ the $\mathcal{R}(n)$ problem is answered in the affirmative in Chap. 9. In fact, this is a special case of an older and much more general result of Wall [98]. The question that remains is genuinely problematic:

$\mathcal{R}(2)$: Is every algebraic 2-complex geometrically realizable?

Whilst important in its own right, the $\mathcal{R}(2)$-problem is also of interest via its relation to a notorious and more obviously geometrical problem in low dimensional topology. First make a definition; say that a 3-dimensional cell complex X is *cohomologically 2-dimensional* when $H_3(\widetilde{X}; \mathbf{Z}) = H^3(X; \mathcal{B}) = 0$ for all coefficient systems $\mathcal{B}$ on X. The problem may then be stated as follows:

$\mathcal{D}(2)$: Let X be a finite connected cell complex of geometrical dimension 3 which is cohomologically 2-dimensional. Is X is homotopy equivalent to a finite complex of geometrical dimension 2?

Both $\mathcal{D}(2)$ and $\mathcal{R}(2)$ problems are parametrized by the fundamental group under discussion; each finitely presented group G has its own $\mathcal{D}(2)$ problem and its own $\mathcal{R}(2)$ problem. Moreover, for a given fundamental group G the $\mathcal{D}(2)$ problem is entirely equivalent to the $\mathcal{R}(2)$ problem; to solve one is to solve the other. This equivalence was shown by the present author in [51, 52], subject to a mild condition on G which was subsequently shown to be unnecessary by Mannan [71].

This book is in two parts, Theory and Practice. In this Preface we give a brief outline of the theory; a summary of the practical aspects is given in the Conclusion.

The Method of Syzygies   The basic model in the theory of modules is the theory of vector spaces over a field. However, the modules encountered in this book are defined over more general rings and in dealing with them it is useful to keep in mind how far one is being forced to deviate from the basic paradigm.

Linear algebra over a field is rendered tractable by the fact that every module over a field is free; that is, has a spanning set of linearly independent vectors. General module theory takes as its point of departure the observation that when a module M is not free we may at least make a first approximation to its being free by taking a surjective homomorphism $\varphi : F_0 \rightarrow M$ where F_0 is free to obtain an exact sequence

$$0 \rightarrow K_1 \rightarrow F_0 \overset{\varphi}{\rightarrow} M \rightarrow 0.$$

We find it instructive to regard the kernel K_1 as *a first derivative* of M. Setting aside temporarily the question of uniqueness one may repeat the construction and approximate K_1 in turn by a free module to obtain an exact sequence

$$0 \rightarrow K_2 \rightarrow F_1 \rightarrow K_1 \rightarrow 0.$$

Iterating we obtain a long exact sequence

$$\overset{\partial_{n+1}}{\longrightarrow} F_n \overset{\partial_n}{\longrightarrow} F_{n-1} \overset{\partial_{n-1}}{\longrightarrow} \cdots \overset{\partial_3}{\longrightarrow} F_2 \overset{\partial_2}{\longrightarrow} F_1 \overset{\partial_1}{\longrightarrow} F_0 \rightarrow M \rightarrow 0$$

with K_n, K_2, K_1 as connecting modules.

Thus arises the notion of *free resolution*, made famous by the work of Hilbert on Invariant Theory [43]. The intermediate modules K_n are called the *syzygies of M*. Indeed, the etymology ($\sigma\upsilon\zeta\upsilon\gamma o\zeta$ = yoke) is determined by the conventional view that the K_n are connections in this sense. Nevertheless, we prefer to regard them as objects in their own right, as *derivatives of M*. Before doing this, however, we must first answer the question we have avoided; to what extent are they unique?

At one level the most simple minded considerations show that they *cannot possibly* be unique; given an exact sequence

$$0 \rightarrow K_1 \rightarrow F_0 \overset{\varphi}{\rightarrow} M \rightarrow 0$$

then by stabilizing the middle term thus $0 \rightarrow K_1 \oplus \Lambda \rightarrow F_0 \oplus \Lambda \overset{\varphi}{\rightarrow} M \rightarrow 0$ it is clear that if K_1 is to be considered as a first derivative of M then $K_1 \oplus \Lambda$ must also be so considered. So much must have been apparent to Hilbert. Even so, it is clear that the pioneers of the subject considered that the syzygies *ought*, somehow, to be unique. In the original context of Invariant Theory [28] this can be made to work if the resolution is, in some sense, minimal. In our context, as we shall see, the notion of 'uniqueness via minimality' fails badly. However there is indeed a sense in which the syzygies are uniquely specified, and it is to this we now turn.

Stable Modules and Schanuel's Lemma   According to legend, in the autumn of 1958, during a lecture of Kaplansky at the University of Chicago, Stephen Schanuel, then still an undergraduate, observed that if we are given exact sequences of modules over a ring Λ

$$0 \to K \to \Lambda^n \xrightarrow{\varphi} M \to 0;$$

$$0 \to K' \to \Lambda^m \xrightarrow{\varphi} M \to 0$$

then $K \oplus \Lambda^m \cong K' \oplus \Lambda^n$. In fact, Schanuel proved slightly more than this; however it suggests that given Λ-modules K, K' we should write:

$$K \sim K' \iff K \oplus \Lambda^m \cong K' \oplus \Lambda^n \quad \text{for some positive integers } m, n.$$

When this happens we say that K, K' are *stably equivalent*. The relation '$\sim$' is an equivalence relation on Λ modules and, applied to the above exact sequences, Schanuel's Lemma shows that $K \sim K'$; it is in this sense that syzygies are unique.

Schanuel's Lemma explains neatly why the attempt to force uniqueness of the syzygy modules by minimising the resolution is, in general, doomed to failure. Thus suppose that m is the minimum number of generators of the Λ-module M and suppose given exact sequences

$$0 \to K \to \Lambda^m \xrightarrow{\varphi} M \to 0;$$

$$0 \to K' \to \Lambda^m \xrightarrow{\varphi} M \to 0.$$

Schanuel's Lemma then tells us that $K \oplus \Lambda^m \cong K' \oplus \Lambda^m$. We are left to solve the following:

Cancellation Problem Does $K \oplus \Lambda^m \cong K' \oplus \Lambda^m$ imply that $K \cong K'$?

In dealing with modules over integral group rings the expected answer is 'No'; as we shall see, cancellation is the exception not the rule. The failure of cancellation may be starkly portrayed by representing the stable module $[K]$ as a graph.

When M is a finitely generated Λ-module, the stable module $[M]$ has the structure of a directed graph in which the vertices are the isomorphism classes of modules $N \in [M]$ and where we draw an edge $N_1 \to N_2$ when $N_2 \cong N_1 \oplus \Lambda$. We will show, in Chap. 1, that $[M]$ is a 'tree with roots that do not extend infinitely downwards'. This graphical method of representing stable modules is due to Dyer and Sieradski [24].

The extent to which cancellation fails in $[M]$ is captured by the amount of branching. We illustrate the point with some examples; **A** below represents a tree with a single root and no branching above level two; **B** represents a tree with two roots but with no branching above level one; **C** represents a tree with a single root and no branching whatsoever. Cancellation holds in **C** but fails in both **A** and **B**.

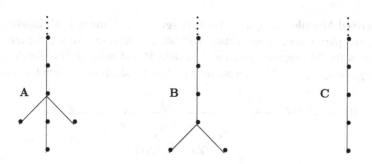

A significant difference between finite and infinite groups is the extent of our knowledge of the branching behaviour in stable modules over $\mathbf{Z}[G]$. When G is finite, the Swan-Jacobinski Theorem [46, 93] imposes severe restrictions on the type of branching that may occur; for example, the odd syzygies $\Omega_{2n+1}(\mathbf{Z})$ can behave only like $\mathbf{B}$ and $\mathbf{C}$ with possibly multiple roots but with no branching above level one; the even syzygies $\Omega_{2n}(\mathbf{Z})$ may resemble any of the three types but nothing worse. By contrast, when G is infinite very little is known in detail about the levels at which a stable module over $\mathbf{Z}[G]$ may branch.[1] We explore this question for some familiar infinite groups starting with the most basic case, namely the stable class of 0.

Iterated Fibre Squares and Stably Free Modules   In passing from finite groups to infinite groups the first point of difference is the increased incidence of non-cancellation. For finite Φ non-cancellation over $\mathbf{Z}[\Phi]$ is comparatively rare. By the theorem of Swan and Jacobinski, it can only occur when the real group ring

$$\mathbf{R}[\Phi] \cong \prod_{i=1}^{m} M_{d_i}(\mathcal{D}_i)$$

fails the *Eichler condition*; that is when for some i, $d_i = 1$ and $\mathcal{D}_i = \mathbf{H}$ is the division ring of Hamiltonian quaternions. However, the proof of the Swan-Jacobinski theorem does not survive the passage to infinite groups and so we are forced to fall back on other methods.

The approach which has proved profitable is the method of iterated fibre squares which was used by Swan in [94] to consider the *extent* to which non-cancellation fails in finite groups which fail the Eichler condition. We elaborate the necessary theory of fibre squares in Chap. 3. As a working method it proceeds like this; take a convenient finite group Φ and establish the cancellation properties of $\mathbf{Z}[\Phi]$ from first principles by using the method of fibre squares. Now generalize the statement, replacing $\mathbf{Z}[\Phi]$ by $R[\Phi]$; on taking $R = \mathbf{Z}[G]$ where G is infinite one hopes to analyze the cancellation properties of $R[\Phi] \cong \mathbf{Z}[G \times \Phi]$. Some successful attempts are exhibited in Chaps. 10 through 12.

---

[1]Although over more general rings, for example the coordinate rings of spheres, the pattern of branching away from the main stem may be very complicated.

The Derived Module Category We have set ourselves the task of classifying algebraic complexes and, in particular, algebraic 2-complexes. To see the relevance of syzygies for this, suppose given a Λ-module M and write $\Omega_n(M)$ for the stable class any nth-syzygy of M; then we may portray an algebraic 2-complex formally as

$$0 \longrightarrow \Omega_3(\mathbf{Z}) \longrightarrow F_2 \xrightarrow{\partial_2} F_1 \xrightarrow{\partial_1} F_0 \longrightarrow \mathbf{Z} \longrightarrow 0$$
$$\searrow \quad \nearrow \quad \searrow \quad \nearrow$$
$$\Omega_2(\mathbf{Z}) \qquad \Omega_1(\mathbf{Z})$$

showing, in particular, that when X is a connected geometric 2-complex with $\pi_1(X) = G$ the $\mathbf{Z}[G]$-module $\pi_2(\widetilde{X})$ is constrained to lie in the third syzygy $\Omega_3(\mathbf{Z})$.

The Ω_n formalism was first introduced by Heller in the context of modular representations of finite groups [39]. In that restricted setting it is relatively easy, with suitable interpretations, to regard the correspondence $M \mapsto \Omega_n(M)$ as a functor. In more general contexts attempting to make Ω_n functorial involves additional technical complications.

The first question to be answered is '*In what category is $\Omega_n(M)$ supposed to live?*' As a first approximation we take the quotient of the category $\mathcal{M}\mathrm{od}_\Lambda$ of Λ-modules obtained by ignoring morphisms which factorize through a free module; more precisely, we equate morphisms whose difference factorizes through a free module; that is if $f, g : M \to N$ are Λ-homomorphisms we write '$f \approx g$' when $f - g$ can be written as a composite $f - g = \xi \circ \eta$ as below where F is a free module:

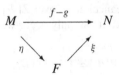

The quotient category $\mathcal{D}\mathrm{er}(\Lambda) = \mathcal{M}\mathrm{od}_\Lambda / \approx$ is called the *derived module category*. It is too crude an approximation, if only on the basis of size for, as we have imposed no size restrictions, our modules can be arbitrarily large. We can attempt to restrict all definitions to apply only to finitely generated modules; thus if N is a module we say that its stable class $[N]$ is finitely generated when N is finitely generated; in that case, *any* module in $[N]$ is also finitely generated. In the original context of modular representation theory, such size restriction causes no difficulty. In our more general context however, the difficulty arises that if M is finitely generated then $\Omega_n(M)$ need not be. To restrict attention to rings where this behaviour does not occur would exclude the integral group rings $\mathbf{Z}[G]$ of many interesting groups [53] (See Appendix D).

However, under a mild restriction on the ring,[2] if M is countably generated so also is $\Omega_n(M)$; then restricting all definitions to apply only to countably generated modules yields a derived module category $\mathcal{D}\mathrm{er}_\infty(\Lambda)$ of realistic size.

---

[2]Weak coherence. See Chap. 1.

There is, however, a complication more subtle than mere size. Recall that any projective module is a direct summand of a free module. Thus the above condition '$f \approx g$' is equivalent to the requirement that $f - g$ factors through a projective. This has the eventual consequence for modules K, K' over Λ that

$$K \cong_{\mathcal{D}er} K' \iff K \oplus P \cong_{\Lambda} K' \oplus P'$$

for some projective modules P, P'; that is, isomorphism classes in $\mathcal{D}er$ correspond not to stability classes of modules but, in MacLane's terminology, to *projective equivalence classes*[3] ([68], p. 101). Moreover, this applies even when all modules under consideration are finitely generated. In the original context of modular representation theory all projective modules are free, there is no distinction between stability and projective equivalence and Ω_n defines a functor on the derived module category. However, in general, to obtain functoriality one must consider not Ω_n but rather its analogue using the appropriate notion of *generalized syzygy*; disregarding finiteness restrictions and taking the successive kernels in a projective resolution $\mathcal{P}$

$$\xrightarrow{\partial_{n+1}} P_n \xrightarrow{\partial_n} P_{n-1} \xrightarrow{\partial_{n-1}} \cdots \xrightarrow{\partial_3} P_2 \xrightarrow{\partial_2} P_1 \xrightarrow{\partial_1} P_0 \to M \to 0$$

with D_n, D_2, D_1 branching below.

the correspondence $M \mapsto D_n$ gives a functor $D_n : \mathcal{D}er_\infty \to \mathcal{D}er_\infty$. As classes of modules $\Omega_n(M) \subset D_n(M)$ and we may regard $\Omega_n(M)$ as a sort of *polarization state* of $D_n(M)$. We note that for most computational purposes we may legitimately revert to $\Omega_n(M)$ as $\mathrm{Hom}_{\mathcal{D}er}(\Omega_n(M), N) \equiv \mathrm{Hom}_{\mathcal{D}er}(D_n(M), N)$.

Eliminating Injectives   In the late 1940s the introduction of Eilenberg-Maclane cohomology as the *derived functors* of Hom completely transformed module theory. The indeterminate nature of syzygies was replaced by the definiteness of computable invariants. In the aftermath the syzygetic method, insofar as it was still pursued, was regarded as an unwelcome reminder of a more primitive past. For us now, however, its rehabilitation via the derived module category raises the question of relating syzygies directly to cohomology.

Here we encounter a difficulty which is inherent in the cohomological method itself. In the standard treatments it is shown that one may compute the derived functor of $\mathrm{Hom}(-, -)$ *either* by taking a projective resolution in the first variable *or*, *equally*, by taking an injective co-resolution in the second. Moreover, this symmetry is not a point of esoteric scholarship, or at least, not merely so. With each variable one has a long exact sequence obtained by systematic appeal to the properties of the appropriate type of module. Which leads us back to the two sorts of modules themselves.

---

[3]For countably generated modules it is technically more convenient to replace the relation of projective equivalence by the equivalent notion of *hyperstable equivalence*, which is to say that $K \oplus \Lambda^\infty \cong_{\Lambda} K' \oplus \Lambda^\infty$. But again, see Chap. 1.

Projective modules, as direct summands of free modules, were in common use[4] before the name was ever applied to them; however the history and nature of injective modules is entirely different. Whereas projective modules are unavoidable, injective modules are a deliberate contrivance, only introduced to have arrow-theoretic properties dual to those of projectives [6]. Whereas projective modules are natural, injective modules are formal. Whereas projective modules are constructible (and we shall show how to construct some of them) injective modules are essentially non-constructible. One needs a theorem to show they exist. Except in the most elementary cases, where the point is irrelevant, they are not describable by any *effective* process. In our context this last point is the most pressing; injectives are so different from the objects with which we must deal that, arguments of formal simplicity notwithstanding, the need to dispense with them becomes insistent.[5]

The elimination of magic from homological algebra, in this case the avoidance of injective modules, forces us in every case to use projective resolutions. Whilst dispensing with the dualising services of injectives it is nevertheless essential to employ some form of homological duality which, however weak, can be confined entirely within the 'projective quotient' category. In fact, this requirement has a precedent as does the remedy; in the cohomology of lattices over finite groups the dual arrow theoretic properties of projectives are possessed by projectives themselves. Thus one may dispense with injectives entirely and describe the theory solely in terms of projectives. This is *Tate cohomology*, a point to which we will return. Our solution is comparable but not quite so convenient.

Corepresentability of Cohomology   The appropriate notion, which we shall use systematically, is that of 'coprojectivity'; a module M is said to be *coprojective* when $\operatorname{Ext}^1(M, \Lambda) = 0$. To see how coprojectivity works take an exact sequence $\mathcal{E} = (0 \to K \xrightarrow{i} F \xrightarrow{\varphi} M \to 0)$ where F is free so that K is a first syzygy of M; if $\alpha : K \to N$ is a Λ-homomorphism one may form the pushout diagram

$$
\begin{array}{c}
\mathcal{E} \\
\downarrow \quad c = \\
\alpha_*(\mathcal{E})
\end{array}
\left(
\begin{array}{ccc}
0 \to K \xrightarrow{i} & F \xrightarrow{\varphi} & M \to 0 \\
\downarrow \alpha & \downarrow \nu & \downarrow \mathrm{Id} \\
0 \to N \to \varinjlim(\alpha, i) \to & M \to 0
\end{array}
\right)
$$

from which we obtain the *connecting homomorphism* $\delta : \operatorname{Hom}_\Lambda(K, N) \to \operatorname{Ext}^1(M, N)$ by means of $\delta([\mathcal{E}]) = [\alpha_*(\mathcal{E})]$. When M is coprojective (and not otherwise) δ descends to give a natural equivalence $\delta : \operatorname{Hom}_{\mathcal{D}\mathrm{er}}(K, -) \to \operatorname{Ext}^1(M, -)$ so that we may write

$$
\operatorname{Ext}^1(M, -) \cong \operatorname{Hom}_{\mathcal{D}\mathrm{er}}(\Omega_1(M), -).
$$

---

[4]For example in Wedderburn theory.

[5]The disadvantages, *for any practical purpose*, of an object about which one has to think hard before even being able to admit its existence ought to be obvious. Doubtless some will regret this as yet another instance of a depressing but universal trend; in Weber's succinct phrase 'The elimination of Magic from the World' ([99], p. 105).

In other-words, when M is coprojective, $\Omega_1(M)$ is a *corepresenting object* for $\text{Ext}^1(M, -)$[6] considered as a functor on the derived module category. More generally, in higher dimensions there is a corresponding corepresentation theorem

$$H^n(M, -) \cong \text{Hom}_{\mathcal{D}\text{er}}(\Omega_n(M), -)$$

which holds provided that $H^n(M, \Lambda) = 0$. That is, we have replaced the *derived functor* H^n by the *derived object* Ω_n. Corepresenting cohomology in this way is the first step towards geometrizing extension theory so as to be able to apply it to the question of realizing algebraic complexes. Moreover, the groups $\text{Hom}_{\mathcal{D}\text{er}}(\Omega_n(M), N)$ are then natural generalizations of the Tate cohomology groups defined for modules over finite groups.

Homotopy Classification and the Swan Homomorphism   The problem of classifying algebraic complexes up to homotopy equivalence may be compared with the simpler Yoneda theory of module extensions up to congruence [68, 101]. For a specified fundamental group G let $\mathbf{Alg}_n(\mathbf{Z})$ denote the set of homotopy types of algebraic n-complexes of the form

$$A_* = (0 \to J \to A_n \to A_{n-1} \to \cdots \to A_0 \to \mathbf{Z} \to 0).$$

The stabilization $\Sigma_+(A_*)$ is obtained by adding $\Lambda = \mathbf{Z}[G]$ to the final two terms thus

$$\Sigma_+(A_*) = (0 \to J \oplus \Lambda \to A_n \oplus \Lambda \to A_{n-1} \to \cdots \to A_0 \to \mathbf{Z} \to 0)$$

and $\mathbf{Alg}_n(\mathbf{Z})$ also acquires a tree structure by drawing arrows $A_* \to \Sigma_+(A_*)$. Moreover the correspondence $A_* \mapsto J$ defines a mapping of trees, 'algebraic π_n',

$$\pi_n : \mathbf{Alg}_n(\mathbf{Z}) \to \Omega_{n+1}(\mathbf{Z}).$$

In his unpublished paper [12] Browning described the fibres $\pi_2 : \mathbf{Alg}_2(\mathbf{Z}) \to \Omega_3(\mathbf{Z})$ for those finite groups G which satisfy the Eichler condition. In [52], generalizing a criterion of Swan [91], we showed, still within the confines of finite groups, how to circumvent dependence on the Eichler condition and gave a rather different description of the fibres of π_2. Here we show how to extend the description of [52] to a much wider class of rings.[7]

A significant difficulty lies in being able to generalize the Swan mapping. In the original version [91] the homomorphism property of the Swan mapping is an easy consequence of special circumstances; in the wider context it is less obvious. Again

---

[6]Notice that the blank space would normally have to be co-resolved by means of injectives; the coprojectivity hypothesis removes this necessity.

[7]We note that a very special case of our classification theorem, for algebraic n-complexes over the group rings of n-dimensional Poincaré Duality groups ($n \geq 4$), was given by Dyer in [23].

take an exact sequence $\mathcal{E} = (0 \to J \xrightarrow{i} F \xrightarrow{\varphi} M \to 0)$ where F is free; if $\alpha : J \to J$ is a Λ-homomorphism one may again form the pushout diagram

$$
\begin{array}{ccc}
J & \xrightarrow{i} & F \\
\downarrow \alpha & & \downarrow \nu \\
J & \to & \varinjlim(\alpha, i)
\end{array}
$$

It turns out (*Swan's projectivity criterion*) that $\varinjlim(\alpha, i)$ is projective precisely when α is an isomorphism in $\mathcal{D}$er. When M and J are finitely generated one obtains a mapping

$$S : \mathrm{Aut}_{\mathcal{D}\mathrm{er}}(J) \to \widetilde{K}_0(\Lambda)$$

to the reduced projective class group of Λ. This is the generalized Swan mapping and is, nontrivially, a homomorphism. This result was first shown in [56]. Moreover, despite the apparent dependence upon J, when M is coprojective it depends only upon M and is independent of the sequence $\mathcal{E}$ used to produce it. More generally, if

$$0 \to J \to A_n \to A_{n-1} \to \cdots \to A_0 \to \mathbf{Z} \to 0$$

is an algebraic n-complex and $H^{n+1}(M, \Lambda) = 0$ the same mapping $S : \mathrm{Aut}_{\mathcal{D}\mathrm{er}}(J) \to \widetilde{K}_0(\Lambda)$ again reappears independently of the sequence used to produce it. By contrast, however, the natural mapping $\nu_J : \mathrm{Aut}_\Lambda(J) \to \mathrm{Aut}_{\mathcal{D}\mathrm{er}}(J)$ is heavily dependent on J. The detailed homotopy classification of algebraic n-complexes over M requires a knowledge of the cosets $\mathrm{Ker}(S)/\mathrm{Im}(\nu_J)$ as J runs through $\Omega_{n+1}(M)$.

Imposing the coprojectivity condition or its higher dimensional analogues does, of course, restrict the range of applicability of the theory. In practice it is not too serious; for example, the classification of algebraic 2-complexes over $\mathbf{Z}[G]$ requires us to impose the condition

$$H^3(\mathbf{Z}, \mathbf{Z}[G]) = 0.$$

This condition is satisfied in many familiar cases; in particular, when G is a virtual duality group of virtual dimension n it is satisfied whenever $n \neq 3$.

Parametrizing the First Syzygy   In applying the classification theorem to our original problem one needs specific information about the syzygies $\Omega_n(\mathbf{Z})$. In practice, this is a matter of severe computational difficulty. At the time of writing, the only finite fundamental groups for which there are complete descriptions for *all* $\Omega_n(\mathbf{Z})$ are certain groups of periodic cohomology. For infinite fundamental groups the situation is far worse.

In the first instance we are content to study $\Omega_1(\mathbf{Z})$. Here we find that the branching properties at the minimal level are intimately related to the existence of stably free modules; that is, to the stable class of the zero module. When G is infinite and $\mathrm{Ext}^1(\mathbf{Z}, \mathbf{Z}[G]) = 0$ we show that the stably free modules describe a lower bound for the branching behaviour in $\Omega_1(\mathbf{Z})$ and give a complete description of the minimal level $\Omega_1^{\min}(\mathbf{Z})$. This is done in Chap. 13.

Finally, in the most familiar case where $\mathrm{Ext}^1(\mathbf{Z}, \mathbf{Z}[G]) \neq 0$, namely when $G \cong F_n \times C_m$, we give a complete description of all the *odd* syzygies $\Omega_{2n+1}(\mathbf{Z})$. By way of illustration we conclude the book with Edwards' solution [25, 26] of the $\mathcal{R}(2)$ problem for the groups $C_\infty \times C_m$.

Acknowledgments   The author wishes to express his thanks to his colleagues Dr. R.M. Hill and Dr. M.L. Roberts; the former for his insights into cyclotomic fields; the latter for some helpful discussions on free ideal rings.

The author has had the advantage of being able to rehearse the theory presented here over a number of years to the captive audience of his students; in alphabetical order: Tim Edwards, Susanne Gollek, Jodie Humphreys, Pouya Kamali, Daniel Laydon, Wajid Mannan, Dominique Miranda, Jamil Nadim, Seamus O'Shea, Jonathan Remez, Isidoros Strouthos. The theory has gained thereby, not only in clarity from their perceptive comments but also, as will be seen in the text, in substance from some original and significant contributions.

London, England   F.E.A. Johnson

Contents

Part I
Theory

Chapter 1
Preliminaries

Many of the arguments in this book are formulated in terms of modules over the group ring $\mathbf{Z}[G]$ where G is a specified fundamental group. Thus, in part, this book is concerned with the general theory of modules and so, by association, with the general theory of rings. Given the pathology of which the subject is capable there is a tendency, frequently indulged in the literature, to present Ring Theory as a menagerie of wild beasts with strange and terrifying properties. Regardless of appearances that is not our aim here. The rings we consider are comparatively well behaved. However, in order to explain quite how well behaved we are forced to discuss a small amount of pathology if only to say what delinquencies we need not tolerate.

1.1 Restrictions on Rings and Modules

The rings we encounter are typically, though not exclusively, integral group rings. In principle we would prefer simply to say that the rings we meet will have properties which are no worse than the worst behaviour one can expect from $\mathbf{Z}[G]$ where G is a finitely presented group; but of course we must be more precise than that. The first restriction we impose is the *invariant basis number property* (= IBN); that is, for positive integers a, b:

$$\Lambda^a \cong \Lambda^b \implies a = b. \tag{IBN}$$

Although this condition is a definite restriction it is too weak for many purposes and there are two progressively stronger notions which are more useful; the first is the *surjective rank property* (= SR):

If $\varphi : \Lambda^N \to \Lambda^n$ is a surjective Λ-homomorphism then, $n \leq N$. (SR)

Finally we have the so-called *weak finiteness* property (= WF).

If $\varphi : \Lambda^a \to \Lambda^a$ is a surjective Λ-homomorphism then φ is bijective. (WF)

F.E.A. Johnson, *Syzygies and Homotopy Theory*, Algebra and Applications 17, DOI 10.1007/978-1-4471-2294-4_1, © Springer-Verlag London Limited 2012

It is straightforward to see that WF $\Longrightarrow$ SR $\Longrightarrow$ IBN. In [15] Cohn shows that if there exists a ring homomorphism $\Lambda \to \mathbf{F}$ to a field then Λ has the SR property. Thus if A is a commutative ring then *any group ring* $A[G]$ satisfies SR. Furthermore, in addition to possessing the SR property, for any group G the integral group ring $\mathbf{Z}[G]$ also satisfies WF. The main details of a proof of this last were outlined in a paper of Montgomery [75].

For reasons explained below, we also impose the following very mild restriction:

Weak Coherence If M is a countably generated Λ-module and $N \subset M$ is a Λ-submodule then N is also countably generated.

We denote by $\mathcal{M}od_\Lambda$ the category of right Λ-modules and by $\mathcal{M}od_\infty$ the full subcategory of countably generated modules; $\mathcal{M}od_\infty$ is then equivalent to a small category. The force of imposing the weak coherence condition is that $\mathcal{M}od_\infty$ becomes an abelian category in the formal sense of [74].

There is a stronger notion; let $\mathcal{M}od_{fp}(= \mathcal{M}od_{fp}(\Lambda))$ denote the category of *finitely presented* right Λ-modules; Λ is said to be *coherent* when $\mathcal{M}od_{fp}$ is an abelian category. Ideally one would like to impose this stronger condition. However, to do so would exclude too many significant examples.

Clearly every countable ring is weakly coherent. Hence, the integral group ring $\mathbf{Z}[G]$ of any countable group G is weakly coherent. By contrast, coherence is a far less common property. Admittedly, if G is finite then $\mathbf{Z}[G]$ is coherent; however, there are many finitely presented infinite groups G where $\mathbf{Z}[G]$ fails to be coherent, even some which satisfy otherwise strong geometrical finiteness conditions. For example, if G contains a direct product of two nonabelian free groups then $\mathbf{Z}[G]$ fails to be coherent. The topic is considered further in Appendix D.

Finally, we need to mention duality. We set out with the intention of always working with right modules. Over general rings, this is not possible if one wants also to deal with duality, for if M is a right Λ-module then the dual module $\text{Hom}_\Lambda(M, \Lambda)$ is naturally a left module via the action

$$\bullet : \Lambda \times \text{Hom}_\Lambda(M, \Lambda) \to \text{Hom}_\Lambda(M, \Lambda)$$
$$(\lambda \bullet f)(x) \qquad = \qquad \lambda f(x)$$

In general there is no way around this; there exist rings in which the category of left modules is not equivalent to the category of right modules. However, in the case of group rings $\Lambda = \mathbf{Z}[G]$ we can circumvent this difficulty by the familiar device of converting left modules back to right modules

$$* : \text{Hom}_\Lambda(M, \Lambda) \times \Lambda \to \text{Hom}_\Lambda(M, \Lambda)$$
$$f * \lambda \qquad = \qquad \overline{\lambda} \bullet f$$

via the canonical (anti)-involution $\overline{g} = g^{-1}$. More generally one may do this whenever the ring Λ has a distinguished (anti)-involution. With this convention the dual module $\text{Hom}_\Lambda(M, \Lambda)$ so equipped as a right module is denoted by M^*.

1.2 Stable Modules and Tree Structures

Let Λ be a ring with the surjective rank property SR of Sect. 1.1. We denote by '$\sim$' the stability relation on Λ modules; that is

$$M_1 \sim M_2 \quad \Longleftrightarrow \quad M_1 \oplus \Lambda^{n_1} \cong M_2 \oplus \Lambda^{n_2}$$

for some integers $n_1, n_2 \geq 0$; the relation '$\sim$' is an equivalence on isomorphism classes of Λ-modules. For any Λ-module M, we denote by $[M]$ the corresponding *stable module*; that is, the set of isomorphism classes of modules N such that $N \sim M$. One sees easily that:

M is finitely generated if and only if each $N \in [M]$ is finitely generated.   (1.1)

When M is a *nonzero* finitely generated Λ-module we define the Λ-rank of M by

$$\mathrm{rk}_\Lambda(M) = \min\{a \in \mathbf{Z}_+ \text{ for which there is a surjective } \Lambda\text{-homomorphism}$$

$$\varphi : \Lambda^a \to M\}.$$

Proposition 1.2 *If $N \in [M]$ then for each integer $a > 0$, $N \oplus \Lambda^a \ncong N$.*

Proof Put $\mu = \mathrm{rk}_\Lambda(N)$ and let $\varphi : \Lambda^\mu \to N$ be a surjective homomorphism. If $N \cong N \oplus \Lambda^a$ for some $a \geq 1$ then for all $k \geq 1$, $N \cong N \oplus \Lambda^{ka}$. Choose $k \geq 1$ such that $\mu < ka$. Let $h_k : N \to N \oplus \Lambda^{ka}$ be an isomorphism and let $\pi_k : N \oplus \Lambda^{ka} \to \Lambda^{ka}$ be the projection. Then $\pi_k \circ h_k \circ \mu : \Lambda^\mu \to \Lambda^{ka}$ is a surjective homomorphism and $\mu < ka$. This is a contradiction, hence $N \ncong N \oplus \Lambda^a$ when $a \geq 1$.   □

We define a function $g : [M] \times [M] \to \mathbf{Z}$, the 'gap function' as follows

$$g(N_1, N_2) = g \quad \Longleftrightarrow \quad N_1 \oplus \Lambda^{a+g} \cong N_2 \oplus \Lambda^a,$$

where both a and $a + g$ are positive integers. We must first show that:

Proposition 1.3 *g is a well defined function.*

Proof Suppose that $N_1 \oplus \Lambda^p \cong N_2 \oplus \Lambda^q$ and also that $N_1 \oplus \Lambda^r \cong N_2 \oplus \Lambda^s$. We will show

$$p - q = r - s.\qquad\qquad (*)$$

To see this, observe that $N_1 \oplus \Lambda^{p+r} \cong N_2 \oplus \Lambda^{q+r}$ and that $N_1 \oplus \Lambda^{p+r} \cong N_2 \oplus \Lambda^{p+s}$. Thus

$$N_2 \oplus \Lambda^{q+r} \cong N_2 \oplus \Lambda^{p+s}.$$

Suppose that $q + r \neq p + s$. Then without loss of generality we may suppose that $p + s < q + r$. Putting $N_3 = N_2 \oplus \Lambda^{p+s}$ and $\alpha = q + r - (p + s)$ we see that

$N_3 \oplus \Lambda^\alpha \cong N_3$ where $\alpha > 0$. This contradicts Proposition 1.2 above. Hence $q + r = p + s$ and so $p - q = r - s$ as claimed. $\qquad \square$

It is straightforward to check that

$$g(N, N \oplus \Lambda^b) = b, \tag{1.4}$$

$$g(N_2, N_1) = -g(N_1, N_2), \tag{1.5}$$

$$g(N_1, N_3) = g(N_1, N_2) + g(N_2, N_3). \tag{1.6}$$

Lemma 1.7 *Let Λ be a ring with the surjective rank property and let M be a finitely generated Λ-module; if $K \in [M]$ is such that $0 \leq g(K, M)$ then $g(K, M) \leq \mathrm{rk}_\Lambda(M)$.*

Proof Put $m = \mathrm{rk}_\Lambda(M)$ and let $\varphi : \Lambda^m \to M$ be a surjective Λ-homomorphism. Suppose that $K \in [M]$ is such that $0 \leq g(K, M) = k$ and let $h : M \oplus \Lambda^a \to K \oplus \Lambda^{a+k}$ be an isomorphism. If $\pi : K \oplus \Lambda^{a+k} \to \Lambda^{a+k}$ is the projection then $\pi \circ h \circ (\varphi \oplus \mathrm{Id}) : \Lambda^{m+a} \to \Lambda^{a+k}$ is also a surjective homomorphism. Hence by the surjective rank property for Λ, $a + k \leq a + m$ and so $k \leq m$ as claimed. $\qquad \square$

We say that a module $M_0 \in [M]$ is a *root module* for $[M]$ when $0 \leq g(M_0, K)$ for all $K \in [M]$. We show:

Theorem 1.8 *Let Λ be a ring with the surjective rank property and let M be a finitely generated Λ-module; then $[M]$ contains a root module.*

Proof If $K \in [M]$, either $g(K, M) < 0$ or, by above, $0 \leq g(K, M)$ and $g(K, M) \leq \mathrm{rk}_\Lambda(M)$. Either way

$$g(K, M) \leq \mathrm{rk}_\Lambda(M),$$

and the mapping $K \mapsto g(K, M)$ gives a function $[M] \to \mathbf{Z}$ which is bounded above by $\mathrm{rk}_\Lambda(M)$. Thus there exists $M_0 \in [M]$ which maximises this function; that is,

$$g(M_0, M) = \max\{g(K, M) : K \in [M]\}.$$

We claim that for all $K \in [M]$, $0 \leq g(M_0, K)$. Otherwise, if there exists $K \in [M]$ such that $g(M_0, K) < 0$ then $g(K, M_0) < 0$ and so

$$g(K, N) = g(K, M_0) + g(M_0, N) > g(M_0, N)$$

which contradicts the choice of M_0. Thus $0 \leq g(M_0, K)$ for all $K \in [M]$, and M_0 is a root module as claimed. $\qquad \square$

If M_0 is a root module for $[M]$ we may define a height function $h : [M] \to \mathbf{N}$ by

$$h(L) = g(M_0, L).$$

Whilst ostensibly the height function depends upon M_0, in fact it is intrinsic to the stable module $[M]$; to see this, suppose that M_0 and M_0' are both root modules for $[M]$ and consider the respective height functions $h(L) = g(M_0, L)$ and $h'(L) = g(M_0', L)$. From (1.6) above $g(M_0, L) = g(M_0, M_0') + g(M_0', L)$ so that

$$h(L) = g(M_0, M_0') + h'(L).$$

However $g(M_0, M_0') = h(M_0') \geq 0$ whilst $g(M_0, M_0') = -g(M_0', M_0) = -h'(M_0) \leq 0$. Thus $g(M_0, M_0') = 0$ and so

$$h(L) = h'(L).$$

When the ring Λ has the surjective rank property and M is a finitely generated Λ-module we may speak unequivocally of *the height function* $h : [M] \to \mathbf{N}$ on the stable module $[M]$.

When M is a finitely generated Λ-module, the stable module $[M]$ has the structure of a graph in which the vertices are the isomorphism classes of modules $N \in [M]$ and where we draw an edge $N_1 \to N_2$ when $N_2 \cong N_1 \oplus \Lambda$. Recall that a graph is said to be a *tree* when it contains no nontrivial loop. Since each module $N \in [M]$ has a unique arrow which exits the vertex represented by N, namely the arrow $N \to N \oplus \Lambda$, it follows that the only way of having a non trivial loop in $[M]$ would be if $N \cong N \oplus \Lambda^a$ for some $a > 0$. However, this possibility is precluded by Proposition 1.2, so that we have:

Proposition 1.9 *Let Λ be a ring having the surjective rank property; if M is a finitely generated module over Λ then $[M]$ is an infinite (directed) tree.*

Without attempting any more precise characterization of the (directed) tree structures which may arise in this way, it is evident that they are good deal more specialised than indicated by the statement of Proposition 1.9. For example, we have already observed that a unique arrow exits any vertex. Furthermore, the existence of root modules and the associated existence of a height function $h : [M] \to \mathbf{N}$ implies that $[M]$ may be represented as a 'tree with roots'. In particular if we regard the integers $\mathbf{Z}$ as a directed tree in the obvious way, namely:

$$\mathbf{Z} = (\cdots \to -(\mathbf{n}+1) \to -\mathbf{n} \to \cdots \to -1 \to 0 \to 1 \to \cdots \to \mathbf{n} \to (\mathbf{n}+1) \to \cdots)$$

then it is an easy deduction from the height function, as constructed on $[M]$, that $\mathbf{Z}$ does not imbed in $[M]$. We may paraphrase this by saying that the roots of $[M]$ *do not extend infinitely downwards*. To illustrate the point consider again the tree diagrams noted in the Introduction; **A** below represents a tree with a single root and no branching above level two; **B** represents a tree with two roots but with no branching above level one; **C** represents a tree with a single root and no branching whatsoever.

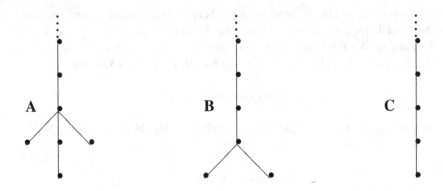

These examples all actually arise; denoting the quaternion group of order $4n$ by $Q(4n)$ then **A** represents the stable class of 0 over the integral group ring $\mathbf{Z}[Q(24)]$ whilst **B** represents the stable class $\Omega_3(\mathbf{Z})$ over $\mathbf{Z}[Q(32)]$. Any stable module in which cancellation holds is represented by **C**; for example (as we shall see in Chap. 15) the stable class $\Omega_3(\mathbf{Z})$ over the group ring $\mathbf{Z}[C_\infty \times C_m]$ for any integer $m \geq 2$.

1.3 Stably Free Modules and Gabel's Theorem

The most basic cancellation problem arises when one considers [0], the stable class of the zero module; evidently a module S belongs to [0] when, for some integers a, $b \geq 1$

$$S \oplus \Lambda^a \cong \Lambda^b.$$

Any such module S is finitely generated. More generally, one says that a module S is *stably free* when $S \oplus \Lambda^a$ is a free module of unspecified rank, finite or infinite. Clearly any free module is stably free; the issue is whether a stably free module is necessarily free. In fact, nothing new is gained by allowing infinitely generated stably free modules as shown by the following observation of Gabel [32, 65, 67].

Theorem 1.10 *Let S be a stably free Λ module; if S is not finitely generated then S is free.*

Proof Let F_X denote the free Λ module on the set X. The hypotheses may be expressed as follows:

(i) S is not finitely generated;

(ii) for some set X and some *finite set* Y there is a Λ-isomorphism $h : F_X \xrightarrow{\simeq} S \oplus F_Y$.

Note that X is necessarily infinite. Now let $\pi : S \oplus F_Y \to F_Y$ be the projection; putting

$$\widehat{h} = \pi \circ h : F_X \to F_Y$$

then $\widehat{h}$ is surjective. Moreover, h induces an isomorphism $h : \text{Ker}(\widehat{h}) \xrightarrow{\simeq} S$ so that it is enough to show that $\text{Ker}(\widehat{h})$ is free.

As F_Y is free we may choose a right inverse $s : F_Y \to F_X$ for $\widehat{h}$. For each $y \in Y$ there exists a finite subset $\sigma(y) \subset X$ such that $s(y)$ is a linear combination in the elements of $\sigma(y)$. Put $Z = \bigcup_{y \in Y} \sigma(y)$ and $\overline{Z} = X - Z$. Then Z is finite so that $\overline{Z}$ is infinite.

Now $\pi \circ h : F_Z \to F_Y$ is also surjective so that F_X is an internal sum (not necessarily direct) $F_X = \text{Ker}(\widehat{h}) + F_Z$. However $F_Z/(\text{Ker}(\widehat{h}) \cap F_Z) \cong F_Y$ so from the exact sequence

$$0 \to \text{Ker}(\widehat{h}) \cap F_Z \to F_Z \to F_Z/(\text{Ker}(\widehat{h}) \cap F_Z) \to 0$$

we see that

$$(\text{Ker}(\widehat{h}) \cap F_Z) \oplus F_Y \cong F_Z. \tag{1.11}$$

From the exact sequence $0 \to \text{Ker}(\widehat{h}) \cap F_Z \to \text{Ker}(\widehat{h}) \to F_X/F_Z \to 0$ and the isomorphism $F_X/F_Z \cong F_{\overline{Z}}$ we see that

$$\text{Ker}(\widehat{h}) \cong (\text{Ker}(\widehat{h}) \cap F_Z) \oplus F_{\overline{Z}}. \tag{1.12}$$

As $\overline{Z}$ is infinite we may write it as a disjoint union $\overline{Z} = Y_1 \sqcup W$ where $Y_1 \subset \overline{Z}$ is a finite subset such that $|Y_1| = |Y|$. In particular we may write

$$F_{\overline{Z}} \cong F_Y \oplus F_W$$

for some infinite subset $W \subset X$ so from (1.12) we get

$$\text{Ker}(\widehat{h}) \cong (\text{Ker}(\widehat{h}) \cap F_Z) \oplus F_Y \oplus F_W. \tag{1.13}$$

From (1.11) and (1.13) we see that

$$\text{Ker}(\widehat{h}) \cong F_Z \oplus F_W \cong F_{Z \sqcup W} \tag{1.14}$$

so that $\text{Ker}(\widehat{h})$ is free as required. $\qquad\square$

Gabel's Theorem confines the problem of stably free modules to the realm of finitely generated modules. Even so, the subject admits a certain amount of pathology; to avoid this we must impose the strongest of the restrictions of Sect. 1.1.

Given a stably free module S such that $S \oplus \Lambda^a \cong \Lambda^b$ one is tempted to make a definition of the *rank* $\text{rk}(S)$ of S by

$$\text{rk}(S) = b - a.$$

This is not a definition for all rings Λ as in general the rank of a free module is not well defined. It becomes a definition if Λ satisfies the (IBN) condition; however, there are still problems. Cohn [15] has given an example of a ring Λ with the IBN property which possesses a nonzero module S satisfying $S \oplus \Lambda^2 \cong \Lambda$ so that in this case $\mathrm{rk}(S) = -1$.

The surjective rank property by itself still leaves open the possibility that there exist nonzero stably free modules S of rank 0; that is, which satisfy $S \oplus \Lambda \cong \Lambda$. To avoid this we need something stronger still; we note:

Proposition 1.15 *Let S be a nonzero finitely generated stably free module over a weakly finite ring Λ. Then $0 < \mathrm{rk}(S)$.*

When discussing the rank of stably free modules we shall, without further comment, assume that the ring Λ is weakly finite. We shall say that such a ring Λ has the *stably free cancellation property* ($=$ SFC) when

$$S \oplus \Lambda^a \cong \Lambda^b \quad \Longrightarrow \quad S \cong \Lambda^{b-a}.$$

1.4 Projective Modules and Hyperstability

A module P over Λ is said to be *projective* when it is a direct summand of a free module. Projective modules arise inevitably once cohomology is introduced. When P is finitely generated projective then, for some positive integer n and some Λ-module Q

$$P \oplus Q \cong \Lambda^n.$$

Clearly stably free modules are projective but the converse is frequently false. Whilst our interest is directed towards the existence or nonexistence of nontrivial stably free modules, a discussion, however brief, of the general case is unavoidable.

Let $\mathcal{P}(\Lambda)$ denote the set of isomorphism classes of finitely generated projective Λ-modules. $\mathcal{P}(\Lambda)$ becomes an abelian monoid under direct sum;

$$[P] + [Q] = [P \oplus Q].$$

The Grothendieck group $K_0(\Lambda)$ is the universal abelian group obtained from $\mathcal{P}(\Lambda)$. The reduced Grothendieck group $\widetilde{K}_0(\Lambda)$ is the quotient

$$\widetilde{K}_0(\Lambda) = K_0(\Lambda)/[\Lambda]$$

by the subgroup generated by the class of Λ and there is a surjective monoid homomorphism

$$\mathcal{P}(\Lambda) \to \widetilde{K}_0(\Lambda).$$

Whereas standard K-theory typically concerns itself with computing $\widetilde{K}_0(\Lambda)$ we shall be more concerned with the dual problem, calculating what is lost under

$\mathcal{P}(\Lambda) \to \widetilde{K}_0(\Lambda)$. At a number of points we will need to use the celebrated result of Grothendieck [2, 4].

If R is a coherent ring of finite global dimension then $\widetilde{K}_0(R[t, t^{-1}]) \cong \widetilde{K}_0(R)$.

(1.16)

Gabel's Theorem does not extend to general projective modules. There are many interesting rings which possess infinitely generated projective modules which are not free. However, they cannot be 'too big' as the following result of Kaplansky [62] shows:

Every projective module is a direct sum of countably generated modules. (1.17)

In recent years there has been renewed research interest on the topic of infinitely generated projective modules. However, the question does not impact significantly upon our considerations. We do, however, need to consider the famous 'conjuring trick' of Eilenberg.

We denote by Λ^∞ the countable coproduct $\Lambda^\infty = \Lambda \oplus \Lambda \oplus \cdots \oplus \Lambda \oplus \Lambda \oplus \cdots$ or, alternatively, as the direct limit $\Lambda^\infty = \varinjlim \Lambda^n$ where $\Lambda^n \subset \Lambda^n \oplus \Lambda = \Lambda^{n+1}$ under the inclusion $x \mapsto (x, 0)$. Evidently a countable coproduct of copies of Λ^∞ is isomorphic to Λ^∞;

$$\Lambda^\infty \oplus \Lambda^\infty \oplus \cdots \oplus \Lambda^\infty \oplus \Lambda^\infty \oplus \cdots \cong \Lambda^\infty. \tag{1.18}$$

In this context one has Eilenberg's trick.

Proposition 1.19 *If $P \in \mathcal{M}od_\infty$ is projective then $P \oplus \Lambda^\infty \cong \Lambda^\infty$.*

Proof As P is countably generated choose a surjective Λ-homomorphism

$$\eta : \Lambda^\infty \to P.$$

Putting $Q = \mathrm{Ker}(\eta)$ gives an exact sequence $0 \to Q \to \Lambda^\infty \to P \to 0$ which splits since P is projective. We have

$$P \oplus Q \cong \Lambda^\infty \tag{I}$$

so that from (1.18), $(P \oplus Q) \oplus (P \oplus Q) \oplus \cdots \oplus (P \oplus Q) \oplus (P \oplus Q) \oplus \cdots \cong \Lambda^\infty$ which we may re-bracket as

$$P \oplus (Q \oplus P) \oplus (Q \oplus P) \oplus \cdots \oplus (Q \oplus P) \oplus (Q \oplus P) \oplus \cdots \cong \Lambda^\infty. \tag{II}$$

However from (I) it also follows that $Q \oplus P \cong \Lambda^\infty$; substitution in (II) gives

$$P \oplus \Lambda^\infty \oplus \Lambda^\infty \oplus \cdots \oplus \Lambda^\infty \oplus \Lambda^\infty \oplus \cdots \cong \Lambda^\infty$$

so that, from (1.18), $P \oplus \Lambda^\infty \cong \Lambda^\infty$ as stated. $\qquad\square$

We shall say that a module $M \in \mathcal{M}\mathrm{od}_\infty$ is *hyperstable* when $M \cong M_0 \oplus \Lambda^\infty$ for some module $M_0 \in \mathcal{M}\mathrm{od}_\infty$. Then $M \oplus \Lambda^\infty \cong M_0 \oplus \Lambda^\infty \oplus \Lambda^\infty \cong M_0 \oplus \Lambda^\infty \cong M$ so that we have:

A module $M \in \mathcal{M}\mathrm{od}_\infty$ is hyperstable if and only if $M \cong M \oplus \Lambda^\infty$. \hfill (1.20)

The *hyperstabilization* $\widehat{M}$ of the module $M \in \mathcal{M}\mathrm{od}_\Lambda^\infty$ is defined to be

$$\widehat{M} = M \oplus \Lambda^\infty.$$

Clearly the operation is idempotent; $\widehat{\widehat{M}} \cong \widehat{M}$. Moreover, if $P \in \mathcal{M}\mathrm{od}_\infty$ is a projective module it follows easily from Eilenberg's Trick that:

$$\widehat{M} \cong \widehat{M} \oplus P \cong \widehat{M \oplus P}. \hfill (1.21)$$

Chapter 2
The Restricted Linear Group

A celebrated result of H.J.S. Smith [47, 86] shows that when Λ is a commutative integral domain which possesses a Euclidean algorithm then an arbitrary $m \times m$ matrix X over Λ can be expressed as a product $X = E_+ D E_-$ where D is diagonal and E_+, E_- are products of elementary unimodular matrices. This chapter is a general study of rings whose matrices possess an analogue of such a Smith Normal Form.

2.1 Some Identities Between Elementary Matrices

Given a ring Λ we denote by $M_n(\Lambda)$ the ring of $(n \times n)$-matrices over Λ and by $GL_n(\Lambda)$ the group of invertible $n \times n$-matrices over Λ. For each $n \geq 2$, $M_n(\Lambda)$ has the canonical Λ-basis $\epsilon(i,j)_{1 \leq i,j \leq n}$ given by $\epsilon(i,j)_{r,s} = \delta_{ir}\delta_{js}$. The elementary invertible matrices $E(i,j;\lambda)$ ($\lambda \in \Lambda$) and $D(i,\delta)$ ($\delta \in \Lambda^*$) which perform row and column operations are expressed in terms of the basic matrices as follows;

$$E(i,j;\lambda) = I_n + \lambda\epsilon(i,j) \quad (i \neq j);$$
$$D(i,\delta) = I_n + (\delta - 1)\epsilon(i,i).$$

There are a number of familiar identities between these matrices:

$$E(i,j;\lambda)E(i,j;\mu) = E(i,j;\lambda+\mu); \tag{2.1}$$

$$E(i,j;\lambda)^{-1} = E(i,j;-\lambda); \tag{2.2}$$

$$[E(i,j;\lambda), E(j,k;\mu)] = E(i,k;\lambda\mu) \quad (i \neq k); \tag{2.3}$$

$$[E(i,j;\lambda), E(k,l;\mu)] = 1 \quad (\{i,j\} \cap \{k,l\} = \emptyset). \tag{2.4}$$

Here we are taking the commutator $[X,Y]$ to be $[X,Y] = XYX^{-1}Y^{-1}$.

$$D(i,\lambda)D(i,\mu) = D(i,\lambda\mu); \tag{2.5}$$

$$D(i,\lambda)^{-1} = D(i,\lambda^{-1}); \tag{2.6}$$

F.E.A. Johnson, *Syzygies and Homotopy Theory*, Algebra and Applications 17, DOI 10.1007/978-1-4471-2294-4_2, © Springer-Verlag London Limited 2012

$$D(i, \lambda)E(i, j; \mu) = E(i, j; \lambda\mu)D(i, \lambda); \tag{2.7}$$

$$D(i, \lambda)E(j, k; \mu) = E(j, k; \mu)D(i, \lambda) \quad (i \notin \{j, k\}). \tag{2.8}$$

For 2×2 matrices we have the following identity where $u \in \Lambda^*$;

$$\begin{bmatrix} u & 0 \\ 0 & u^{-1} \end{bmatrix} = \begin{bmatrix} 1 & 0 \\ 1 & 1 \end{bmatrix} \begin{bmatrix} 1 & -1 \\ 0 & 1 \end{bmatrix} \begin{bmatrix} 1 & 0 \\ 1 & 1 \end{bmatrix} \begin{bmatrix} 1 & u^{-1} \\ 0 & 1 \end{bmatrix} \begin{bmatrix} 1 & 0 \\ -u & 1 \end{bmatrix} \begin{bmatrix} 1 & u^{-1} \\ 0 & 1 \end{bmatrix}.$$

It generalises to the following identity with $i \neq j$:

$$\Delta(i, u)\Delta(j, u^{-1}) = E(j, i; 1)E(i, j; -1)E(j, i; 1)E(i, j; u^{-1})$$

$$\times E(j, i; -u)E(i, j; u^{-1}). \tag{2.9}$$

Let Σ_n denote the group of permutations on $\{1, \ldots, n\}$. For each $\sigma \in \Sigma_n$ there is an $n \times n$ permutation matrix $P(\sigma)$ defined by

$$P(\sigma)_{r,s} = \delta_{r, \sigma(s)}.$$

It is straightforward to see that:

$$P \text{ defines an injective homomorphism } P : \Sigma_n \to GL_n(\Lambda). \tag{2.10}$$

The permutation matrices $P(\sigma)$ can be expressed as products of matrices of the form $D(i, -1)$ and $E(i, j; \pm 1)$. As Σ_n is generated by the transpositions (i, j) it suffices to express each $P(i, j)$ as a product of this type. In fact, we have:

$$P(\sigma) = D(j, -1)E(i, j; 1)E(j, i; -1)E(i, j; 1). \tag{2.11}$$

It is useful to record how the permutation matrices $P(\sigma)$ interact with the $E(i, j; \lambda)$ and $D(i, \delta)$. First the action on the basic matrices $\epsilon(i, j)$;

$$P(\sigma)\epsilon(i, j) = \epsilon(\sigma(i), \sigma(j))P(\sigma). \tag{2.12}$$

This easily implies

$$P(\sigma)E(i, j; \lambda) = E(\sigma(i), \sigma(j); \lambda)P(\sigma). \tag{2.13}$$

Alternatively expressed:

$$E(i, j; \lambda)P(\sigma) = P(\sigma)E(\sigma^{-1}(i), \sigma^{-1}(j); \lambda). \tag{2.14}$$

Whilst

$$P(\sigma)D(i, \delta) = D(\sigma(i), \delta)P(\sigma). \tag{2.15}$$

In the special case where $\lambda \in \Lambda^*$ we have the matrix equation

$$\begin{pmatrix} 0 & 1 \\ 1 & 0 \end{pmatrix} \begin{pmatrix} 1 & \lambda \\ 0 & 1 \end{pmatrix} = \begin{pmatrix} 1 & \lambda^{-1} \\ 0 & 1 \end{pmatrix} \begin{pmatrix} -\lambda^{-1} & 0 \\ 0 & \lambda \end{pmatrix} \begin{pmatrix} 1 & 0 \\ \lambda^{-1} & 1 \end{pmatrix}$$

which generalises to

$$P(i,j)E(i,j;\lambda) = E(i,j;\lambda^{-1})D(i,-\lambda^{-1})D(j,\lambda)E(j,i;\lambda^{-1}). \qquad (2.16)$$

2.2 The Restricted Linear Group

For $n \geq 2$ we denote by $D_n(\Lambda)$ the subgroup of $GL_n(\Lambda)$ defined by

$$D_n(\Lambda) = \{D(1,\delta) : \delta \in \Lambda^*\};$$

that is

$$D_n(\Lambda) = \left\{ \begin{bmatrix} \delta & & & 0 \\ & 1 & & \\ & & \ddots & \\ 0 & & & 1 \end{bmatrix} \right\}$$

and by $E_n(\Lambda)$ the subgroup of $GL_n(\Lambda)$ generated by the matrices $E(i,j;\lambda)$ ($\lambda \in \Lambda$). From (2.7) and (2.8)

$$D(1,\delta)E(i,j;\lambda)D(1,\delta)^{-1} = \begin{cases} E(1,j;\delta\lambda) & i=1, \\ E(i,1;\lambda\delta^{-1}) & j=1, \\ E(i,j;\lambda) & i \notin \{i,j\}. \end{cases}$$

We see that:

$$D_n(\Lambda) \text{ normalises } E_n(\Lambda). \qquad (2.17)$$

We define the *restricted linear group* $GE_n(\Lambda)$ to be the subgroup of $GL_n(\Lambda)$ given as the internal product

$$GE_n(\Lambda) = D_n(\Lambda) \cdot E_n(\Lambda).$$

In general $GE_n(\Lambda)$ is a proper subgroup of $GL_n(\Lambda)$ and Λ is said to be *weakly Euclidean* when $GE_n(\Lambda) = GL_n(\Lambda)$ for all $n \geq 2$. We shall examine this notion at greater length in Sects. 2.4 and 2.5. From (2.17) we get:

$$E_n(\Lambda) \text{ is a normal subgroup of } GE_n(\Lambda). \qquad (2.18)$$

We put $\widehat{\Sigma}_n = \{P(\sigma) : \sigma \in \Sigma_n\}$.

Proposition 2.19 $\widehat{\Sigma}_n \subset GE_n(\Lambda)$.

Proof It follows from (2.11) that $P(i,j) \in E_n(\Lambda)$ for each transposition (i,j). The conclusion follows as Σ_n is generated by transpositions. $\qquad \square$

It is useful to have different descriptions of $GE_n(\Lambda)$. For $\delta_1, \ldots, \delta_n \in \Lambda^*$ let $\Delta(\delta_1, \ldots, \delta_n)$ denote the diagonal matrix

$$\Delta(\delta_1, \ldots, \delta_n) = \begin{bmatrix} \delta_1 & & & 0 \\ & \delta_2 & & \\ & & \ddots & \\ 0 & & & \delta_n \end{bmatrix}$$

and put $\Delta_n(\Lambda) = \{\Delta(\delta_1, \ldots, \delta_n) : \delta_i \in \Lambda^*\}$.

Proposition 2.20 $\Delta_n(\Lambda)$ *is a subgroup of* $GE_n(\Lambda)$.

Proof It is straightforward to see that $\Delta_n(\Lambda)$ is a subgroup of $GL_n(\Lambda)$. Note that $D(j, \delta_j) \in GE_n(\Lambda)$ as, by (2.15),

$$D(j, \delta) = P(1, j)D(1, \delta)P(1, j)$$

and $D(i, \delta), P(1, j) \in GE_n(\Lambda)$. The conclusion follows as $\Delta(\delta_1, \ldots, \delta_n) = \prod_{j=1}^n D(j, \delta_j)$. $\qquad\qquad\qquad\qquad\qquad\qquad\qquad\qquad\qquad\qquad\qquad\qquad\qquad\quad\square$

By (2.7) we have $\Delta(\delta_1, \ldots, \delta_n)E(i, j; \lambda)\Delta(\delta_1, \ldots, \delta_n)^{-1} = E(i, j; \delta_i \lambda \delta_j^{-1})$ from which we see that:

$$\Delta_n(\Lambda) \text{ normalises } E_n(\Lambda). \qquad\qquad\qquad\qquad\qquad (2.21)$$

Now $\Delta_n(\Lambda)$, $E_n(\Lambda)$ are subgroups of $GE_n(\Lambda) = D_n(\Lambda)E_n(\Lambda)$ and $D_n(\Lambda) \subset \Delta_n(\Lambda)$. We obtain another description of $GE_n(\Lambda)$.

$$GE_n(\Lambda) = \Delta_n(\Lambda) \cdot E_n(\Lambda). \qquad\qquad\qquad\qquad\qquad (2.22)$$

The constructions GE_n, E_n are functorial under ring homomorphisms. In particular, given a surjective ring homomorphism $\pi : A \to B$ the induced map $\pi_* : E_n(A) \to E_n(B)$ is also surjective. However unless the induced map on units $\pi_* : A^* \to B^*$ is also surjective then the induced homomorphism $\pi_* : GE_n(A) \to GE_n(B)$, need not be surjective. We say that ring homomorphism $\pi : A \to B$ has the *lifting property for units* when the induced map on units $\pi_* : A^* \to B^*$ is surjective. Then we have:

Proposition 2.23 *Let* $\pi : A \to B$ *be a surjective ring homomorphism with the lifting property for units; then* $\pi_* : GE_n(A) \to GE_n(B)$ *is surjective.*

2.3 Matrices with a Smith Normal Form

If Λ is a ring and $\alpha_1, \ldots, \alpha_n \in \Lambda$ write $\Delta(\alpha_1, \ldots, \alpha_n)$ for the diagonal matrix

$$\Delta(\alpha_1, \ldots, \alpha_n) = \begin{pmatrix} \alpha_1 & & 0 \\ & \ddots & \\ 0 & & \alpha_n \end{pmatrix}.$$

We shall say that $X \in M_n(\Lambda)$ *has a Smith Normal Form* when

$$X = E_+ \Delta(\alpha_1, \ldots, \alpha_n) E_-$$

for some $E_+, E_- \in E_n(\Lambda)$ and some $\alpha_1, \ldots, \alpha_n \in \Lambda$. More generally, if $X, Y \in M_n(\Lambda)$ we write $X \sim Y$ when $X = E_+ Y E_-$ for some $E_+, E_- \in E_n(\Lambda)$. Evidently we have:

Proposition 2.24 *Let* $X \sim Y \in M_n(\Lambda)$; *if* Λ *is commutative then* $\det(X) = \det(Y)$.

The identities of Sect. 2.1 allow us to move units around; let $\alpha_1, \ldots, \alpha_n \in \Lambda$ and $u_1, \ldots, u_n \in \Lambda^*$. Writing $D(r) = \Delta(1, u_r)\Delta(r, u_r^{-1})$, $D = D(2)D(3) \cdots D(n)$ and $\mathbf{u} = u_1 \cdots u_n$ we see that $\Delta(\alpha_1 u_1, \ldots, \alpha_n u_n) D = \Delta(\alpha_1 \mathbf{u}, \alpha_2, \ldots, \alpha_n)$. However, each $D(r) \in E_n(\Lambda)$ by (2.9); thus for $u_1, \ldots, u_n \in \Lambda^*$:

$$\Delta(\alpha_1 u_1, \ldots, \alpha_n u_n) \sim \Delta(\alpha_1 \mathbf{u}, \alpha_2, \ldots, \alpha_n) \quad \text{where } \mathbf{u} = u_1 \cdots u_n. \tag{2.25}$$

Similarly for $v_1, \ldots, v_n \in \Lambda^*$:

$$\Delta(v_1 \alpha_1, \ldots, v_n \alpha_n) \sim \Delta(\mathbf{v}\alpha_1, \alpha_2, \ldots, \alpha_n) \quad \text{where } \mathbf{v} = v_n \cdots v_1. \tag{2.26}$$

Write $T(i, j) = E(i, j; 1)E(j, i; -1)E(i, j; 1) \in E_n(\Lambda)$; when τ is the transposition which interchanges the indices i and j we have

$$\Delta(\alpha_{\tau(1)}, \ldots, \alpha_{\tau(n)}) = T(i, j)\Delta(\alpha_1, \ldots, \alpha_n)T(j, i).$$

Writing an arbitrary permutation σ as a product of transpositions we see that

$$\Delta(\alpha_{\sigma(1)}, \ldots, \alpha_{\sigma(n)}) \sim \Delta(\alpha_1, \ldots, \alpha_n). \tag{2.27}$$

The following is useful:

Proposition 2.28 *Let* Λ *be a commutative ring and suppose that* $X \in M_n(\Lambda)$ *has a Smith Normal Form where* $n \geq 2$; *if* $\det(X)$ *is indecomposable in* Λ *then*

$$X \sim \Delta(\det(X), 1, \ldots, 1).$$

Proof As X has a Smith Normal Form then $X \sim \Delta(\alpha_1, \ldots, \alpha_n)$ for some $\alpha_1, \ldots, \alpha_n \in \Lambda$. By Proposition 2.24 $\det(X) = \prod_{r=1}^n \alpha_r$. As $\det(X)$ is indecomposable it follows that there is an index t such that $\alpha_r \in \Lambda^*$ for $r \neq t$. Denoting by σ the transposition $(1, t)$ we see from (2.27) that $X \sim \Delta(\alpha_{\sigma(1)}, \ldots, \alpha_{\sigma(n)})$ where $\alpha_{\sigma(r)} \in \Lambda^*$ for $r \geq 2$. By (2.25) $X \sim \Delta(\prod_r^n \alpha_{\sigma(r)}, 1, \ldots, 1)$. However $\prod_r^n \alpha_{\sigma(r)} = \prod_{r=1}^n \alpha_r = \det(X)$. $\square$

The Smith Normal Form is compatible with products of rings as we now proceed to show. Suppose $\Lambda = \Lambda_1 \times \Lambda_2$ is a direct product of rings. Then the projections

$\pi_r : \Lambda \to \Lambda_r$ induce ring homomorphisms $\pi_r : M_m(\Lambda) \to M_m(\Lambda_r)$ so that

$$(\pi_1, \pi_2) : M_m(\Lambda) \to M_m(\Lambda_1) \times M_m(\Lambda_2) \text{ is a ring isomorphism.} \qquad (2.29)$$

In consequence:

$$(\pi_1, \pi_2) : GL_m(\Lambda) \to GL_m(\Lambda_1) \times GL_m(\Lambda_2) \text{ is an isomorphism of groups.} \quad (2.30)$$

Requiring slightly more care is:

Proposition 2.31 (π_1, π_2) *induces an isomorphism* $E_m(\Lambda) \xrightarrow{\simeq} E_m(\Lambda_1) \times E_m(\Lambda_2)$.

Proof $\pi_r(E(k, l; (\lambda_1, \lambda_2))) = E(k, l; \lambda_r)$ so that (π_1, π_2) induces a group homomorphism $(\pi_1, \pi_2) : E_m(\Lambda) \to E_m(\Lambda_1) \times E_m(\Lambda_2)$. Evidently (π_1, π_2) is injective since it is already the restriction of a ring isomorphism. To show that (π_1, π_2) is onto $E_m(\Lambda_1) \times E_m(\Lambda_2)$ we first show that:

If $X \in E_m(\Lambda_1)$ then there exists $\widetilde{X} \in E_m(\Lambda)$ such that $(\pi_1, \pi_2)(\widetilde{X}) = (X, \mathrm{Id})$.

$$(*)$$

To prove $(*)$ write $X \in E_m(\Lambda_1)$ as a product $X = E(k_1, l_1; \lambda_1) \cdots E(k_N, l_N; \lambda_N)$ and write $\widetilde{X} = E(k_1, l_1; (\lambda_1, 0)) \cdots E(k_N, l_N; (\lambda_N, 0))$. One sees easily that $(\pi_1, \pi_2)(\widetilde{X}) = (X, \mathrm{Id})$. Similarly:

If $Y \in E_m(\Lambda_2)$ then there exists $\widetilde{Y} \in E_m(\Lambda)$ such that $(\pi_1, \pi_2)(\widetilde{Y}) = (\mathrm{Id}, Y)$.

$$(**)$$

Surjectivity of (π_1, π_2) now follows, for if $(X, Y) \in E_m(\Lambda_1) \times E_m(\Lambda_2)$ then

$$(X, Y) = (X, \mathrm{Id})(\mathrm{Id}, Y) = (\pi_1, \pi_2)(\widetilde{X})(\pi_1, \pi_2)(\widetilde{Y}) = (\pi_1, \pi_2)(\widetilde{X}\widetilde{Y}). \qquad \square$$

The existence of Smith Normal Forms is compatible with products in the strongest sense; let $X \in M_m(\Lambda)$ and put $X_r = \pi_r(X) \in M_m(\Lambda_r)$ for $r = 1, 2$. Suppose given sequences $\alpha_s^r \in \Lambda_r$ $(s = 1, \ldots, m, r = 1, 2)$; with this notation:

Theorem 2.32 *For any permutations* σ, τ *we have:*

$$X \sim \begin{pmatrix} (\alpha_{\sigma(1)}^1, \alpha_{\tau(1)}^2) & & & \\ & (\alpha_{\sigma(2)}^1, \alpha_{\tau(2)}^2) & & \\ & & \ddots & \\ & & & (\alpha_{\sigma(n)}^1, \alpha_{\tau(n)}^2) \end{pmatrix}$$

$$\Longleftrightarrow \quad X_r \sim \begin{pmatrix} \alpha_1^r & & & \\ & \alpha_2^r & & \\ & & \ddots & \\ & & & \alpha_n^r \end{pmatrix}.$$

We shall say that the ring Λ is *generalized Euclidean* when for all $m \geq 2$,

$$M_m(\Lambda) = E_m(\Lambda)\mathcal{D}_m(\Lambda)E_m(\Lambda). \tag{2.33}$$

The terminology *generalized Euclidean* is taken from the classical sufficient conditions for this to occur, namely those of the Smith Normal Form, which may be re-expressed thus:

Proposition 2.34 (Smith [86]) *Let Λ be a commutative integral domain with a Euclidean algorithm; then Λ is generalized Euclidean.*

It follows from Theorem 2.32 that:

Proposition 2.35 *Let Λ_1, Λ_2 be generalized Euclidean rings; then $\Lambda_1 \times \Lambda_2$ is also generalized Euclidean.*

To consider an example, we denote by $\mathbf{Z}_n$ the ring of residues mod n; that is, $\mathbf{Z}_n = \mathbf{Z}/n$. If n is square free we may write it as a product of distinct primes $n = p_1 p_2 \cdots p_k$. Then $\mathbf{Z}_n \cong \mathbf{F}_{p_1} \times \cdots \times \mathbf{F}_{p_k}$ where $\mathbf{F}_p$ is the field with p elements and so

$$\mathbf{Z}_n[t, t^{-1}] \cong \mathbf{F}_{p_1}[t, t^{-1}] \times \cdots \times \mathbf{F}_{p_k}[t, t^{-1}].$$

However, each $\mathbf{F}_{p_r}[t, t^{-1}]$ is a Euclidean ring and so is generalized Euclidean by Proposition 2.34; from Proposition 2.35 we see that:

If n is square free then $\mathbf{Z}_n[t, t^{-1}]$ is generalized Euclidean.   (2.36)

Another useful result is:

Proposition 2.37 *Let Λ_1 be a generalized Euclidean ring; if $\varphi : \Lambda_1 \to \Lambda_2$ is a surjective ring homomorphism then Λ_2 is also generalized Euclidean ring.*

Proof Let $A \in M_m(\Lambda_2)$. First choose $\widetilde{A} \in M_m(\Lambda_1)$ such that $\varphi_*(\widetilde{A}) = A$. Since Λ_1 is generalized Euclidean, we may write $\widetilde{A} = P_1 D P_2$ where $P_i \in E_m(\Lambda_1)$ and $D \in \mathcal{D}_m(\Lambda_1)$. Then $\varphi_*(P_i) \in E_m(\Lambda_2)$, $\varphi_*(D) \in \mathcal{D}_m(\Lambda_2)$ and the required decomposition is given by

$$A = \varphi_*(P_1)\varphi_*(D)\varphi_*(P_2). \qquad \square$$

Next consider diagonal matrices

$$\Delta = \begin{pmatrix} \delta_1 & & & 0 \\ & \delta_2 & & \\ & & \ddots & \\ 0 & & & \delta_m \end{pmatrix} \in M_m(\Lambda).$$

We say that Δ is of *restricted type* when $\delta_2, \ldots, \delta_m \in \Lambda^*$ and that $\Delta \in M_m(\Lambda)$ is of *very restricted type* when $\delta_2 = \cdots = \delta_m = 1$. Likewise we say that a matrix $X \in M_m(\Lambda)$ has a Smith Normal Form of *restricted type* (resp. *very restricted type*) when $X \sim \Delta$ where Δ is a diagonal matrix of *restricted type* (resp. *very restricted type*). From (2.25) and (2.27) it follows easily that if $X \in M_m(\Lambda)$ then:

$$\left\{ \begin{matrix} X \text{ has a Smith Normal} \\ \text{Form of restricted type} \end{matrix} \right\} \quad \Longleftrightarrow \quad \left\{ \begin{matrix} X \text{ has a Smith Normal} \\ \text{Form of very restricted type} \end{matrix} \right\}. \qquad (2.38)$$

Given a ring Λ a two sided ideal J in Λ is of *radical type* when $u + j \in \Lambda^*$ for any $j \in J$ and $u \in \Lambda^*$. We say that a ring homomorphism $\varphi : A \to B$ has the *strong lifting property for units* when, in addition to the lifting property for units, φ satisfies

$$\alpha \in A^* \quad \Longleftrightarrow \quad \pi(\alpha) \in B^*. \qquad (2.39)$$

It is straightforward to see that if $\varphi : A \to B$ is a surjective ring homomorphism with the lifting property for units then:

φ has the strong lifting property for units $\quad \Longleftrightarrow \quad$ Ker(φ) is of radical type. (2.40)

If J is a two-sided ideal in Λ we shall say that a matrix $U = (u_{ij}) \in M_m(\Lambda)$ has *restricted form* with respect to J when $u_{22}, \ldots, u_{mm} \in \Lambda^*$ whilst $u_{ij} \in J$ when $i \neq j$ and $i, j \geq 2$; otherwise expressed, the image $\overline{U}$ in Λ/J then takes the form

$$\overline{U} = \begin{pmatrix} * & * & * & * \\ * & \overline{u_{22}} & 0 & 0 \\ * & 0 & \ddots & 0 \\ * & 0 & 0 & \overline{u_{mm}} \end{pmatrix}.$$

Now suppose that J is a two sided ideal of radical type in Λ and that $U \in M_m(\Lambda)$ has restricted form with respect to J. As r descends from m to 2 we may, by means of suitable row and column operations, successively kill the rth row and column leaving units in the (r, r)th places; we obtain:

Proposition 2.41 *Let J be a two sided ideal of radical type in Λ; if $U \in M_m(\Lambda)$ has restricted form with respect to J then U has a Smith Normal Form of restricted type.*

Lemma 2.42 (Lifting Lemma) *Let $\varphi : \Lambda \to \check{\Lambda}$ be a surjective ring homomorphism with the strong lifting property for units; if $X \in M_m(\Lambda)$ is such that $\varphi(X)$ has a Smith Normal Form of restricted type then X also has a Smith Normal Form of restricted type.*

Proof By hypothesis, $\varphi(X)$ has a Smith Normal Form of restricted type which, by (2.38), we may take to be of the form

$$\begin{pmatrix} \xi & & & & 0 \\ & 1 & & & \\ & & \ddots & & \\ 0 & & & & 1 \end{pmatrix}.$$

Formally, there exist $E_+, E_- \in E_m(\check{\Lambda})$ such that $E_+\varphi(X)E_- = \Delta(\xi, 1, \ldots, 1)$. As φ is surjective we may now choose $E'_+, E'_- \in E_m(\Lambda)$ such that $\varphi(E'_\sigma) = E_\sigma$. One sees easily that $E'_+ X E'_-$ has restricted form with respect to the two sided ideal $\mathrm{Ker}(\varphi)$. However, as φ has the strong lifting property for units then $\mathrm{Ker}(\varphi)$ is of radical type. By Proposition 2.41, $E'_+ X E'_-$ has a Smith Normal Form of restricted type. Hence X also has a Smith Normal Form of restricted type. $\qquad\square$

2.4 Weakly Euclidean Rings

The ring Λ is said to be *weakly Euclidean* when $GL_n(\Lambda) = GE_n(\Lambda)$ for all $n \geq 2$. As $\Delta_m(\Lambda)$ normalises $E_m(\Lambda)$ this is equivalent to requiring that each $X \in GL_n(\Lambda)$ has a Smith Normal Form. We may now establish a Recognition Criterion for this property:

Proposition 2.43 *Let* $\varphi : A \to B$ *be a surjective ring homomorphism where* B *is weakly Euclidean; if* φ *has the strong lifting property for units then* A *is also weakly Euclidean.*

Proof Let $X \in GL_m(A)$; then $\varphi(X) \in GL_m(B)$. By hypothesis on B, $\varphi(X)$ is a product

$$\varphi(X) = D(\check{\delta}, 1)E,$$

where $\check{\delta} \in B^*$ and $E \in E_m(B)$. In particular, $\varphi(X)$ has a Smith Normal Form of very restricted type. By Lemma 2.42 and (2.38), X also has a Smith Normal Form of very restricted type so we may write $X = E_+ D(\delta, 1)E_-$ for some $E_+, E_- \in E_m(A)$. As X is invertible then $\delta \in A^*$. It follows from (2.17) that $E'_+ = D(\delta^{-1}, 1)E_+ D(\delta, 1) \in E_m(A)$ and so $X = D(\delta, 1)E$ where $E = E'_+ E_- \in E_m(A)$. $\qquad\square$

There is a useful generalization of the notion of weakly Euclidean ring; for any ring Λ the group $GL_m(\Lambda)$ imbeds via stabilization in $GL_n(\Lambda)$ for $m \leq n$; moreover $GL_m(\Lambda) \cdot E_n(\Lambda)$ is a subgroup of $GL_n(\Lambda)$ and is normalised by $GL_m(\Lambda)$. We say that Λ is *m-weakly Euclidean* when $GL_n(\Lambda) = GL_m(\Lambda) \cdot E_n(\Lambda) =$ for $m \leq n$. Evidently if Λ is m-weakly Euclidean then it is Λ is n-weakly Euclidean when $m \leq n$. Moreover, Λ is 1-weakly Euclidean precisely when it is weakly Euclidean.

2.5 Examples of Weakly Euclidean Rings

The standard theory of reduction by row operations shows that:

$$\text{Any division ring is weakly Euclidean.} \qquad (2.44)$$

Moreover, from Smith's Theorem Proposition 2.34, it follows a fortiori that:

$$\text{Any commutative Euclidean domain is weakly Euclidean.} \qquad (2.45)$$

Let L be a (possibly noncommutative) local ring; it is straightforward to see that the canonical homomorphism $\pi : L \to L/\mathrm{rad}(L)$ has the strong lifting property for units. Moreover, as $L/\mathrm{rad}(L)$ is a division ring it is weakly Euclidean. Applying Proposition 2.43 we get the following which first seems to have been observed by Klingenberg [64].

Corollary 2.46 *If L is a (possibly noncommutative) local ring then L is weakly Euclidean.*

Note that weakly Euclidean rings are closed under products; that is:

Proposition 2.47 *If $R_1, \ldots, R_m$ are weakly Euclidean rings then $R_1 \times \cdots \times R_m$ is also weakly Euclidean.*

We note also the following adjunct to Morita theory:

Theorem 2.48 *Let R be a weakly Euclidean ring; then for each $n \geq 1$, the ring $M_n(R)$ of $n \times n$ matrices over R is also weakly Euclidean.*

Proof If V is an R-module then for each $m \geq 1$ put $V^{(m)} = \underbrace{V \oplus \cdots \oplus V}_{m}$; for each $m \geq 1$ there is a ring isomorphism $\Psi : \mathrm{End}_R(V^{(m)}) \xrightarrow{\simeq} M_m(\mathrm{End}_R(V))$ given by

$$\Psi(\alpha)_{rs} = \pi_s \alpha i_r,$$

where $i_s : V \to V^{(m)}$ is the inclusion of the sth summand and $\pi_r : V^{(m)} \to V$ is projection onto the rth-factor. When V is a free module of rank n with basis $\{\epsilon_k\}_{1 \leq k \leq n}$ we identify $\mathrm{End}_R(V)$ with $M_n(R)$. Moreover $V^{(m)}$ then has the basis $\{E_t\}_{1 \leq t \leq mn}$ where, on writing $t = n(a-1) + b$ with $1 \leq a \leq m$ and $1 \leq b \leq n$,

$$E_t = i_a(\epsilon_b)$$

enabling us to identify $\mathrm{End}_R(V^{(m)})$ with $M_{mn}(R)$. Ψ then becomes the 'block decomposition' isomorphism

$$\Psi : M_{mn}(R) \xrightarrow{\simeq} M_m(M_n(R))$$

inducing a group isomorphism $\Psi : GL_{mn}(R) \xrightarrow{\simeq} GL_m(M_n(R))$. It will suffice to show that

$$\Psi(GE_{mn}(R)) \subset GE_m(M_n(R)).$$

For then, by the weakly Euclidean hypothesis on R, $GE_{mn}(R) = GL_{mn}(R)$ and so

$$GL_m(M_n(R)) = \Psi(GL_{mn}(R)) \subset GE_m(M_n(R))$$

and hence $GL_m(M_n(R)) = GE_m(M_n(R))$ showing that $M_n(R)$ is weakly Euclidean. To show that $\Psi(GE_{mn}(R)) \subset GE_m(M_n(R))$ first observe that $\Psi(\Delta_{mn}(R)) \subset \Delta_m(M_n(R))$. As $GE_{mn}(R) = \Delta_{mn}(R) \cdot E_{mn}(R)$ it therefore suffices to show that

$$\Psi(E(s,t;\lambda)) \in GE_m(M_n(R))$$

for $1 \leq s, t \leq mn$ and $s \neq t$, $\lambda \in R$. Write $s = n(a-1) + b$, $t = n(a'-1) + b'$ with $1 \leq a, a' \leq m$ and $1 \leq b, b' \leq n$. If $a = a'$ then $\Psi(E(s,t;\lambda)) \in \Delta_m(M_n(R))$ whilst if $a \neq a'$ $\Psi(E(s,t;\lambda)) \in E_m(M_n(R))$. $\square$

For $n \geq 2$ we denote by F_n the free *nonabelian* group of rank n, whilst F_1 will denote the infinite cyclic group C_∞. We have the following theorem of Cohn [17]:

Theorem 2.49 *For any division ring D the group ring $D[F_n]$ is weakly Euclidean.*

Let R be a ring and G a group; we say that the group ring $R[G]$ has *only trivial units* when each $\lambda \in R[G]^*$ has the form $\lambda = ug$ for $u \in R^*$ and $g \in G$.

Proposition 2.50 *Let $\pi : A \to B$ be a surjective ring homomorphism with the strong lifting property for units and suppose that the ideal $\mathrm{Ker}(\pi)$ is nilpotent; if G is a group for which $B[G]$ has only trivial units then the induced homomorphism $\pi_* : A[G] \to B[G]$ has the strong lifting property for units.*

Proof Putting $J = \mathrm{Ker}(\pi)$ observe that $\mathrm{Ker}(\pi_*) = J[G] = \{\sum_{g \in G} \xi_g g : \xi_g \in J\}$. Now, by hypothesis, $J^M = \{0\}$ for some $M \geq 1$. It follows that if $X \in \mathrm{Ker}(\pi_*)$ then $X^M = 0$. Hence $1 - X \in A[G]^*$ as

$$(1 - X)(1 + X + X^2 + \cdots + X^{M-1}) = 1.$$

Now let $\alpha \in A[G]$ satisfy $\pi_*(\alpha) \in B[G]^*$. We must show that $\alpha \in A[G]^*$. By hypothesis, $B[G]^*$ has only trivial units; that is $\pi_*(\alpha) = ug$ for some $u \in B^*$ and $g \in G$. As π is surjective we may choose $v \in A$ such that $\pi(v) = u$. It then follows that $v \in A^*$ as π has the strong lifting property for units. Put $\gamma = v^{-1}\alpha g^{-1}$; and $X = 1 - \gamma$; then $\gamma = 1 - X$ where $X \in \mathrm{Ker}(\pi_*)$, so that $\gamma \in A[G]^*$ by the above and $\alpha = v\gamma g \in A[G]^*$ as required. $\square$

If G is a group and X, Y are subsets of G we say that $g \in G$ is represented as a product in X, Y when $g = xy$ for some $x \in X$ and $y \in Y$. We say that $g \in G$ is *uniquely represented* as a product in X, Y when, in addition, if $g = x'y'$ with $x' \in X$ and $y' \in Y$ then $x = x'$ and $y = y'$. We say that G satisfies the *two unique products* condition (abbreviated to $\mathcal{TUP}$) when, given finite subsets X, Y of G with $2 \leq |X|$ and $2 \leq |Y|$, at least two elements of G are uniquely represented as products in X, Y. In Appendix C we give a fuller account of this topic; in particular we will see:

Theorem 2.51 *Let G be a $\mathcal{TUP}$ group; then for any (possibly noncommutative) integral domain A, $A[G]$ has only trivial units.*

It is known (cf. Appendix C) that the free group F_n satisfies the $\mathcal{TUP}$ condition; thus for any division ring D the group ring $D[F_n]$ has only trivial units. Suppose that L is a local ring in which $\mathrm{rad}(L)$ is nilpotent and put $D = L/\mathrm{rad}(L)$; by applying Proposition 2.50 to the induced homomorphism $L[F_n] \to D[F_n]$ we may extend Theorem 2.49 as follows:

Corollary 2.52 *If L is a local ring for which $\mathrm{rad}(L)$ is nilpotent then the group ring $L[F_n]$ is weakly Euclidean.*

With very little change the requirement that $\mathrm{rad}(L)$ be nilpotent may be replaced by the hypothesis that L is complete to obtain:

Proposition 2.53 *If $\widehat{L}$ is a complete local ring then $\widehat{L}[F_n]$ is weakly Euclidean for any $n \geq 1$.*

We may generate a useful class of examples by the triangular algebra construction. If R is a ring and n is an integer ≥ 2 we define

$$\mathcal{T}_n(R) = \{X \in M_n(R) : X_{ij} = 0 \text{ for } i < j\};$$

that is $\mathcal{T}_n(R)$ consists of elements of the form

$$\begin{pmatrix} X_{11} & & & \\ & X_{22} & & 0 \\ & & \ddots & \\ & * & & \ddots \\ & & & X_{nn} \end{pmatrix}.$$

Observe that $\mathcal{T}_n(R)$ is a subring of $M_n(R)$. Moreover, there is an obvious surjective ring homomorphism

$$\delta : \mathcal{T}_n(R) \to \underbrace{R \times \cdots \times R}_{n}; \quad \delta(X) = (X_{11}, X_{22}, \ldots, X_{nn}).$$

It is straightforward to check that δ has the strong lifting property for units. From this observation together with Propositions 2.43 and 2.47 we see that:

Proposition 2.54 *If R is weakly Euclidean then $\mathcal{T}_n(R)$ is also weakly Euclidean.*

2.6 The Dieudonné Determinant

If $P(\sigma)$ is a permutation matrix then by (2.15)

$$P(\sigma)\Delta(\delta_1,\ldots,\delta_n)P(\sigma)^{-1} = \Delta(\delta_{\sigma(1)},\ldots,\delta_{\sigma(n)}).$$

In particular, in $GE_n(\Lambda)$ the subgroup $\widehat{\Sigma}_n$ of permutation matrices normalises the subgroup $\Delta_n(\Lambda)$ of diagonal matrices. We define the *core subgroup* $C_n(\Lambda)$ of $GE_n(\Lambda)$ to be the internal product

$$C_n(\Lambda) = \Delta_n(\Lambda) \cdot \widehat{\Sigma}_n. \tag{2.55}$$

We have seen, in (2.21), that $\Delta_n(\Lambda)$ normalises $E_n(\Lambda)$. It follows from (2.13) that $\widehat{\Sigma}_n$ also normalises $E_n(\Lambda)$. Thus $C_n(\Lambda)$ normalises $E_n(\Lambda)$. However, $GE_n(\Lambda) = \Delta_n(\Lambda) \cdot E_n(\Lambda)$ so that, a fortiori,

$$GE_n(\Lambda) = C_n(\Lambda) \cdot E_n(\Lambda). \tag{2.56}$$

It is clear that $\Delta_n(\Lambda) \cap \widehat{\Sigma}_n = \{1\}$ so that the decomposition (2.55) is actually a semi-direct product. Alternatively we can regard $C_n(\Lambda)$ as a semidirect product

$$C_n(\Lambda) \cong (\Lambda^*)^n \bullet \Sigma_n, \tag{2.57}$$

where Σ_n acts on $(\Lambda^*)^n$ via $\sigma \cdot (\delta_1,\ldots,\delta_n) = (\delta_{\sigma(1)},\ldots,\delta_{\sigma(n)})$. Let $(\Lambda^*)^{ab}$ denote the abelianization of the unit group $(\Lambda^*)^{ab}$; that is

$$(\Lambda^*)^{ab} = \Lambda^*/[\Lambda^*,\Lambda^*]. \tag{2.58}$$

If $\delta \in \Lambda^*$ we denote by $[\delta]$ its class in $(\Lambda^*)^{ab}$. One sees easily that the multiplication map

$$\mu : (\Lambda^*)^n \to (\Lambda^*)^{ab}; \quad \mu(\delta_1,\ldots,\delta_n) = [\delta_1],\ldots,[\delta_n]$$

is a homomorphism. Moreover, μ extends to a homomorphism from the semidirect product

$$\widehat{\mu} : (\Lambda^*)^n \bullet \Sigma_n \to (\Lambda^*)^{ab} \quad \text{by } \widehat{\mu}(\delta_1,\ldots,\delta_n;\sigma) = [\text{sign}(\sigma)][\delta_1],\ldots,[\delta_n].$$

We define the *proto-determinant* $\text{prot}_n : C_n(\Lambda) \to (\Lambda^*)^{ab}$ to be the homomorphism which corresponds to $\widehat{\mu}$ under the isomorphism (2.57); that is:

$$\text{prot}_n (\Delta(\delta_1,\ldots,\delta_n) \cdot P(\sigma)) = [\text{sign}(\sigma)][\delta_1],\ldots,[\delta_n]. \tag{2.59}$$

The proto-determinant is compatible with stabilization. For any integers k,n with $1 \leq k$ the stabilization inclusion $s_{n,k} : GL_n(\Lambda) \to GL_{n+k}(\Lambda)$

$$s_{n,k}(A) = \begin{pmatrix} A & 0 \\ 0 & I_k \end{pmatrix}$$

induces the subsidiary inclusions $s_{n,k} : E_n(\Lambda) \to E_{n+k}(\Lambda)$, $s_{n,k} : \Delta_n(\Lambda) \to \Delta_{n+k}(\Lambda)$, $s_{n,k} : GE_n(\Lambda) \to GE_{n+k}(\Lambda)$, $s_{n,k} : \widehat{\Sigma}_n \to \widehat{\Sigma}_{n+k}$ and hence also $s_{n,k} : \mathcal{C}_n(\Lambda) \to \mathcal{C}_{n+k}(\Lambda)$ and the following diagram commutes for each k, n with $1 \leq k$

$$
\begin{array}{ccc}
\mathcal{C}_{n+k}(\Lambda) & \overset{\mathrm{prot}_{n+k}}{\to} & (\Lambda^*)^{ab} \\
\uparrow s_{n,k} & & \uparrow \mathrm{Id} \\
\mathcal{C}_n(\Lambda) & \overset{\mathrm{prot}_n}{\to} & (\Lambda^*)^{ab}
\end{array}
$$

By a *weak determinant* for Λ we mean a family $\{\det_n\}_{2 \leq n}$ of group homomorphisms

$$\det_n : GE_n(\Lambda) \to (\Lambda^*)^{ab}$$

such that

(i) $\det_n$ extends prot_n; that is $\det_{n \mid \mathcal{C}_n(\Lambda)} = \mathrm{prot}_n$;
(ii) the family $\{\det_n\}_{2 \leq n}$ is compatible with stabilization; that is, the diagram below commutes for each k, n with $1 \leq k$:

$$
\begin{array}{ccc}
GE_{n+k}(\Lambda) & \overset{\det_{n+k}}{\to} & (\Lambda^*)^{ab} \\
\uparrow s_{n,k} & & \uparrow \mathrm{Id} \\
GE_n(\Lambda) & \overset{\det_n}{\to} & (\Lambda^*)^{ab}
\end{array}
$$

Observe that:

Proposition 2.60 *If $\{\det_n\}_{2 \leq n}$ is a weak determinant for Λ then $\det_n(E) = 1$ for each $E \in E_n(\Lambda)$.*

Proof It suffices to show that $\det_n(E(i, j; \lambda)) = 1$ for each generator $E(i, j; \lambda)$. When $n \geq 3$ then, by (2.3), for some $r \leq n$, $E(i, j; \lambda) = [E(i, r; \lambda), E(r, j; 1)]$ so that

$$\det_n(E(i, j; \lambda)) = \det_n(E(i, r; \lambda))\det_n(E(r, j; 1))$$
$$\times \det_n(E(i, r; \lambda))^{-1}\det_n(E(r, j; 1))^{-1}$$

and the conclusion follows as $\det_n$ takes values in the abelian group $(\Lambda^*)^{ab}$. For $n = 2$, the result follows by stabilization as

$$\det_2(E(i, j; \lambda)) = \det_3(s_{3,2}(E(i, j; \lambda))) = 1. \qquad \square$$

It follows that a weak determinant for Λ is unique; that is:

Proposition 2.61 *If $\{\det_n\}_{2 \leq n}$ and $\{\det'_n\}_{2 \leq n}$ are weak determinants for Λ then $\det_n = \det'_n$ for each $n \geq 2$.*

Proof Write $X \in GE_n(\Lambda)$ in the form $X = C \cdot E$ where $C \in \mathcal{C}_n(\Lambda)$ and $E \in E_n(\Lambda)$. Then $\det_n(C) = \mathrm{prot}_n(C)$ and $\det_n(E) = 1$ so that

$$\det_n(X) = \det_n(C) \cdot \det_n(E) = \text{prot}_n(C).$$

Repeating the calculation gives $\det'_n(X) = \text{prot}_n(C)$ so that $\det'_n = \det_n$.   □

2.7 Equivalent Formulations of the Dieudonné Condition

We say that Λ is a *Dieudonné ring* when Λ possesses a weak determinant. In this section we derive a simple necessary and sufficient condition for Λ to be Dieudonné, namely that for each $n \geq 2$

$$D_n(\Lambda) \cap E_n(\Lambda) = D_n([\Lambda^*, \Lambda^*]),$$

where

$$D_n([\Lambda^*, \Lambda^*]) = \{D(1, h) : h \in [\Lambda^*, \Lambda^*]\}.$$

We shall also show that when $\{\det_n\}_{2 \leq n}$ is a weak determinant for Λ then the sequence

$$1 \to E_n(\Lambda) \subset GE_n(\Lambda) \xrightarrow{\det_n} (\Lambda^*)^{ab} \to 1$$

is exact. We begin by observing:

Proposition 2.62 $D_n([\Lambda^*, \Lambda^*]) \subset E_n(\Lambda)$ *for each* $n \geq 2$.

Proof First consider the case $n = 2$. Then for $\alpha, \beta \in \Lambda^*$,

$$\begin{bmatrix} \alpha\beta\alpha^{-1} & 0 \\ 0 & \beta^{-1} \end{bmatrix} = \begin{bmatrix} 1 & 0 \\ \alpha^{-1} & 1 \end{bmatrix} \begin{bmatrix} 1 & -\alpha \\ 0 & 1 \end{bmatrix} \begin{bmatrix} 1 & 0 \\ \alpha^{-1} & 1 \end{bmatrix} \begin{bmatrix} 1 & \alpha\beta^{-1} \\ 0 & 1 \end{bmatrix}$$
$$\times \begin{bmatrix} 1 & 0 \\ -\beta\alpha^{-1} & 1 \end{bmatrix} \begin{bmatrix} 1 & \alpha\beta^{-1} \\ 0 & 1 \end{bmatrix}$$

from which we see that:

$$\begin{bmatrix} \alpha\beta\alpha^{-1} & 0 \\ 0 & \beta^{-1} \end{bmatrix} \in E_2(\Lambda). \tag{I}$$

Replacing β by β^{-1} and taking $\alpha = 1$ we obtain the following, which is also a consequence of (2.9)

$$\begin{bmatrix} \beta^{-1} & 0 \\ 0 & \beta \end{bmatrix} \in E_2(\Lambda). \tag{II}$$

Taking the product $\begin{bmatrix} \alpha\beta\alpha^{-1}\beta^{-1} & 0 \\ 0 & 1 \end{bmatrix} = \begin{bmatrix} \alpha\beta\alpha^{-1} & 0 \\ 0 & \beta^{-1} \end{bmatrix} \begin{bmatrix} \beta^{-1} & 0 \\ 0 & \beta \end{bmatrix}$ we conclude that

$$\begin{bmatrix} \alpha\beta\alpha^{-1}\beta^{-1} & 0 \\ 0 & 1 \end{bmatrix} \in E_2(\Lambda). \tag{III}$$

The statement for $n = 2$ now follows as $[\Lambda^*, \Lambda^*]$ is generated by the basic commutators $[\alpha, \beta] = \alpha\beta\alpha^{-1}\beta^{-1}$ whilst for $n > 2$ the conclusion follows by stabilization as

$$D_n([\Lambda^*, \Lambda^*]) = s_{n,2}(D_2([\Lambda^*, \Lambda^*])) \subset s_{n,2}(E_2(\Lambda) \subset E_n(\Lambda). \qquad \square$$

It should cause no confusion to identify Λ^* with $D_n(\Lambda)$ and $[\Lambda^*, \Lambda^*]$ with $D_n([\Lambda^*, \Lambda^*])$. In particular we shall write

$$GE_n(\Lambda) = \Lambda^* \cdot E_n(\Lambda) \quad \text{and} \quad GE(\Lambda) = \Lambda^* \cdot E(\Lambda),$$

where

$$E(\Lambda) = \varinjlim E_n(\Lambda) \quad \text{and} \quad GE(\Lambda) = \varinjlim GE_n(\Lambda).$$

Whenever $2 \le k < n$ we have, by Proposition 2.62, a sequence of inclusions

$$[\Lambda^*, \Lambda^*] \subset \Lambda^* \cap E_k(\Lambda) \subset \Lambda^* \cap E_n(\Lambda) \subset \Lambda^* \cap E(\Lambda).$$

Let $\natural_n : (\Lambda^*)^{ab} \to GE_n(\Lambda)/E_n(\Lambda)$ be the composition $\natural_n = \nu_n \circ [\,]_n$ where

$$[\,]_n : \Lambda^*/[\Lambda^*, \Lambda^*] \to \Lambda^*/\Lambda^* \cap E_n(\Lambda)$$

is the natural surjection and

$$\nu_n : \Lambda^*/\Lambda^* \cap E_n(\Lambda) \to \Lambda^* E_n(\Lambda)/E_n(\Lambda) = GE_n(\Lambda)/E_n(\Lambda)$$

is the Noether isomorphism $\nu_n([\delta]_n) = \delta \cdot E_n(\Lambda)$. Let $\natural_\infty$ be the composition $\natural_\infty = \nu_\infty \circ [\,]_\infty : (\Lambda^*)^{ab} \to GE(\Lambda)/E(\Lambda)$ where $\nu_\infty = \varinjlim \nu_n$ and $[\,]_\infty = \varinjlim [\,]_n$. We get a commutative diagram with exact rows

$$\begin{cases} 1 \to \Lambda^* \cap E(\Lambda)/[\Lambda^*, \Lambda^*] \to (\Lambda^*)^{ab} \xrightarrow{\natural_\infty} GE(\Lambda)/E(\Lambda) \to 1 \\ \qquad\qquad \uparrow i_{\infty,n} \qquad\qquad\quad \|\mathrm{Id} \qquad\qquad \uparrow s_{\infty,n} \\ 1 \to \Lambda^* \cap E_n(\Lambda)/[\Lambda^*, \Lambda^*] \to (\Lambda^*)^{ab} \xrightarrow{\natural_n} GE_n(\Lambda)/E_n(\Lambda) \to 1 \qquad (2.66) \\ \qquad\qquad \uparrow i_{n,k} \qquad\qquad\quad \|\mathrm{Id} \qquad\qquad \uparrow s_{n,k} \\ 1 \to \Lambda^* \cap E_k(\Lambda)/[\Lambda^*, \Lambda^*] \to (\Lambda^*)^{ab} \xrightarrow{\natural_k} GE_k(\Lambda)/E_k(\Lambda) \to 1 \end{cases}$$

where the mappings $i_{\infty,n}$, $i_{n,k}$, $s_{\infty,n}$, $s_{n,k}$ are induced by stabilization. Note that $i_{\infty,n}$, $i_{n,k}$ are injective whilst $s_{\infty,n}$, $s_{n,k}$ are surjective.

The homomorphism $\Lambda^* \to K_1(\Lambda) = GL(\Lambda)/E(\Lambda)$; $\delta \mapsto D(1, \delta) \cdot E(\Lambda)$ induces a canonical mapping $i : (\Lambda^*)^{ab} \to K_1(\Lambda)$. As we now see, the Dieudonné condition on a ring Λ can be expressed in a number of ways. The conditions below are comprehensive without being exhaustive.

Theorem 2.64 *For any ring Λ the conditions below are equivalent:*

(i) *Λ admits a weak determinant;*
(ii) *$\Lambda^* \cap E_n(\Lambda) = [\Lambda^*, \Lambda^*]$ for each $n \ge 2$;*

(iii) $\Lambda^* \cap E(\Lambda) = [\Lambda^*, \Lambda^*]$;
(iv) *the canonical mapping* $\natural_n : (\Lambda^*)^{ab} \to GE_n(\Lambda)/E_n(\Lambda)$ *is an isomorphism for each* $n \geq 2$;
(v) *the canonical mapping* $\natural_\infty : (\Lambda^*)^{ab} \to GE(\Lambda)/E(\Lambda)$ *is an isomorphism*;
(vi) *the canonical mapping* $i : (\Lambda^*)^{ab} \to K_1(\Lambda)$ *is injective*.

Proof The equivalence of (ii) and (iii) follows directly from the definition of $E(\Lambda)$ as $\varinjlim E_n(\Lambda)$ whilst the equivalence of (ii) with (iv) and (iii) with (v) follows directly from the exactness of the rows in (2.66). Thus it suffices to show that (i) $\iff$ (ii) and (iii) $\iff$ (vi).

(i) $\implies$ (ii) Let $\{\det_n\}_{2 \leq n}$ be a weak determinant for Λ. As $\det_n(E) = 1$ for any $E \in E_n(\Lambda)$ then $\det_n$ induces a homomorphism $(\det_n)_* : GE_n(\Lambda)/E_n(\Lambda) \to (\Lambda^*)^{ab}$. Thus consider the diagram

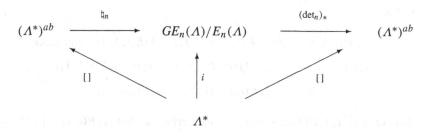

where $\natural_n$ is the homomorphism of (2.66) and $i : \Lambda^* \to GE_n(\Lambda)/E_n(\Lambda)$ is the homomorphism $i(\delta) = D_n(1, \delta) \cdot E_n(\Lambda)$. Computing we see that

$$\natural_n([\delta]) = D_n(1, \delta) \cdot E_n(\Lambda) \quad \text{and} \quad (\det_n)_*(D_n(1\delta) \cdot E_n(\Lambda)) = [\delta]$$

so that $(\det_n)_* \circ \natural_n = \mathrm{Id}_{|(\Lambda^*)^{ab}}$. In particular, $\natural_n$ is injective. However, by (2.66),

$$\mathrm{Ker}(\natural_n) = \Lambda^* \cap E_n(\Lambda)/[\Lambda^*, \Lambda^*]$$

and so $\Lambda^* \cap E_n(\Lambda) = [\Lambda^*, \Lambda^*]$. This proves (i) $\implies$ (ii).

(ii) $\implies$ (i) From the filtration

$$[\Lambda^*, \Lambda^*] \subset \Lambda^* \cap E_k(\Lambda) \subset \Lambda^* \cap E_n(\Lambda) \subset \Lambda^* \tag{I}$$

we obtain a commutative diagram with exact rows

$$\begin{cases} 1 \to E_n(\Lambda) \to GE_n(\Lambda) \xrightarrow{\gamma_n} \Lambda^*/\Lambda^* \cap E_n(\Lambda) \to 1 \\ \quad\quad \uparrow s_{n,k} \quad\quad \uparrow s_{n,k} \quad\quad\quad \uparrow \sigma_{n,k} \\ 1 \to E_k(\Lambda) \to GE_k(\Lambda) \xrightarrow{\gamma_k} \Lambda^*/\Lambda^* \cap E_k(\Lambda) \to 1 \\ \quad\quad \uparrow s_{k,1} \quad\quad \uparrow s_{k,1} \quad\quad\quad \uparrow \sigma_{k,1} \\ 1 \to [\Lambda^*, \Lambda^*] \to \quad \Lambda^* \quad \xrightarrow{[]} \Lambda^*/[\Lambda^*, \Lambda^*] \to 1 \end{cases} \tag{II}$$

in which γ_n is the composition $\gamma_n = \nu^{-1} \circ \langle \rangle_n$ where $\langle \rangle_n : GE_n(\Lambda) \to GE_n(\Lambda)/E_n(\Lambda)$ is the canonical mapping and $\nu^{-1} : GE_n(\Lambda)/E_n(\Lambda) \to \Lambda^*/\Lambda^* \cap E_n(\Lambda)$

is the inverse of the Noether isomorphism already considered. Here the mappings $s_{n,k}$ are induced by stabilization and the $\sigma_{n,k}$ are the appropriate quotient mappings from the filtration (i). By hypothesis $\Lambda^* \cap E_n(\Lambda) = [\Lambda^*, \Lambda^*]$ so that each $\sigma_{n,k}$ is the identity and (II) becomes

$$
\begin{cases}
1 \to & E_n(\Lambda) \to GE_n(\Lambda) \xrightarrow{\gamma_n} (\Lambda^*)^{ab} \to 1 \\
& \uparrow s_{n,k} \qquad \uparrow s_{n,k} \qquad \| \mathrm{Id} \\
1 \to & E_k(\Lambda) \to GE_k(\Lambda) \xrightarrow{\gamma_k} (\Lambda^*)^{ab} \to 1 \\
& \uparrow s_{k,1} \qquad \uparrow s_{k,1} \qquad \| \mathrm{Id} \\
1 \to [\Lambda^*, \Lambda^*] \to & \Lambda^* \xrightarrow{[\,]} (\Lambda^*)^{ab} \to 1
\end{cases}
\qquad \text{(III)}
$$

Evidently γ_k is a homomorphism and we claim it extends prot_k. To see this, first note that $s_{k,1}(\delta) = D_k(1, \delta)$ so that, by commutativity, $\gamma_k(s_{k,1}(\delta)) = [\delta]$. By (2.11) we have

$$
P(i, 1) = s_{k,1}(-1)E(1, 1; 1)E(1, i; -1)E(i, 1; 1) \quad \text{and hence}
$$

$$
\gamma_k(P(i, 1)) = \gamma_k(s_{k,1}(-1))\gamma_k(E(i, 1; 1)E(1, i; -1)E(i, 1; 1))
$$

$$
= [-1]\gamma_k(E(i, 1; 1)E(1, i; -1)E(i, 1; 1)).
$$

However $E(i, 1; 1)E(1, i; -1)E(i, 1; 1) \in \mathrm{Ker}(\gamma_k)$ so that $\gamma_k(P(i, 1) = [-1]$. Now $P(i, j) = P(i, 1)P(j, 1)P(i, 1)$ so that .

$$
\gamma_k(P(i, j)) = \gamma_k(P(i, 1))\gamma_k(P(j, 1))\gamma_k(P(i, 1)) = [-1][-1][-1] = [-1].
$$

It follows by induction that $\gamma_k(P(\sigma) = [\mathrm{sign}\,\sigma]$. Also $D_k(r, \delta) = P(1, r)D_k(1, \delta) \times P(1, r)$ so that

$$
\gamma_k(D_k(r, \delta)) = \gamma_k(P(1, r))\gamma_k(D_k(1, \delta))\gamma_k(P(1, r)) = [-1][\delta][-1] = [\delta].
$$

However $\Delta(\delta_1, \ldots, \delta_k) = \prod_{r=1}^{k} D_k(r, \delta_r)$, hence $\gamma_k(\Delta(\delta_1, \ldots, \delta_k)) = [\delta_1][\delta_2] \cdots [\delta_k]$. Thus

$$
\gamma_k(\Delta(\delta_1, \ldots, \delta_n)P(\sigma)) = [\mathrm{sign}(\sigma)][\delta_1][\delta_2] \cdots [\delta_k]
$$

and so γ_k is a homomorphism extending prot_k. Moreover, as is clear from (III) above, $\{\gamma_n\}_{2 \leq n}$ is compatible with stabilization; that is $\{\gamma_n\}_{2 \leq n}$ is a weak determinant for Λ and so (ii) $\Longrightarrow$ (i).

(iii) $\Longleftrightarrow$ (vi) The canonical mapping $i : (\Lambda^*)^{ab} \to K_1(\Lambda)$ is simply the composition

$$
(\Lambda^*)^{ab} \xrightarrow{\natural_\infty} GE(\Lambda)/E(\Lambda) \subset GL(\Lambda)/E(\Lambda)
$$

so that, by exactness of the rows in (2.66), $\mathrm{Ker}(i) = \mathrm{Ker}(\natural_\infty) = \Lambda^* \cap E(\Lambda)/[\Lambda^*, \Lambda^*]$. In particular, $\Lambda^* \cap E(\Lambda) = [\Lambda^*, \Lambda^*] \Longleftrightarrow i$ is injective. This proves (iii) $\Longleftrightarrow$ (vi) and completes the proof. $\qquad \square$

The proof of Theorem 2.64 shows more than the formal statement. As there is at most one weak determinant for Λ, the proof that (ii) $\Longrightarrow$ (i) and, in particular, the diagram (III) shows that:

Corollary 2.65 *If* $\{\det_n\}_{2 \geq n}$ *is a weak determinant for* Λ *then for each* n *the sequence* $1 \to E_n(\Lambda) \to GE_n(\Lambda) \overset{\det_n}{\to} (\Lambda^*)^{ab} \to 1$ *is exact.*

We note that the condition '$\Lambda^* \cap E_2(\Lambda) = [\Lambda^*, \Lambda^*]$' occurs, albeit obliquely, in Cohn's study of GL_2 [16]. Cohn shows that for those rings Λ which are 'universal for GE_2' there is an isomorphism $GE_2(\Lambda)/E_2(\Lambda) \cong (\Lambda^*)^{ab}$ ([16], Theorem (9.1)). Although not expressed as such, his proof may be re-arranged to give the condition '$\Lambda^* \cap E_2(\Lambda) = [\Lambda^*, \Lambda^*]$' directly.

2.8 A Recognition Criterion for Dieudonné Rings

We begin with a straightforward group-theoretic observation:

Proposition 2.66 *Let* $\varphi : G \to H$ *be a surjective group homomorphism; then*

$$\varphi^{ab} : G^{ab} \to H^{ab} \text{ is an isomorphism} \quad \Longleftrightarrow \quad \mathrm{Ker}(\varphi) \subset [G, G].$$

Proof ($\Longrightarrow$) First note the inclusions

$$[G, G] \subset \varphi^{-1}([H, H]) \subset G, \tag{$*$}$$

$$\mathrm{Ker}(\varphi) \subset \varphi^{-1}([H, H]) \subset G. \tag{$**$}$$

We have a Noether isomorphism $\widehat{\varphi} : G/\varphi^{-1}([H, H]) \to H/[H, H] = H^{ab}$ whilst ($**$) gives an exact sequence

$$1 \to \varphi^{-1}([H, H])/[G, G] \to G/[G, G] \overset{\nu}{\to} G/\varphi^{-1}([H, H]) \to 1.$$

Moreover, the induced map $\varphi^{ab} : G^{ab} \to H^{ab}$ is the composition $\varphi^{ab} = \widehat{\varphi} \circ \nu$. Thus we have an exact sequence

$$1 \to \varphi^{-1}([H, H])/[G, G] \to G^{ab} \overset{\varphi^{ab}}{\longrightarrow} H^{ab} \to 1$$

in which, by hypothesis, φ^{ab} is an isomorphism. Hence $[G, G] = \varphi^{-1}([H, H])$ and so $\mathrm{Ker}(\varphi) \subset [G, G]$ by ($**$). This proves ($\Longrightarrow$).

($\Longleftarrow$) Conversely suppose that $\mathrm{Ker}(\varphi) \subset [G, G]$. Then we have an exact sequence

$$1 \to [G, G]/\mathrm{Ker}(\varphi) \to G/\mathrm{Ker}(\varphi) \overset{\mu}{\longrightarrow} G^{ab} \to 1.$$

As φ is surjective there is a Noether isomorphism $\varphi_* : G/\mathrm{Ker}(\varphi) \to H$. Define

$$v = \mu \circ \varphi_*^{-1} : H \to G^{ab}.$$

Then v is surjective and the diagram below commutes where $\natural_G, \natural_H$ are the canonical homomorphisms:

$$
\begin{array}{ccc}
G & \overset{\natural_G}{\to} & G^{ab} \\
\varphi \downarrow \ v \nearrow & & \downarrow \varphi^{ab} \\
H & \overset{\natural_H}{\to} & H^{ab}
\end{array}
$$

Let $v^{ab} : H^{ab} \to G^{ab}$ be the homomorphism induced from v via the universal property for abelianization. Then v^{ab} is surjective. Moreover $\varphi^{ab} \circ v = \natural_H \implies \varphi^{ab} \circ v^{ab} = \mathrm{Id}$ so that v^{ab} is also injective and hence is an isomorphism with $(v^{ab})^{-1} = \varphi^{ab}$. In particular, φ^{ab} is an isomorphism. This proves ($\Longleftarrow$) and completes the proof.   □

In what follows we adopt the convention that $\prod_{r=1}^{N} \lambda_r$ means $\lambda_1 \cdots \lambda_N$; moreover, for a ring homomorphism $\varphi : A \to B$, φ_u will denote the induced map on units

$$\varphi_u = \varphi_{|A^*} : A^* \to B^*.$$

We have a Recognition Criterion for Dieudonné rings.

Theorem 2.67 *Let $\varphi : A \to B$ be a ring homomorphism such that*

(i) *$\varphi_u : A^* \to B^*$ is surjective and*
(ii) *$\varphi_u^{ab} : (A^*)^{ab} \to (B^*)^{ab}$ is an isomorphism.*

If B is a Dieudonné ring then so also is A.

Proof By Theorem 2.64 we must show $[A^*, A^*] = A^* \cap E(A)$. However, $[A^*, A^*] \subset A^* \cap E(A)$ by Proposition 2.62 so it suffices to show that $A^* \cap E(A) \subset [A^*, A^*]$. Thus let $\delta \in A^* \cap E(A)$. Then $\varphi(\delta) \in B^* \cap E(B)$. However, B is a Dieudonné ring so that, by Theorem 2.64, we may write

$$\varphi(\delta) = \prod_{r=1}^{N} [\alpha_r, \beta_r]$$

with $\alpha_r, \beta_r \in B^*$. Now $\varphi_u : A^* \to B^*$ is surjective so that we may choose $\widehat{\alpha_r}, \widehat{\beta_b} \in A^*$ such that $\varphi(\widehat{\alpha_r}) = \alpha_r$ and $\varphi(\widehat{\beta_r}) = \beta_r$. Put

$$\gamma = \prod_{r=1}^{N} [\widehat{\alpha_r}, \widehat{\beta_r}] \in [A^*, A^*].$$

Then $\varphi_u(\gamma^{-1}\delta) = 1$; that is, $\gamma^{-1}\delta \in \text{Ker}(\varphi_u)$. However, φ_u^{ab} is an isomorphism so that, by Proposition 2.66, $\text{Ker}(\varphi_u) \subset [A^*, A^*]$. Hence we may write $\delta = \gamma\eta$ for some $\eta \in [A^*, A^*]$. Thus $\delta \in [A^*, A^*]$ as $\gamma, \eta \in [A^*, A^*]$. □

The standard theory of the determinant over a commutative ring shows that:

$$\text{Any commutative ring is a Dieudonné ring.} \qquad (2.68)$$

The theorem of Dieudonné [21] shows:

$$\text{Any division ring is a Dieudonné ring.} \qquad (2.69)$$

We give a complete proof of (2.69) in Appendix A. More interestingly, the group rings of certain infinite groups satisfy the Dieudonné condition. For example:

Proposition 2.70 *Let G be a finitely generated $\mathcal{TUP}$ group such that $H_1(G, \mathbf{Z})$ is torsion free. Then for any commutative integral domain A the group ring $A[G]$ satisfies the Dieudonné condition.*

Proof As noted in Theorem 2.51, the $\mathcal{TUP}$ condition guarantees that $A[G]$ has only trivial units; that is

$$A[G]^* \cong A^* \times G.$$

Let $\nu : A[G] \to A[G^{ab}]$ be the canonical mapping. As $G^{ab} \cong H_1(G; \mathbf{Z})$ is finitely generated and torsion free then G^{ab} is a free abelian group of finite rank and so also satisfies the $\mathcal{TUP}$ condition (see Appendix C). In particular, $A[G^{ab}]$ also has only trivial units; that is

$$A[G^{ab}]^* \cong A^* \times G^{ab}.$$

The canonical mapping ν thus induces a surjection $\nu : A[G]^* \to A[G^{ab}]^*$ and an isomorphism

$$\nu : (A[G]^*)^{ab} \xrightarrow{\sim} A[G^{ab}]^*.$$

Moreover, $A[G^{ab}]^*$ is its own abelianization. As $A[G^{ab}]$ is commutative it satisfies the Dieudonné condition. Thus $A[G]$ also satisfies the Dieudonné condition by Theorem 2.67. □

As examples of groups G satisfying the hypotheses of Proposition 2.70 we may take, for example, any finite product $G = G_1 \times \cdots \times G_N$ where G_i is either a free group or the fundamental group of an orientable surface (cf. Appendix C).

2.9 Fully Determinantal Rings

To a ring Λ is associated a canonical homomorphism $i : (\Lambda^*)^{ab} \to K_1(\Lambda)$. We shall say that Λ is *fully determinantal* when i admits a left inverse. Given such a left inverse $\delta : K_1(\Lambda) \to (\Lambda^*)^{ab}$ then composition with the canonical mappings

$v_n : GL_n(\Lambda) \to K_1(\Lambda)$ gives a family $\{\delta_n\}_{2 \leq n}$ of group homomorphisms

$$\delta_n = \delta \circ v_n : GL_n(\Lambda) \to (\Lambda^*)^{ab}$$

making the following diagram commute for each k, n with $1 \leq k$:

$$GL_{n+k}(\Lambda) \overset{\delta_{n+k}}{\to} (\Lambda^*)^{ab}$$
$$\uparrow s_{n,k} \qquad\qquad \uparrow \text{Id}$$
$$GL_n(\Lambda) \overset{\delta_n}{\to} (\Lambda^*)^{ab}$$

The family $\{\delta_n\}_{2 \leq n}$ is then said to be a *full determinant* for Λ. As i has a left inverse it is injective. The restriction $\delta_n : GE_n(\Lambda) \to (\Lambda^*)^{ab}$ is then the (unique) weak determinant of Λ which exists by virtue of the injectivity of i; hence δ_n extends $\text{prot}_n : C_n(\Lambda) \to (\Lambda^*)^{ab}$. Clearly one has:

If Λ is Dieudonné and weakly Euclidean then Λ is fully determinantal.   (2.71)

In particular, as division rings are both Dieudonné and weakly Euclidean then

Any division ring is fully determinantal.   (2.72)

We note that a commutative ring Λ is fully determinantal as there is a natural choice of left inverse for i induced from the standard determinant construction. However, in contrast to weak determinants, full determinants, where they exist, need not necessarily be unique. This may be true even in the commutative case. In general, if $i : (\Lambda^*)^{ab} \to K_1(\Lambda)$ has a left inverse then any full determinant of Λ is unique only when there is no nontrivial group homomorphism $\text{Coker}(i) \to (\Lambda^*)^{ab}$.

In conjunction with the Dieudonné condition the additional condition of being weakly Euclidean is sufficient but not necessary for the possession of a full determinant. Every commutative ring is fully determinantal but very few are weakly Euclidean. Less trivially, as the following shows, there are examples of noncommutative rings which are fully determinantal and, in general, very far from being weakly Euclidean.

Theorem 2.73 *If A is a commutative integral domain then the group ring $A[F_N]$ is fully determinantal.*

Proof Let k be the field of fractions of A. As both F_N and C_∞^N satisfy the $\mathcal{TUP}$ condition then $k[F_N]^* \cong k^* \times F_N$ and $k[C_\infty^N]^* \cong k^* \times C_\infty^N$ so that $(k[F_N]^*)^{ab} \cong k^* \times C_\infty^N$. We saw in Proposition 2.70 that $k[F_N]$ satisfies the Dieudonné condition. Moreover, by the theorem of Cohn Theorem 2.48 $k[F_N]$ is weakly Euclidean. Thus

$k[F_N]$ is fully determinantal; moreover, Proposition 2.61 and the weakly Euclidean condition guarantee the determinant is unique.

Let $\delta_m : GL_m(k[F_N]) \to (k[F_N]^*)^{ab}$ be the determinant; there is a commutative diagram

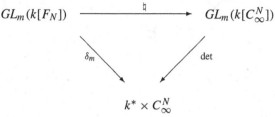

where $\natural : GL_m(k[F_N]) \to GL_m(k[C_\infty^N])$ be the homomorphism induced from the abelianization $F_N \to C_\infty^N$. Evidently this imbeds in a larger commutative diagram

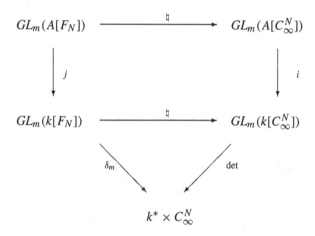

As $A[C_\infty^N]$ is commutative, if $X \in GL_m(A[C_\infty^N])$ then $\det(X) \in A[C_\infty^N]^*$. However, again by the $\mathcal{TUP}$ condition, $A[C_\infty^N]^* \cong A^* \times C_\infty^N$. From the commutativity of the above diagram, if $X \in GL_m(A[F_N])$ then $\delta_m(j(X)) \in A[C_\infty^N]^* = A^* \times C_\infty^N$. Once more by the $\mathcal{TUP}$ condition $(A[F_N]^*)^{ab} \cong A^* \times C_\infty^N$. We define a homomorphism $\delta'_m : GL_m(A[F_N]) \to (A[F_N]^*)^{ab}$ by

$$\delta'_m = \delta_m \circ j.$$

Again by Proposition 2.70, $A[F_N]$ satisfies the Dieudonné condition and it straightforward to see that δ'_m extends the weak determinant $d_m : GE_m(A[F_N]) \to (A[F_N]^*)^{ab}$. Finally the diagram below commutes

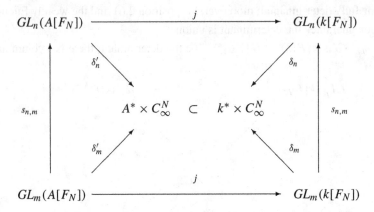

in consequence of which $A[F_N]$ is fully determinantal as claimed.   □

Chapter 3
The Calculus of Corners and Squares

This chapter is a systematic study of projective modules via the decomposition of rings into fibre squares. The genesis of this approach was geometric, namely 'Mayer-Vietoris patching' for vector bundles. Its translation into pure algebra was effected by Milnor in [73] and subsequently extended, notably by Karoubi [63]. In its original formulation, the method aimed to describe the stable structure of projective modules. However, our treatment is heavily influenced by that of Swan in [94], wherein the method is adapted to the dual 'unstable' theory; that is, the study of what is lost on stabilization.

3.1 The Category of Corners

By a *corner* A we mean a quintuple $A = (A_+, A_-, A_0, \varphi_+, \varphi_-)$ where A_+, A_-, A_0 are rings and $\varphi_+ : A_+ \to A_0$, $\varphi_- : A_- \to A_0$ are ring homomorphisms. We portray a corner conventionally as

$$
A = \left\{
\begin{array}{cc}
 & A_- \\
\cdot & \downarrow \varphi_- \\
A_+ & \overset{\varphi_+}{\to} A_0
\end{array}
\right.
$$

If $B = (B_+, B_-, B_0, \psi_+, \psi_-)$ is also a corner then by a corner morphism $h : A \to B$ we mean a triple $h = (h_+, h_-, h_0)$ of ring homomorphisms $h_\sigma : A_\sigma \to B_\sigma$ making the following diagram commute.

F.E.A. Johnson, *Syzygies and Homotopy Theory*, Algebra and Applications 17, DOI 10.1007/978-1-4471-2294-4_3, © Springer-Verlag London Limited 2012

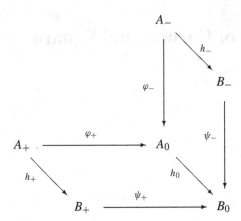

There is evidently a category **Corners** whose objects are corners and whose morphisms are as described above. A group valued functor $\mathcal{G} : \textbf{Rings} \rightarrow \textbf{Groups}$ applied to a corner $\mathcal{A}$ gives a diagram of group homomorphisms

$$
\mathcal{G}(\mathcal{A}) = \begin{cases}
& \mathcal{G}(A_-) \\
& \downarrow \mathcal{G}(\varphi_-) \\
\mathcal{G}(A_+) & \overset{\mathcal{G}(\varphi_+)}{\rightarrow} \mathcal{G}(A_0)
\end{cases}
$$

We write

$$\overline{\mathcal{G}}(\mathcal{A}) = \operatorname{Im} \mathcal{G}(\varphi_+)\backslash \mathcal{G}(A_0)/\operatorname{Im} \mathcal{G}(\varphi_-).$$

The set $\overline{\mathcal{G}}(\mathcal{A})$ has a distinguished point, denoted by $*$, namely the class of the identity from $\mathcal{G}(A_0)$. The correspondence $\mathcal{A} \mapsto \overline{\mathcal{G}}(\mathcal{A})$ defines a functor $\overline{\mathcal{G}} : \textbf{Corners} \rightarrow \textbf{Sets}_*$ to the category $\textbf{Sets}_*$ of based sets and basepoint preserving maps. When φ_+, φ_- are understood we write this as $\overline{\mathcal{G}}(\mathcal{A}) = \mathcal{G}(A_+)\backslash\mathcal{G}(A_0)/\mathcal{G}(A_-)$.

The functors we use most often are $\mathcal{G} = GL_n, E_n, GL \, (= \varinjlim GL_n), E \, (= \varinjlim E_n)$. However, over any ring Λ

$$GL_n(\Lambda) \cdot E_{n+k}(\Lambda) = E_{n+k}(\Lambda) \cdot GL_n(\Lambda);$$

thus writing $GE_{n,k}(\Lambda) = GL_n(\Lambda) \cdot E_{n+k}(\Lambda)$ we see that $GE_{n,k}(\Lambda)$ is a subgroup of $GL_{n+k}(\Lambda)$ and we shall also employ the functor $\Lambda \mapsto GE_{n,k}(\Lambda)$. For $k \geq 1$ the stabilization operator

$$X \mapsto \begin{pmatrix} X & 0 \\ 0 & I_k \end{pmatrix}$$

induces mappings $s_{n,k} : \overline{GL_n}(\mathcal{A}) \rightarrow \overline{GL_{n+k}}(\mathcal{A})$ and $s_{n,k} : \overline{E_n}(\mathcal{A}) \rightarrow \overline{E_{n+k}}(\mathcal{A})$. These satisfy $s_{n,m} = s_{n+k,m-k} \circ s_{n,k}$ for $1 \leq k < m$. In addition, there is stabilization map $s'_{n,k} : \overline{GL_n}(\mathcal{A}) \rightarrow \overline{GE_{n,k}}(\mathcal{A})$.

3.2 Modules over a Corner

By a module over a corner $\mathcal{A}$ we mean a triple $\mathcal{M} = (M_+, M_-, \alpha)$ where M_+, M_- are modules over A_+ A_- respectively with the property that $M_+ \otimes A_0 \cong M_- \otimes A_0$ and where $\alpha : M_- \otimes A_0 \to M_+ \otimes A_0$ is a specific A_0-isomorphism. If $\mathcal{N} = (N_+, N_-, \beta)$ is also a module over $\mathcal{A}$ then by an $\mathcal{A}$-morphism $f : \mathcal{M} \to \mathcal{N}$ we mean a pair $f = (f_+, f_-)$ where $f_\sigma : M_\sigma \to N_\sigma$ is a homomorphism over A_σ such that the following commutes:

$$
\begin{array}{ccc}
M_- \otimes A_0 & \xrightarrow{f_- \otimes \mathrm{Id}} & N_- \otimes A_0 \\
\downarrow \alpha & & \downarrow \beta \\
M_+ \otimes A_0 & \xrightarrow{f_+ \otimes \mathrm{Id}} & N_+ \otimes A_0
\end{array}
$$

There is a category $\mathcal{M}od_{\mathcal{A}}$ whose objects are modules over the corner $\mathcal{A}$ and whose morphisms are as described above. Observe that $\mathcal{M}od_{\mathcal{A}}$ has coproducts given by

$$
\mathcal{M} \oplus \mathcal{N} = (M_+ \oplus N_+, M_- \oplus N_-; \alpha \oplus \beta); \qquad f \oplus g = (f_+ \oplus g_+, f_- \oplus g_-).
$$

We may transfer properties of A_σ-modules to $\mathcal{A}$-modules in an obvious way; thus say that an $\mathcal{A}$-module $\mathcal{M} = (M_+, M_-; \alpha)$ is *finitely generated* when M_σ is finitely generated over A_σ for $\sigma = +, -$. Similarly we may transfer the notion of exactness. Firstly observe that $(0, 0; \mathrm{Id})$ is a zero object in the category of $\mathcal{A}$-modules in the formal sense of category theory. Now suppose given a sequence $\mathcal{E}$ of $\mathcal{A}$-modules thus:

$$
\mathcal{E} = (0 \to (K_+, K_-; \gamma) \overset{(i_+, i_-)}{\to} (M_+, M_-; \alpha) \overset{(p_+, p_-)}{\to} (Q_+, Q_-; \beta) \to 0);
$$

then we say that $\mathcal{E}$ is a short exact sequence of $\mathcal{A}$-modules when, for $\sigma = +, -$ the sequences

$$
0 \to K_\sigma \overset{i_\sigma}{\to} M_\sigma \overset{p_\sigma}{\to} Q_\sigma \to 0 \quad \text{and} \quad 0 \to K_\sigma \otimes A_0 \overset{i_\sigma}{\to} M_\sigma \otimes A_0 \overset{p_\sigma}{\to} Q_\sigma \otimes A_0 \to 0
$$

are all exact. Here a certain degree of circumspection is called for as we cannot, a priori, guarantee that $- \otimes A_0$ is an exact functor. The following condition will be sufficient for our purposes.

Proposition 3.1 *If the sequences* $0 \to K_\sigma \overset{i_\sigma}{\to} M_\sigma \overset{p_\sigma}{\to} Q_\sigma \to 0$ *are split exact for* $\sigma = +, -$ *then* $\mathcal{E}$ *is a short exact sequence of $\mathcal{A}$-modules.*

The corner $\mathcal{A}$ gives rise to a family of finitely generated modules over itself as follows; note that there is a canonical isomorphism $\natural : A_+ \otimes A_0 \overset{\cong}{\longrightarrow} A_- \otimes A_0$ which, by (a slight) abuse of notation, we may confuse with $\mathrm{Id} : A_0 \to A_0$; then for $n \geq 1$ write $\mathbf{F}_n(\mathcal{A}) = (A_+^n, A_-^n; \mathrm{Id}_n)$; we refer to $\mathbf{F}_n(\mathcal{A})$ as the *global free module* of rank n over $\mathcal{A}$. A finitely generated $\mathcal{A}$ module $\mathcal{P}$ will be said to be *globally projective* when there is an isomorphism $\mathcal{P} \oplus \mathcal{Q} \cong \mathbf{F}_n(\mathcal{A})$ for some n and some $\mathcal{A}$ module $\mathcal{Q}$.

3.3 Classification of Modules Within a Local Type

For an $\mathcal{A}$-module $\mathcal{M} = (M_+, M_-; \alpha)$ the pair (M_+, M_-) is called the *local type* of $\mathcal{M}$. The following shows that local type is an isomorphism invariant.

Proposition 3.2 *Let* $\mathcal{M} = (M_+, M_-; \alpha)$, $\mathcal{N} = (N_+, N_-; \beta)$ *be modules over the corner* $\mathcal{A}$; *then* $\mathcal{M} \cong_{\mathcal{A}} \mathcal{N} \Longrightarrow M_\sigma \cong_{A_\sigma} N_\sigma$ *for* $\sigma = +, -$.

Proof Let $f = (f_+, f_-) : (M_+, M_-; \alpha) \to (N_+, N_-; \beta)$ be an $\mathcal{A}$-isomorphism with inverse $g = (g_+, g_-) : (N_+, N_-; \beta) \to (M_+, M_-; \alpha)$. Then $f_\sigma : M_\sigma \to N_\sigma$ is an A_σ-isomorphism with inverse $g_\sigma : N_\sigma \to M_\sigma$.   □

For a local type (M_+, M_-) put

$$\mathrm{Iso}(M_- \otimes A_0, M_+ \otimes A_0) = \left\{ \begin{array}{l} \alpha : M_- \otimes A_0 \to M_+ \otimes A_0 \text{ such} \\ \text{that } \alpha \text{ is an } A_0 \text{ isomorphism} \end{array} \right\}.$$

There is a two-sided action

$$\mathrm{Aut}_{A_+}(M_+) \times \mathrm{Iso}(M_- \otimes A_0, M_+ \otimes A_0) \times \mathrm{Aut}_{A_-}(M_-) \to \mathrm{Iso}(M_- \otimes A_0, M_+ \otimes A_0)$$
$$(h_+, \alpha, h_-) \qquad\qquad\qquad \mapsto \qquad [h_+] \circ \alpha \circ [h_-]$$

where we write $[h_\sigma] = h_\sigma \otimes 1 : M_\sigma \otimes A_0 \to M_\sigma \otimes A_0$. Then if $\mathcal{L}(M_+, M_-)$ denotes the set of isomorphism classes of $\mathcal{A}$-modules of local type (M_+, M_-) there is evidently a surjective mapping

$$\natural : \mathrm{Iso}(M_- \otimes A_0, M_+ \otimes A_0) \to \quad \mathcal{L}(M_+, M_-)$$
$$\natural(\alpha) \qquad\qquad = \quad [(M_+, M_-, \alpha)]$$

It is straightforward to see that:

Proposition 3.3 $\natural$ *induces a bijection*

$$\natural : \mathrm{Aut}_{A_+}(M_+) \backslash \mathrm{Iso}(M_- \otimes A_0, M_+ \otimes A_0) / \mathrm{Aut}_{A_-}(M_-) \longrightarrow \mathcal{L}(M_+, M_-).$$

Evidently Proposition 3.3 gives a complete classification of $\mathcal{A}$-modules within a local type.

3.4 Locally Projective Modules and the Patching Condition

We say that an $\mathcal{A}$-module $\mathcal{L}$ is *locally free* when $\mathcal{L} \cong (A_+^n, A_-^n, \alpha)$ for some $\alpha \in GL_n(A_0)$. Likewise an $\mathcal{A}$-module $\mathcal{M} = (M_+, M_-; \alpha)$ is *locally projective* when M_σ is projective over A_σ for $\sigma = +, -$. Recall that a finitely generated $\mathcal{A}$ module $\mathcal{P}$ is *globally projective* when there is an isomorphism $\mathcal{P} \oplus \mathcal{Q} \cong (A_+^n, A_-^n, I_n)$ for some n and some $\mathcal{A}$ module $\mathcal{Q}$. Clearly we have:

A globally projective $\mathcal{A}$-module is locally projective.   (3.4)

Although the converse is false in general, it is, as we will show later in this chapter, true under favourable conditions which are relatively easy to satisfy. We begin with a useful simplification:

Proposition 3.5 Let $\mathcal{P} = (P_+, P_-; \alpha)$ be a finitely generated locally projective $\mathcal{A}$-module; then there exists a finitely generated locally projective $\mathcal{A}$-module $\mathcal{Q} = (Q_+, Q_-; \beta)$ such that $\mathcal{P} \oplus \mathcal{Q}$ is locally free.

Proof Suppose that $\mathcal{P} = (P_+, P_-; \alpha)$ is a finitely generated locally projective $\mathcal{A}$-module. We may choose projective modules $\widetilde{K}_+, \widetilde{K}_-$ over A_+, A_- respectively so that for some positive integers a, b,

$$P_+ \oplus \widetilde{K}_+ \cong A_+^a; \qquad P_- \oplus \widetilde{K}_- \cong A_-^b.$$

Putting $K_+ = \widetilde{K}_+ \oplus A_+^b$ and $K_- = \widetilde{K}_- \oplus A_-^a$ we see that $P_\sigma \oplus K_\sigma \cong A_\sigma^{a+b}$ for $\sigma = +, -$. So also $(P_\sigma \otimes A_0) \oplus (K_\sigma \otimes A_0) \cong A_0^{a+b}$ for $\sigma = +, -$. However $P_+ \otimes A_0 \cong P_- \otimes A_0$; putting $Q_\sigma = K_\sigma \oplus A_\sigma^{a+b}$ we see by Schanuel's Lemma that $Q_+ \otimes A_0 \cong Q_- \otimes A_0$. Choose an isomorphism $\beta : Q_- \otimes A_0 \xrightarrow{\simeq} Q_+ \otimes A_0$; then $\mathcal{Q} = (Q_+, Q_-; \beta)$ is a locally projective $\mathcal{A}$-module and

$$\mathcal{P} \oplus \mathcal{Q} \cong (P_+ \oplus Q_+, P_- \oplus Q_-; \alpha \oplus \beta).$$

However, $P_\sigma \oplus Q_\sigma \cong A_\sigma^{2(a+b)}$ so that $(P_+ \oplus Q_+, P_- \oplus Q_-; \alpha \oplus \beta)$ is locally free. $\square$

The 'finite generation' hypotheses in Proposition 3.5 may be dropped with suitable modifications which we leave to the reader. Next we observe that the classification of modules within a local type given by Proposition 3.3 becomes slightly more concrete when we classify locally free modules. Given a corner $\mathcal{A}$ we denote by

$$\mathcal{LF}_n(\mathcal{A}) = \left\{ \begin{matrix} \text{Isomorphism classes of } \mathcal{A}\text{-modules of the form} \\ (A_+^n, A_-^n; \alpha) \text{ where } \alpha \in \text{Iso}(A_-^n \otimes A_0, A_+^n \otimes A_0) \end{matrix} \right\}.$$

Identifying $A_+ \otimes A_0 = A_0 = A_- \otimes A_0$ we have $\text{Iso}(A_-^n \otimes A_0, A_+^n \otimes A_0) = GL_n(A_0)$; then Proposition 3.3 reduces to a bijective correspondence:

$$\nu_n : \overline{GL_n}(\mathcal{A}) = GL_n(A_+)\backslash GL_n(A_0)/GL_n(A_-) \xrightarrow{\simeq} \mathcal{LF}_n(\mathcal{A}). \qquad (3.6)$$

Likewise there are stabilization operators $\sigma_{n,k} : \mathcal{LF}_n(\mathcal{A}) \to \mathcal{LF}_{n+k}(\mathcal{A})$ induced from the correspondence $\mathcal{P} \mapsto \mathcal{P} \oplus (A_+^k, A_-^k; I_k)$. Moreover the diagram below commutes:

$$\begin{matrix} \overline{GL_n}(\mathcal{A}) & \xrightarrow{s_{n,k}} & \overline{GL_{n+k}}(\mathcal{A}) \\ \nu_n \downarrow & & \nu_{n+k} \downarrow \\ \mathcal{LF}_n(\mathcal{A}) & \xrightarrow{\sigma_{n,k}} & \mathcal{LF}_{n+k}(\mathcal{A}) \end{matrix} \qquad (3.7)$$

Now consider the following condition on corners $\mathcal{A}$;

$Patch$: For each integer $n \geq 1$ and each $\alpha \in GL_n(A_0)$ there exists $k \geq 1$ and $\beta \in GL_k(A_0)$ such that $\alpha \oplus \beta = [h_+][h_-]$ for some $h_+ \in GL_{n+k}(A_+)$ and $h_- \in GL_{n+k}(A_-)$.

Theorem 3.8 *For any corner A the following conditions are equivalent*:

(i) *every finitely generated locally projective A-module is globally projective*;

(ii) *every finitely generated locally free A-module is globally projective*;

(iii) *A satisfies $Patch$*.

Proof (i) $\Longrightarrow$ (ii) is clear since a (finitely generated) locally free A-module is locally projective.

(ii) $\Longrightarrow$ (i). Let $\mathcal{P} = (P_+, P_-; \alpha)$ be a finitely generated locally projective A-module; by Proposition 3.5 there exists a finitely generated locally projective A-module $\mathcal{Q}' = (Q'_+, Q'_-; \beta)$ such that $\mathcal{P} \oplus \mathcal{Q}'$ is locally free. By hypothesis, $\mathcal{P} \oplus \mathcal{Q}$ is now globally projective over A; that is, there exists a finitely generated A-module $\mathcal{Q}''$ such that $\mathcal{P} \oplus \mathcal{Q}' \oplus \mathcal{Q}'' \cong (A_+^n, A_-^n; I_n)$ and so $\mathcal{P}$ is globally projective over A.

(iii) $\Longrightarrow$ (ii). Let $(A_+^n, A_-^n; \alpha)$ be a locally free A-module. By hypothesis, A satisfies $Patch$ so that there exists a positive integer k and an element $\beta \in GL_k(A_0)$ such that for some $h_+ \in GL_{n+k}(A_+)$, $h_- \in GL_{n+k}(A_-)$

$$\alpha \oplus \beta = [h_+][h_-].$$

It follows from (3.6) that $(A_+^n, A_-^n; \alpha) \oplus (A_+^k, A_-^k; \beta) \cong (A_+^{n+k}, A_-^{n+k}; I_{n+k})$. Thus $(A_+^n, A_-^n; \alpha)$ being a direct summand of a globally free module is globally projective.

(ii) $\Longrightarrow$ (iii). Let $\alpha \in GL_n(A_0)$; then $\mathcal{L} = (A_+^n, A_-^n; \alpha)$ is a locally free A-module. By hypothesis, there exists an A-module $\mathcal{Q} = (Q_+, Q_-; \gamma)$ such that

$$\mathcal{L} \oplus \mathcal{Q} \cong (A_+^k, A_-^k; I_k)$$

for some $k > n$. Write $k = m + n$ where $m \geq 1$; in particular, $A_\sigma^{m+n} \cong Q_\sigma \oplus A_\sigma^n$. Hence putting $\mathcal{K} = (Q_+ \oplus A_+^n, Q_- \oplus A_-^n; \gamma \oplus I_n)$ we see that

$$\mathcal{L} \oplus \mathcal{K} \cong (A_+^{m+2n}, A_-^{m+2n}; I_{m+2n}).$$

However $\mathcal{L} \oplus \mathcal{K} \cong (A_+^{m+2n}, A_-^{m+2n}; \alpha \oplus \beta)$ where $\beta = \gamma \oplus I_n \in GL_{n+m}(A_0) = GL_k(A_0)$. That is, given $\alpha \in GL_n(A_0)$ there exists $\beta \in GL_k(A_0)$ such that

$$(A_+^n, A_-^n; \alpha) \oplus (A_+^k, A_-^k; \beta) \cong (A_+^{n+k}, A_-^{n+k}; I_{n+k}).$$

By (3.6) there exist $h_+ \in GL_{n+k}(A_+)$, $h_- \in GL_{n+k}(A_-)$ such that $\alpha \oplus \beta = [h_+][h_-]$ so that A satisfies $Patch$. This proves (ii) $\Longrightarrow$ (iii) and completes the proof. $\square$

3.5 Completing the Square

By a square $\mathbf{A}$ we mean an octuple $\mathbf{A} = (\widehat{A}, A_+, A_-, A_0; \eta_+, \eta_-, \varphi_+, \varphi_-)$ where $\widehat{A}$, A_+, A_-, A_0 are rings and $\eta_+, \eta_-, \varphi_+, \varphi_-$ are ring homomorphisms making the diagram below commute:

$$\mathbf{A} = \left\{ \begin{array}{ccc} \widehat{A} & \xrightarrow{\eta_-} & A_- \\ \downarrow \eta_+ & & \downarrow \varphi_- \\ A_+ & \xrightarrow{\varphi_+} & A_0 \end{array} \right.$$

If $\mathbf{B} = (\widehat{B}, B_+, B_-, B_0; \xi_+, \xi_-, \psi_+, \psi_-)$ is also a square then by a morphism of squares $f : \mathbf{A} \to \mathbf{B}$ we mean a 4-tuple $f = (\widehat{f}, f_+, f_-, f_0)$ of ring homomorphisms such that, for $\sigma = +, -$, $\xi_\sigma f = f_\sigma \eta_\sigma$ and $\psi_\sigma f_\sigma = f_0 \varphi_\sigma$. Evidently there is a category {Squares} whose objects are squares and whose morphisms are as described. We say that the above square $\mathbf{A}$ is a *fibre square* when $\eta_+ \times \eta_-$ maps $\widehat{A}$ isomorphically onto the fibre product $A_+ \times_\varphi A_- = \{(x_+, x_-) \in A_+ \times A_- : \varphi_+(x_+) = \varphi_-(x_-)\}$. Given a corner $\mathcal{A} = (A_+, A_-, A_0, \varphi_+, \varphi_-)$ we can construct a canonical fibre square

$$\widehat{\mathcal{A}} = \left\{ \begin{array}{ccc} \widehat{A} & \xrightarrow{\pi_-} & A_- \\ \downarrow \pi_+ & & \downarrow \varphi_- \\ A_+ & \xrightarrow{\varphi_+} & A_0 \end{array} \right.$$

where $\widehat{A} = A_+ \times_\varphi A_-$ and π_+, π_- are projections. The construction $\mathcal{A} \mapsto \widehat{\mathcal{A}}$ defines a functor

$$\widehat{\quad} : \{\text{Corners}\} \to \{\text{Fibre squares}\}.$$

In this section we show that, for any corner $\mathcal{A}$, there is a natural 1–1 correspondence

$$\left\{ \begin{array}{c} \text{Isomorphism classes of} \\ \text{finitely generated globally} \\ \text{projective } \mathcal{A}\text{-modules} \end{array} \right\} \longleftrightarrow \left\{ \begin{array}{c} \text{Isomorphism classes of} \\ \text{finitely generated} \\ \text{projective } \widehat{A}\text{-modules} \end{array} \right\}.$$

Let M be a module over $\widehat{A}$ and observe that $M \otimes_{\varphi_- \pi_-} A_0 \equiv M \otimes_{\varphi_+ \pi_+} A_0$. Thus there is a canonical isomorphism $\natural : (M \otimes_{\pi_-} A_-) \otimes_{\varphi_-} A_0 \to (M \otimes_{\pi_+} A_+) \otimes_{\varphi_+} A_0$ making the following commute

$$\begin{array}{ccc} (M \otimes_{\pi_-} A_-) \otimes_{\varphi_-} A_0 & \xrightarrow{\natural} & (M \otimes_{\pi_+} A_+) \otimes_{\varphi_+} A_0 \\ \downarrow v_- & & \downarrow v_+ \\ M \otimes_{\varphi_- \pi_-} A_0 & \xrightarrow{\text{Id}} & M \otimes_{\varphi_+ \pi_+} A_0 \end{array}$$

where $v_\sigma : (M \otimes_{\pi_\sigma} A_-) \otimes_{\varphi_\sigma} A_0 \to M \otimes_{\varphi_\sigma \pi_\sigma} A_0$ is the canonical isomorphism. Thus an $\widehat{A}$-module M gives rise to a module $(M \otimes_{\pi_+} A_+, M \otimes_{\pi_+} A_+; \natural)$ over $\mathcal{A}$. If $f : M \to N$ is a homomorphism of $\widehat{A}$ modules then putting $f_\sigma =$

$f \otimes \text{Id} : M \otimes_{\pi_\sigma} A_\sigma \to N \otimes_{\pi_\sigma} A_\sigma$, the correspondences $M \mapsto (M \otimes_{\pi_+} A_+, M \otimes_{\pi_+} A_+; \natural)$; $f \mapsto (f_+, f_-)$ determine a functor

$$r : \mathcal{M}od_{\widehat{A}} \to \mathcal{M}od_{\mathcal{A}}.$$

Note that r is additive functor; that is, there is a natural equivalence:

$$r(M \oplus N) \approx r(M) \oplus r(N). \tag{3.9}$$

Note that we have a literal equality $\widehat{A} = \langle A_+, A_-, \text{Id} \rangle$; moreover, making the identifications $A_\sigma = \widehat{A} \otimes_{\pi_\sigma} A_\sigma$ and $A_0 = A_\sigma \otimes_{\varphi_\sigma} A_0$ we may write $r(\widehat{A}) = (A_+, A_-, \text{Id})$. By additivity, we have $r(\widehat{A}^n) \cong (A_+^n, A_-^n, I_n)$. If P is a finitely generated projective $\widehat{A}$-module then writing $P \oplus Q \cong \widehat{A}^n$ we see that $r(P) \oplus r(Q) \cong (A_+^n, A_-^n, I_n)$ so that $r(P)$ is a finitely generated globally projective $\mathcal{A}$-module. Thus r induces a mapping

$$r : \left\{ \begin{array}{c} \text{Isomorphism classes of} \\ \text{finitely generated} \\ \text{projective } \widehat{A}\text{-modules} \end{array} \right\} \to \left\{ \begin{array}{c} \text{Isomorphism classes of} \\ \text{finitely generated globally} \\ \text{projective } \mathcal{A}\text{-modules} \end{array} \right\}.$$

We will show that r is a bijection with inverse induced by an additive functor

$$\langle , \rangle : \mathcal{M}od_{\mathcal{A}} \to \mathcal{M}od_{\widehat{A}}.$$

Thus when $\mathcal{M} = (M_+, M_-, \alpha)$ is an $\mathcal{A}$ module define

$$\langle M_+, M_-, \alpha \rangle = \{(m_+, m_-) \in M_+ \times M_- : \alpha[m_-] = m_+\},$$

where $[m_\sigma] = m_\sigma \otimes 1 \in M_\sigma \otimes_{\varphi_\sigma} A_0$. If $(\lambda_+, \lambda_-) \in A_+ \times_\varphi A_-$ then $\varphi_+(\lambda_+) = \varphi_-(\lambda_-)$ so that, for $(m_+, m_-) \in \langle M_+, M_-, \alpha \rangle$,

$$\alpha[m_-\lambda_-] = \alpha[m_-]\varphi_-(\lambda_-) = [m_+]\varphi_+(\lambda_+) = [m_+\lambda_+].$$

Hence $\langle M_+, M_-, \alpha \rangle$ acquires the structure of an $\widehat{A}$-module via the action:

$$\langle M_+, M_-, \alpha \rangle \times A_+ \times_\varphi A_- \longrightarrow \langle M_+, M_-, \alpha \rangle$$

$$(m_+, m_-) \bullet (\lambda_+, \lambda_-) \quad = \quad (m_+\lambda_+, m_-\lambda_-)$$

If $f = (f_+, f_-) : (M_+, M_-, \alpha) \to (N_+, N_-, \beta)$ is a morphism of $\mathcal{A}$-modules then

$$\langle f \rangle = f_+ \times f_- : \langle M_+, M_-, \alpha \rangle \to \langle N_+, N_-, \beta \rangle$$

is an $\widehat{A}$-homomorphism; thus the correspondences $(M_+, M_-, \alpha) \mapsto \langle M_+, M_-, \alpha \rangle$, $f \mapsto \langle f \rangle$ define a functor

$$\langle , \rangle : \mathcal{M}od_{\mathcal{A}} \to \mathcal{M}od_{\widehat{A}}.$$

Note that $\langle\,,\,\rangle$ is also additive; there is a natural equivalence:

$$\langle\mathcal{M}\oplus\mathcal{N}\rangle\approx\langle\mathcal{M}\rangle\oplus\langle\mathcal{N}\rangle. \tag{3.10}$$

Next define a natural transformation $\nu : r\circ\langle\rangle\to\mathrm{Id}_{\mathrm{Mod}_{\mathcal{A}}}$; let $\mathcal{M}=(M_+,M_-,\alpha)\in\mathrm{Mod}_{\mathcal{A}}$. For $\sigma=+,-$ the projection $p_\sigma : M_+\times M_-\to M_\sigma$ defines a morphism of modules

$$p_\sigma : \langle M_+,M_-,\alpha\rangle\to M_\sigma$$

over the ring homomorphism $\pi_\sigma : \widehat{A}\to A_\sigma$ and hence induces a homomorphism of A_σ-modules

$$\nu_\sigma = p_\sigma\otimes 1 : \langle M_+,M_-,\alpha\rangle\otimes_{\pi_\sigma} A_\sigma\longrightarrow M_\sigma.$$

Then $\nu_{\mathcal{M}}=(\nu_+,\nu_-) : r\circ\langle M_+,M_-,\alpha\rangle\to(M_+,M_-,\alpha)$ is a homomorphism of $\mathcal{A}$-modules. Relative to the identifications $\widehat{A}=\langle A_+,A_-,\mathrm{Id}\rangle$, $A_\sigma=\widehat{A}\otimes_{\pi_\sigma} A_\sigma$ and $A_0=A_\sigma\otimes_{\varphi_\sigma} A_0$ we may write $r(\widehat{A})=(A_+,A_-,\mathrm{Id})$, so that:

Proposition 3.11 $\nu_{(A_+,A_-,\mathrm{Id})}=\mathrm{Id} : (A_+,A_-,\mathrm{Id})\to(A_+,A_-,\mathrm{Id}).$

We note that the natural transformation ν is additive; that is relative to the above equivalences:

$$\nu_{\mathcal{M}\oplus\mathcal{N}}\approx\begin{pmatrix}\nu_{\mathcal{M}} & 0\\ 0 & \nu_{\mathcal{N}}\end{pmatrix} \tag{3.12}$$

Proposition 3.13 $\nu_{\mathcal{P}} : r\langle\mathcal{P}\rangle\to\mathcal{P}$ *is an isomorphism if* $\mathcal{P}$ *is a finitely generated globally projective* $\mathcal{A}$-*module.*

Proof Write (A_+^n,A_-^n,I_n) as an n-fold direct sum

$$(A_+^n,A_-^n,I_n)\cong\underbrace{(A_+,A_-,\mathrm{Id})\oplus\cdots\oplus(A_+,A_-,\mathrm{Id})}_{n}.$$

From Proposition 3.11, (3.12) we see that $\nu_{(A_+^n,A_-^n,I_n)}$ is an isomorphism. Now suppose that $\mathcal{P}$ is a finitely generated globally projective $\mathcal{A}$-module; then there exists an $\mathcal{A}$-module $\mathcal{Q}$ such that, for some positive integer n, $\mathcal{P}\oplus\mathcal{Q}\cong(A_+^n,A_-^n,I_n)$. Thus, by above, $\nu_{\mathcal{P}\oplus\mathcal{Q}}$ is an isomorphism. However, writing

$$\nu_{\mathcal{P}\oplus\mathcal{Q}}\approx\begin{pmatrix}\nu_{\mathcal{P}} & 0\\ 0 & \nu_{\mathcal{Q}}\end{pmatrix}$$

we see that both $\nu_{\mathcal{P}}$ and $\nu_{\mathcal{Q}}$ are isomorphisms. $\square$

Next we define a natural transformation $\delta : \mathrm{Id}_{\mathrm{Mod}_{\widehat{A}}}\to\langle\rangle\circ r$ thus; let M be a module over $\widehat{A}$; for $\sigma=+,-$ define $M_\sigma=M\otimes_{\pi_\sigma} A_\sigma$ and for $x\in M$ put $\delta_\sigma(x)=x\otimes_{\pi_\sigma} 1$. Then

$$\delta_M : M\to M_+\times M_-,\quad \delta(x)=(\delta_+(x),\delta_-(x))$$

defines an $\widehat{A}$-homomorphism $\delta_M : M \to \langle M_+, M_-, \natural \rangle = \langle r(M) \rangle$. After making the identifications $A_\sigma = \widehat{A} \otimes_{\pi_\sigma} A_\sigma$ and $A_0 = A_\sigma \otimes_{\varphi_\sigma} A_0$ we may write $r(\widehat{A}) = (A_+, A_-, \text{Id})$ so that, from the equality $\widehat{A} = \langle A_+, A_-, \text{Id} \rangle$ and relative to these identifications:

Proposition 3.14 $\delta_{\widehat{A}} = \text{Id} : \widehat{A} \to \widehat{A}$.

Note that δ is also additive; that is relative to the above equivalences

$$\delta_{M \oplus N} \approx \begin{pmatrix} \delta_M & 0 \\ 0 & \delta_N \end{pmatrix}. \tag{3.15}$$

In the manner of the proof of Proposition 3.13 it follows that $\delta_{\widehat{A}^n}$ is an isomorphism. If P is a finitely generated projective $\widehat{A}$-module then writing $P \oplus Q \cong \widehat{A}^n$ for some positive integer n and appealing to (3.15) we see that:

Proposition 3.16 *If P is a finitely generated projective $\widehat{A}$-module then $\delta_P : P \to \langle r(P) \rangle$ is an isomorphism.*

We have the useful consequence that:

Corollary 3.17 *Let $P \cong \langle P_+, P_-; \alpha \rangle$ be a finitely generated projective $\widehat{A}$-module; then for $\sigma = +, -$ there is an A_σ-isomorphism $P_\sigma \cong (\pi_\sigma)_*(P) = P \otimes_{\pi_\sigma} A_\sigma$.*

It follows easily from Propositions 3.13 and 3.16 that:

Theorem 3.18 *For any corner $\mathcal{A}$ the functor r induces a 1–1 correspondence*

$$r : \left\{ \begin{array}{c} \textit{Isomorphism classes of} \\ \textit{finitely generated} \\ \textit{projective } \widehat{A}\textit{-modules} \end{array} \right\} \to \left\{ \begin{array}{c} \textit{Isomorphism classes of} \\ \textit{finitely generated globally} \\ \textit{projective } \mathcal{A}\textit{-modules} \end{array} \right\}.$$

In Theorem 3.18 the inverse to r is induced from $\langle \, , \, \rangle$. Likewise in the following, which is now an immediate consequence of Theorem 3.8 and is a re-interpretion of Milnor's Theorem [74] in our context.

Theorem 3.19 (Milnor) *If the corner $\mathcal{A}$ satisfies the condition $\mathcal{P}$atch then the functor r induces a 1–1 correspondence*

$$r : \left\{ \begin{array}{c} \textit{Isomorphism classes of} \\ \textit{finitely generated} \\ \textit{projective } \widehat{A}\textit{-modules} \end{array} \right\} \to \left\{ \begin{array}{c} \textit{Isomorphism classes of} \\ \textit{finitely generated locally} \\ \textit{projective } \mathcal{A}\textit{-modules} \end{array} \right\}.$$

In the next two sections we consider some commonly occuring conditions under which the patching condition is attained.

3.6 Practical Patching Conditions

We introduce a sequence of progressively stronger conditions which guarantee that $\mathcal{A}$ satisfies $\mathcal{P}atch$.

$\mathcal{W}$: For each $n \geq 1$ and each $\gamma \in E_n(A_0)$ there exist $m \geq 1$, $h_+ \in GL_{n+m}(A_+)$, $h_- \in GL_{n+m}(A_-)$ such that $\gamma \oplus I_m = [h_+][h_-]$.

Proposition 3.20 $\mathcal{A}$ satisfies $\mathcal{W}$ $\implies$ $\mathcal{A}$ satisfies $\mathcal{P}atch$.

Proof Let $\alpha \in GL_n(A_0)$. Observe that, as is well known,

$$\begin{pmatrix} \alpha & 0 \\ 0 & \alpha^{-1} \end{pmatrix} = \begin{pmatrix} 1 & \alpha \\ 0 & 1 \end{pmatrix} \begin{pmatrix} 1 & 0 \\ -\alpha^{-1} & 1 \end{pmatrix} \begin{pmatrix} 1 & \alpha \\ 0 & 1 \end{pmatrix} \begin{pmatrix} 1 & -1 \\ 0 & 1 \end{pmatrix} \begin{pmatrix} 1 & 0 \\ 1 & 1 \end{pmatrix} \begin{pmatrix} 1 & -1 \\ 0 & 1 \end{pmatrix}$$

so that $\alpha \oplus \alpha^{-1} = \begin{pmatrix} \alpha & 0 \\ 0 & \alpha^{-1} \end{pmatrix} \in E_{2n}(A_0)$. By condition $\mathcal{W}$ there exists $m \geq 1$, $h_+ \in GL_{2n+m}(A_+)$ and $h_- \in GL_{2n+m}(A_-)$ such that $\alpha \oplus \alpha^{-1} \oplus I_m = [h_+][h_-]$. Now put $k = n + m$ and $\beta = \alpha^{-1} \oplus I_m$; thus $\alpha \oplus \beta = [h_+][h_-]$ so satisfying $\mathcal{P}atch$. $\square$

We say that $\mathcal{A}$ is $\overline{E}$-*trivial* when $\overline{E}(\mathcal{A}) = E(A_+)\backslash E(A_0)/E(A_-)$ consists of a single point.

Proposition 3.21 $\mathcal{A}$ is $\overline{E}$-trivial $\implies$ $\mathcal{A}$ satisfies $\mathcal{W}$.

Proof Suppose that $\mathcal{A}$ is $\overline{E}$-trivial and let $\gamma \in E_n(A_0)$. We must show that there exist $m \geq 1$, $h_+ \in GL_{n+m}(A_+)$, $h_- \in GL_{n+m}(A_-)$ such that $\gamma \oplus I_m = [h_+][h_-]$. Let $\widehat{\gamma}$ denote the stabilization of γ;

$$\widehat{\gamma} = \begin{pmatrix} \gamma & 0 \\ 0 & I_\infty \end{pmatrix} \in E(A_0) = \lim_{N \to \infty} E_N(A_0).$$

By $\overline{E}$-triviality, $\widehat{\gamma}$ may be written as $\widehat{\gamma} = [\gamma_+][\gamma_-]$ with $\gamma_\sigma \in E(A_\sigma)$. However, from the definition $E(A_\sigma) = \lim_{N \to \infty} E_N(A_\sigma)$ there exists $N > n$ such that

$$(\gamma_\sigma)_{ij} = \begin{cases} 0 & \text{if } i \neq j \text{ and } i > N \text{ or } j > N, \\ 1 & \text{if } i = j > N. \end{cases}$$

Define the $N \times N$ matrix h_σ over A_σ by $(h_\sigma)_{ij} = (\gamma_\sigma)_{ij}$ for $1 \leq i, j \leq N$. Then $h_\sigma \in GL_N(A_\sigma)$ and $\widehat{h_\sigma} = \gamma_\sigma$. Moreover, putting $m = N - n$ we have $\gamma \oplus I_m = [h_+][h_-]$ in satisfaction of condition $\mathcal{W}$. $\square$

For $n \geq 2$ we say that $\mathcal{A}$ is $\overline{E_n}$-trivial when $\overline{E_k}(\mathcal{A}) = 1$ for all $k \geq n$; that is, when $E_k(A_+)\backslash E_k(A_0)/E_k(A_-)$ consists of a single point for $k \geq n$. Evidently if $\mathcal{A}$ is $\overline{E_n}$-trivial and $n \leq N$ then $\mathcal{A}$ is $\overline{E_N}$-trivial. Moreover an easy stabilization argument shows:

Proposition 3.22 $\mathcal{A}$ is $\overline{E_n}$-trivial for some $n \implies \mathcal{A}$ is $\overline{E}$-trivial.

Observe that:

If $\mathcal{A}$ is $\overline{E_{n+1}}$-trivial then $s'_{n,k} : \overline{GL_n}(\mathcal{A}) \to \overline{GE_{n,k}}(\mathcal{A})$ is bijective.   (3.23)

The strongest of these conditions is $\overline{E_2}$ triviality. This in turn is implied by the following condition $\mathcal{M}$ which was the original patching condition introduced by Milnor [74].

$\mathcal{M}$: $\mathcal{A} = (A_+, A_-, A_0, \varphi_+, \varphi_-)$ satisfies $\mathcal{M}$ when either φ_+ or φ_- is surjective.

Suppose that $\varphi_+ : A_+ \to A_0$ is surjective. Then for each $k \geq 2$ the induced homomorphisms $\varphi_+ : E_k(A_+) \to E_k(A_0)$ are surjective and so $E_k(A_+)\backslash E_k(A_0)$ consists of a single point; a fortiori $E_k(A_+)\backslash E_k(A_0)/E_k(A_-)$ consists of a single point. Similarly, $E_k(A_+)\backslash E_k(A_0)/E_k(A_-)$ consists of a single point when $\varphi_- : A_- \to A_0$ is surjective. Thus we see that:

Proposition 3.24 $\mathcal{A}$ satisfies $\mathcal{M} \implies \mathcal{A}$ is $\overline{E_2}$-trivial.

To summarize, we have a chain of implications:

$$\mathcal{M} \implies \overline{E_2}\text{-triviality} \implies \overline{E_n}\text{-triviality} \implies \overline{E}\text{-triviality} \implies \mathcal{W} \implies \mathcal{P}atch.$$

3.7 Karoubi Squares

Let S be a multiplicative submonoid of a ring Λ; we say that S is *regular* when S is central in Λ and contains no zero divisors. In that case we obtain a ring Λ_S by formally inverting the elements of S; thus the elements of Λ_S have the form $(\frac{\lambda}{s})$ with the familiar rules for addition and multiplication of fractions and the mapping $\Lambda \to \Lambda_S$; $\lambda \to (\frac{\lambda}{1})$ is then an injective ring homomorphism. By a *Karoubi homomorphism* we mean a 4-tuple (A, B, φ, S) where

(i) A, B are rings and $\varphi : A \to B$ is a ring homomorphism.
(ii) S is a regular submonoid of A and $\varphi(S)$ is a regular submonoid of B.
(iii) For each $s \in S$, the natural mapping $\varphi_* : A/sA \to B/\varphi(s)B$ is an isomorphism.

Given a Karoubi homomorphism (A, B, φ, S) one may construct a commutative square of ring homomorphisms, the associated *Karoubi square*

$$
\begin{array}{ccc}
A & \xrightarrow{\varphi} & B \\
\downarrow i & & \downarrow v \\
A_S & \xrightarrow{\widehat{\varphi}} & B_S
\end{array}
$$

where i, v are the canonical inclusions. In [63] Karoubi showed:

Proposition 3.25 *The square associated to a Karoubi homomorphism (A, B, φ, S) is a fibre square.*

Proof Suppose that $(\frac{a}{s}) \in A_S$ and $b \in B$ satisfy $\widehat{\varphi}(\frac{a}{s}) = v(b)$. Then $\varphi a = b\varphi(s)$. Thus from the commutativity of the following square

$$
\begin{array}{ccc}
A & \xrightarrow{\varphi} & B \\
\downarrow \natural & & \downarrow \natural \\
A/sA & \xrightarrow{\varphi_*} & B/sB
\end{array}
$$

we deduce that $\varphi_*(\natural(a)) = 0$. However, φ_* is an isomorphism so that $\natural(a) = 0$ and hence $a = sa'$ for some $a' \in A$. Thus $(\frac{a'}{1}) = (\frac{a}{s})$ and so $i(a') = (\frac{a}{s})$. Moreover $\varphi(a's) = b\varphi(s)$ so that $\varphi(a')\varphi(s) = b\varphi(s)$. As $\varphi(s)$ is not a zero divisor then $\varphi(a') = b$. If $a'' \in A$ also satisfies $i(a'') = (\frac{a}{s})$ and $\varphi(a'') = b$ then injectivity of i shows that $a'' = a'$, so verifying the fibre square property. $\qquad\square$

Denote by $E_n(s, B)$: the subgroup of $E_n(B)$ generated by elementary
matrices of the form $E(i, j; \varphi(s)b)$ where $b \in B$;
$\mathcal{E}_n(s, B)$: the image of $E_n(s, B)$ in $E_n(B_S)$.

Let $n \geq 3$ and suppose given $s, v \in S$; then for all $b \in B$ we have the following inclusion:

Lemma 3.26 $E(i, j; -(\frac{b}{\varphi(v)}))\mathcal{E}_n(s^2v^2, B)E(i, j; (\frac{b}{\varphi(v)})) \subset \mathcal{E}_n(s, B)$.

Proof When k, l are indices $1 \leq k, l \leq n$ with $k \neq l$ and $b, \beta \in B$ we denote by $W(k, l, b, \beta)$ the following expression

$$
W(k, l, b, \beta) = E\left(i, j; -\left(\frac{b}{\varphi(v)}\right)\right)E\left(k, l; \left(\frac{\varphi(s^2v^2)\beta}{1}\right)\right)E\left(i, j; \left(\frac{b}{\varphi(v)}\right)\right).
$$

As $\mathcal{E}_n(s^2v^2, B)$ is generated by the matrices $E(k, l; (\frac{\varphi(s^2v^2)\beta}{1}))$ with $\beta \in B$ it suffices to show that each $W(k, l, b, \beta) \in \mathcal{E}_n(s, B)$. Recalling that $i \neq j$, there are four cases:

Case I $i \neq l$ and $j \neq k$; then $W(k, l, b, \beta) = E(k, l; (\frac{\varphi(s)\beta'}{1}))$ where $\beta' = \varphi(sv^2)\beta \in B$.

Case II $i \neq l$ and $j = k$; then $W(k, l, b, \beta) = E(i, l; (\frac{\varphi(s)\beta'}{1}))E(j, l; (\frac{\varphi(s)\beta''}{1}))$
where $\beta' = -\varphi(s^2 v)b\beta$ and $\beta'' = \varphi(s^2 v^2)\beta$. Clearly $\beta', \beta'' \in B$.

Case III $i = l$ and $j \neq k$; then $W(k, l, b, \beta) = E(k, i; (\frac{\varphi(s)\beta'}{1}))E(k, j; (\frac{\varphi(s)\beta''}{1}))$
where $\beta' = \varphi(s^2 v^2)\beta$ and $\beta'' = \varphi(s^2 v)\beta b$. Again $\beta', \beta'' \in B$.

In each of Cases I, II, III it is straightforward to see that $W(k, l, b, \beta) \in \mathcal{E}_n(s, B)$.

Case IV $i = l$ and $j = k$. Here for $n \times n$ matrices X, Y we write $[X, Y]$ for the commutator $[X, Y] = XYX^{-1}Y^{-1}$. As $n \geq 3$ we may choose an index $m \neq k, l$; then

$$
E\left(k, l; \left(\frac{\varphi(s^2 v^2)\beta}{1}\right)\right) = \left[E\left(j, m; \left(\frac{\varphi(sv)}{1}\right)\right), E\left(m, i; \left(\frac{\varphi(sv)\beta}{1}\right)\right)\right].
$$

Now take $X = E(j, m; (\frac{\varphi(sv)}{1}))$, $Y = E(m, i; (\frac{\varphi(sv)\beta}{1}))$ and $Z = E(i, j; (\frac{b}{\varphi(v)}))$.
Similar computations to Cases II and III show that $Z^{-1}XZ \in \mathcal{E}_n(s, B)$ and $Z^{-1}YZ \in \mathcal{E}_n(s, B)$. However,

$$
W(k, l, b, \beta) = Z^{-1}[X, Y]Z = [Z^{-1}XZ, Z^{-1}YZ]
$$

so that $W(k, l, b, \beta) \in \mathcal{E}_n(s, B)$. This completes the proof. $\qquad\square$

Proposition 3.27 *Let $n \geq 3$ and let $s \in S$; then for each $x \in E_n(B_S)$ there exists $t \in S$ such that $x^{-1}\mathcal{E}_n(t, B)x \subset \mathcal{E}_n(s, B)$.*

Proof $E_n(B_S)$ is generated by elements of the form $E(i, j; (\frac{b}{\varphi(v)}))$ where $b \in B$ and $v \in S$. Thus we may write x as a product

$$
x = x_m x_{m-1} \cdots x_1,
$$

where $x_r = E(i_r, j_r; (\frac{b_r}{\varphi(v_r)}))$. It suffices to show that for each r, $1 \leq r \leq m$ there exists $t_r \in S$ such that

$$
(x_r x_{r-1} \cdots x_1)^{-1}\mathcal{E}_n(t_r, B)(x_r x_{r-1} \cdots x_1) \subset \mathcal{E}_n(s, B).
$$

For $r = 1$ put $t_1 = s^2 v_1^2$. Then the result follows from Lemma 3.26. Suppose proved for $r - 1$ and put $t_r = t_{r-1}^2 v_r^2$. Then by Lemma 3.26, $x_r^{-1}\mathcal{E}_n(t_r, B)x_r \subset \mathcal{E}_n(t_{r-1}, B)$. so that

$$
(x_{r-1} \cdots x_1)^{-1}x_r^{-1}\mathcal{E}_n(t_r, B)x_r(x_{r-1} \cdots x_1)
$$
$$
\subset (x_{r-1} \cdots x_1)^{-1}\mathcal{E}_n(t_{r-1}, B)(x_{r-1} \cdots x_1).
$$

By induction $(x_{r-1}x_{r-2} \cdots x_1)^{-1}\mathcal{E}_n(t_{r-1}, B)(x_{r-1}x_{r-2} \cdots x_1) \subset \mathcal{E}_n(s, B)$ so that

$$
(x_r x_{r-1} \cdots x_1)^{-1}\mathcal{E}_n(t_{r+1}, B)(x_r x_{r-1} \cdots x_1) \subset \mathcal{E}_n(s, B).
$$

This completes the induction and the proof. $\qquad\square$

We denote by $\mathcal{E}_n(A_S)$, $\mathcal{E}_n(B)$ the respective images of $E_n(A_S)$, $E_n(B)$ in $E_n(B_S)$. Observe that $\mathcal{E}_n(B) = \mathcal{E}_n(1, B)$.

Proposition 3.28 $\mathcal{E}_n(B)\mathcal{E}_n(A_S) \subset \mathcal{E}_n(A_S)\mathcal{E}_n(B)$.

Proof We first show that if $b \in B$ and $x \in \mathcal{E}_n(A_S)$ then $E(i, j; (\frac{b}{1}))x \in \mathcal{E}_n(A_S) \times \mathcal{E}_n(B)$. Clearly $x \in E_n(B_S)$ so that, by Proposition 3.28 there exists $t \in S$ such that $x^{-1}\mathcal{E}_n(t, B)x \subset \mathcal{E}_n(1, B)$. Now $A/tA \cong B/\varphi(t)B$ so that we may write $b = \varphi(a) + tb'$ for some $a \in A$ and $b' \in B$. Then $E(i, j; (\frac{b}{1})) = E(i, j; (\frac{\varphi(a)}{1}))E(i, j; (\frac{tb'}{1}))$. Hence

$$E\left(i, j; \left(\frac{b}{1}\right)\right)x = \left(E\left(i, j; \left(\frac{\varphi(a)}{1}\right)\right)x\right)\left(x^{-1}E\left(i, j; \left(\frac{tb'}{1}\right)\right)x\right).$$

However $E(i, j; (\frac{\varphi(a)}{1}))x \in \mathcal{E}_n(A_S)$ whilst $x^{-1}E(i, j; (\frac{tb'}{1}))x \in \mathcal{E}_n(1, B) = \mathcal{E}_n(B)$ so that $E(i, j; (\frac{b}{1}))x \in \mathcal{E}_n(A_S)\mathcal{E}_n(B)$ as claimed.

As $\mathcal{E}_n(B)$ is generated by elements of the form $E(i, j; (\frac{b}{1}))$ with $b \in B$ it follows that $yx \in \mathcal{E}_n(A_S)\mathcal{E}_n(B)$ for an arbitrary element y of $\mathcal{E}_n(B)$. Hence $\mathcal{E}_n(B)\mathcal{E}_n(A_S) \subset \mathcal{E}_n(A_S)\mathcal{E}_n(B)$ as required. $\qquad\square$

It follows immediately that:

Corollary 3.29 $\mathcal{E}_n(A_S)\mathcal{E}_n(B)$ *is a subgroup of* $E_n(B_S)$ *provided* $n \geq 3$.

Corollary 3.30 $E_n(B_S) = \mathcal{E}_n(A_S)\mathcal{E}_n(B)$ *for* $n \geq 3$.

Proof $E_n(B_S)$ is generated by elements of the form $E(i, j; (\frac{\varphi(b)}{\varphi(s)}))$ where $b \in B$ and $s \in S$. By Corollary 3.29 it suffices to show that each such generator can be expressed as a product $E(i, j; (\frac{\varphi(b)}{\varphi(s)})) = xy$ with $x \in \mathcal{E}_n(A_S)$ and $y \in \mathcal{E}_n(B)$. However, for $b \in B$ and $s \in S$ then we may write

$$b = \varphi(a) + \varphi(s)b'$$

so that

$$\left(\frac{b}{\varphi(s)}\right) = \left(\frac{\varphi(a)}{\varphi(s)}\right) + \left(\frac{b'}{1}\right).$$

Then $E(i, j; (\frac{\varphi(b)}{\varphi(s)})) = E(i, j; (\frac{\varphi(a)}{\varphi(s)}))E(i, j; (\frac{b'}{1})) \in \mathcal{E}_n(A_S)\mathcal{E}_n(B)$. $\qquad\square$

We obtain the following which Swan [94] attributes to M.P. Murthy and A. Vorst.

Corollary 3.31 *Let* (A, B, φ, S) *be a Karoubi homomorphism; then the corner associated to* (A, B, φ, S) *is* $\overline{E_n}$*-trivial for all* $n \geq 3$.

3.8 Lifting Stably Free Modules

Let Λ be a ring; if $k \geq 1$ is an integer by $\mathcal{S}(k, n)$ we mean the standard exact sequence $0 \to \Lambda^k \xrightarrow{i} \Lambda^{k+n} \xrightarrow{p} \Lambda^n \to 0$ where

$$i(\mathbf{y}) = \begin{pmatrix} 0 \\ \vdots \\ \mathbf{y} \end{pmatrix} \quad \text{and} \quad p \begin{pmatrix} \mathbf{x} \\ \vdots \\ \mathbf{y} \end{pmatrix} = \mathbf{x}.$$

By $\mathrm{Aut}_+(\mathcal{S}(k, m))$ we mean the subgroup of $GL_{k+m}(\Lambda)$ consisting of Λ-automorphisms $\widehat{\alpha}$ of Λ^{k+m} for which there exists $\alpha \in GL_m(\Lambda)$ making the following diagram commute:

$$
\begin{array}{ccccccccc}
0 \to & \Lambda^k & \xrightarrow{i} & \Lambda^{k+m} & \xrightarrow{p} & \Lambda^m & \to 0 \\
& \mathrm{Id} \downarrow & & \widehat{\alpha} \downarrow & & \alpha \downarrow & \\
0 \to & \Lambda^k & \xrightarrow{i} & \Lambda^{k+m} & \xrightarrow{p} & \Lambda^m & \to 0
\end{array}
$$

The property of being m-weakly Euclidean can be re-phrased thus:

Proposition 3.32 *If Λ is an m-weakly Euclidean ring and $X \in GL_{k+m}(\Lambda)$ then X can be factorised as $X = \widehat{\alpha} \circ Y$ where $\widehat{\alpha} \in \mathrm{Aut}_+(\mathcal{S}(k, m))$ and $Y \in E_{k+m}(\Lambda)$.*

Lemma 3.33 *Let $\widehat{\mathcal{A}}$ be a fibre square satisfying $\mathcal{P}$atch in which the corner ring A_0 is m-weakly Euclidean and let S_+, S_- be stably free modules of rank m over A_+, A_- respectively such that*

$$S_+ \otimes A_0 \cong S_- \otimes A_0 \cong A_0^m.$$

Then for some $k \geq 1$ there exists an A_0-isomorphism $h : S_- \otimes A_0 \to S_+ \otimes A_0$ and an element $J_{k+m} \in E_{k+m}(A_0)$ such that

$$\langle S_+, S_-; h \rangle \oplus \widehat{A}^k \cong \langle A_+^{k+m}, A_-^{k+m}; J_{k+m} \rangle.$$

Proof For $\sigma = +, -$ choose A_0-isomorphisms $\eta_\sigma : S_\sigma \otimes A_0 \to A_0^m$. As S_σ is stably free of rank m over A_σ then for some $k_\sigma \geq 1$ $S_\sigma \oplus A_\sigma^{k_\sigma} \cong A_\sigma^{k_\sigma + m}$. Take $k = \max\{k_+, k_-\}$ so that $S_\sigma \oplus A_\sigma^k \cong A_\sigma^{k+m}$. Make a specific choice of exact sequences $0 \to A_\sigma^k \xrightarrow{j_\sigma} A_\sigma^{k+m} \xrightarrow{p_\sigma} S_\sigma \to 0$ over A_σ. After tensoring with A_0 and composing the final projection with η_σ we obtain exact sequences over A_0 thus:

$$\mathcal{F}_\sigma : 0 \to A_0^k \xrightarrow{j_\sigma} A_0^{k+m} \xrightarrow{\eta_\sigma p_\sigma} A_0^m \to 0 \quad (\sigma = +, -).$$

Choose a left splitting for $\mathcal{F}_\sigma$; that is, an A_0-homomorphism $r_\sigma : A_0^{k+m} \to A_0^k$ such that $r_\sigma \circ j_\sigma = \mathrm{Id}$. Then define $X_\sigma : A_0^{k+m} \to A_0^m \oplus A_0^k$ by

$$X_\sigma(\mathbf{x}) = \begin{pmatrix} \eta_\sigma p_\sigma(\mathbf{x}) \\ r_\sigma(\mathbf{x}) \end{pmatrix}.$$

One verifies easily that X_σ is a congruence

$$
\begin{array}{c}
\mathcal{F}_\sigma \\
X_\sigma \downarrow \\
\mathcal{S}(k,m)
\end{array}
=
\begin{pmatrix}
0 \to A_0^k \xrightarrow{j_\sigma} A_0^{k+m} \xrightarrow{\eta_\sigma\, p_\sigma} A_0^m \to 0 \\
\quad\;\; \downarrow \text{Id} \qquad\;\; \downarrow X_\sigma \qquad\;\; \downarrow \text{Id} \\
0 \to A_0^k \xrightarrow{i} A_0^{k+m} \xrightarrow{\pi} A_0^m \to 0
\end{pmatrix}.
$$

By the Five Lemma $X_\sigma \in GL_{k+m}(A_0)$. By Proposition 3.32 X_σ can be factorised as $X_\sigma = \widehat{\beta}_\sigma Y_\sigma$ where $\widehat{\beta}_\sigma \in \mathrm{Aut}_+(\mathcal{S}(k,m))$ and $Y_\sigma \in E_{k+m}(A_0)$. Write

$$
\begin{array}{c}
\mathcal{S}(k,m) \\
\widehat{\alpha}_\sigma \downarrow \\
\mathcal{S}(k,m)
\end{array}
=
\begin{pmatrix}
0 \to A_0^k \xrightarrow{i} A_0^{k+m} \xrightarrow{\pi} A_0^m \to 0 \\
\quad\;\; \downarrow \text{Id} \qquad\; \downarrow \widehat{\alpha}_\sigma \qquad\; \downarrow \alpha_\sigma \\
0 \to A_0^k \xrightarrow{i} A_0^{k+m} \xrightarrow{\pi} A_0^m \to 0
\end{pmatrix}.
$$

Juxtaposing X_σ and α_σ we obtain

$$
\begin{array}{c}
\mathcal{F}_\sigma \\
X_\sigma \downarrow \\
\mathcal{S}(k,m) = \\
\widehat{\alpha}_\sigma \downarrow \\
\mathcal{S}(k,m)
\end{array}
\begin{pmatrix}
0 \to A_0^k \xrightarrow{j_\sigma} A_0^{k+m} \xrightarrow{\eta_\sigma\, p_\sigma} A_0^m \to 0 \\
\quad\;\; \downarrow \text{Id} \qquad\; \downarrow \widehat{\beta}_\sigma Y_\sigma \qquad\; \downarrow \text{Id} \\
0 \to A_0^k \xrightarrow{i} A_0^{k+m} \xrightarrow{\pi} A_0^m \to 0 \\
\quad\;\; \downarrow \text{Id} \qquad\; \downarrow \widehat{\alpha}_\sigma \qquad\; \downarrow \alpha_\sigma \\
0 \to A_0^k \xrightarrow{i} A_0^{k+m} \xrightarrow{\pi} A_0^m \to 0
\end{pmatrix}.
$$

Composing and noting that $\widehat{\alpha}_\sigma = \widehat{\beta}_\sigma^{-1}$, we obtain a commutative diagram:

$$
\begin{array}{ccccccccc}
0 \to & A_0^k & \xrightarrow{j_\sigma} & A_0^{k+m} & \xrightarrow{\eta_\sigma\, p_\sigma} & S_\sigma \otimes A_0 & \to 0 \\
& \downarrow \text{Id} & & \downarrow Y_\sigma & & \downarrow \alpha_\sigma \eta_\sigma & \\
0 \to & A_0^k & \xrightarrow{i} & A_0^{k+m} & \xrightarrow{\pi} & A_0^m & \to 0
\end{array}
$$

Inverting the diagram for $\sigma = +$ and composing we get

$$
\begin{array}{ccccccccc}
0 \to & A_0^k & \xrightarrow{j_-} & A_0^{k+m} & \xrightarrow{p_-} & S_- \otimes A_0 & \to 0 \\
& \downarrow \text{Id} & & \downarrow Y_+^{-1} Y_- & & \downarrow \eta_+^{-1}\alpha_+^{-1}\alpha_-\eta_- & \\
0 \to & A_0^k & \xrightarrow{j_+} & A_0^{k+m} & \xrightarrow{p_+} & S_+ \otimes A_0 & \to 0
\end{array}
$$

On putting $h = \eta_+^{-1}\alpha_+^{-1}\alpha_-\eta_- : S_- \otimes A_0 \to S_+ \otimes A_0$ and $J_{k+m} = Y_+^{-1} Y_- \in E_{k+m}(A_0)$ then as $\widehat{\mathcal{A}}$ is assumed to satisfy the patching condition we obtain an exact sequence of projective $\widehat{A}$-modules

$$
0 \to \langle A_+^k, A_-^k; \text{Id}\rangle \xrightarrow{(j_+, j_-)} \langle A_+^{k+m}, A_-^{k+m}; J_{k+m}\rangle \xrightarrow{(p_+, p_-)} \langle S_+, S_-; h\rangle \to 0.
$$

Splitting the above exact sequence and observing that $\langle A_+^k, A_-^k; \text{Id}\rangle = \widehat{A}^k$ now gives an isomorphism $\langle S_+, S_-; h\rangle \oplus \widehat{A}^k \cong \langle A_+^{k+m}, A_-^{k+m}; J_{k+m}\rangle$ as claimed. $\qquad\square$

Corollary 3.34 *Let $\widehat{A}$ be a fibre square which is $\overline{E}$-trivial and in which the corner ring A_0 is m-weakly Euclidean; let $m \leq n$ and let S_+, S_- be stably free modules of rank n over A_+, A_- respectively such that*

$$S_+ \otimes A_0 \cong S_- \otimes A_0 \cong A_0^n;$$

then there exists a stably free module $\widehat{S}$ over $\widehat{A}$ such that $\pi_+(\widehat{S}) \cong S_+$ and $\pi_-(\widehat{S}) \cong S_-$.

Proof The hypothesis of $\overline{E}$-triviality guarantees that $\widehat{A}$ satisfies the patching condition. As $m \leq n$ then A_0 is, a fortiori, n-weakly Euclidean so we may apply Lemma 3.33 to obtain an integer $k \geq 1$, an isomorphism $h : S_+ \otimes A_0 \to S_- \otimes A_0$ and an element $J_{k+n} \in E_{k+n}(A_0)$ such that

$$\langle S_+, S_-; h \rangle \oplus \widehat{A}^k \cong \langle A_+^{k+n}, A_-^{k+n}; J_{k+n} \rangle.$$

Put $\widehat{S} = \langle S_+, S_-; h \rangle$; then for all $\mu \geq 1$, $\widehat{S} \oplus \widehat{A}^k \oplus \widehat{A}^\mu \cong \langle A_+^{k+n}, A_-^{k+n}; J_{k+n} \rangle \oplus \widehat{A}^\mu$. Put $N = k + n$; then

$$J_{N+\mu} = \begin{bmatrix} J_N & 0 \\ 0 & I_\mu \end{bmatrix} \in E_{N+\mu}(A_0).$$

Furthermore $\langle A_+^N, A_-^N; J_N \rangle \oplus \widehat{A}^\mu \cong \langle A_+^{N+\mu}, A_-^{N+\mu}; J_{N+\mu} \rangle$ so that

$$\widehat{S} \oplus \widehat{A}^{k+\mu} \cong \langle A_+^{N+\mu}, A_-^{N+\mu}; J_{N+\mu} \rangle.$$

$\overline{E}$-triviality guarantees that for μ sufficiently large $J_{N+\mu}$ defines the trivial class in $\overline{GL_{N+\mu}}(\widehat{A})$ in which case $\langle A_+^{N+\mu}, A_-^{N+\mu}; J_{N+\mu} \rangle \cong \widehat{A}^{N+\mu}$ and

$$\widehat{S} \oplus \widehat{A}^{k+\mu} \cong \widehat{A}^{k+n+\mu}.$$

We have shown that $\widehat{S}$ is stably free and by construction $\pi_+(\widehat{S}) \cong S_+$ and $\pi_-(\widehat{S}) \cong S_-$. This completes the proof.   □

Let $\mathcal{SF}_n(\Lambda)$ denote the set of isomorphism classes of stably free modules of rank n over the ring Λ. A ring homomorphism $\varphi : \Lambda_1 \to \Lambda_2$ induces a mapping $\varphi : \mathcal{SF}_n(\Lambda_1) \to \mathcal{SF}_n(\Lambda_2)$ by $\varphi(S) = S \otimes_\varphi \Lambda_2$. We obtain:

Theorem 3.35 *Let $\widehat{A}$ be an $\overline{E}$-trivial fibre square in which A_0 is m-weakly Euclidean and satisfies SFC; then $\pi_+ \times \pi_- : \mathcal{SF}_n(\widehat{A}) \to \mathcal{SF}_n(A_+) \times \mathcal{SF}_n(A_-)$ is surjective for each $n \geq m$.*

Proof If S_+, S_- are stably free modules of rank n over A_+, A_- respectively then both $S_+ \otimes A_0$ and $S_- \otimes A_0$ are stably free of rank n over A_0. As A_0 has property SFC then $S_+ \otimes A_0 \cong A_0^n \cong S_- \otimes A_0$ and the conclusion is a consequence of Corollary 3.34.   □

Corollary 3.36 *Let $\widehat{A}$ be an $\overline{E}$-trivial fibre square in which A_0 is weakly Euclidean and satisfies SFC; then $\pi_+ \times \pi_- : SF_n(\widehat{A}) \to SF_n(A_+) \times SF_n(A_-)$ is surjective for each $n \geq 1$.*

3.9 Stably Free Modules of Locally Free Type

Let A be a corner which satisfies $Patch$; we set ourselves the task of parametrizing the set $S\mathcal{F}(\widehat{A}) = \bigcup_{n \geq 1} S\mathcal{F}_n(\widehat{A})$ of stably free modules over the fibre completion $\widehat{A}$ in terms of computable invariants of A. In Sect. 3.8 we gave conditions on A whereby nontrivial stably free modules of rank n over A_+, A_- survive to be non-trivial over $\widehat{A}$. We now take the modules over A_+, A_- to be trivial (i.e. free) of rank n.

We saw in (3.6), (3.7) that there are commutative diagrams

$$
\begin{array}{ccc}
\overline{GL_n}(A) & \xrightarrow{s_{n,k}} & \overline{GL_{n+k}}(A) \\
\nu_n \downarrow & & \nu_{n+k} \downarrow \\
\mathcal{LF}_n(A) & \xrightarrow{\sigma_{n,k}} & \mathcal{LF}_{n+k}(A)
\end{array}
$$

in which the maps ν_n, ν_{n+k} are bijective. We may compare our task with a more familiar aspect of algebraic K-theory, the computation of the direct limit

$$
\varinjlim(\mathcal{LF}_n(A), \sigma_{n,k}) \cong \varinjlim(\overline{GL_n}(A), s_{n,k}).
$$

Our problem is dual to this; we wish to compute $\mathrm{Ker}(\varinjlim)$ rather than $\mathrm{Im}(\varinjlim)$. Writing $*$ for the class of $\widehat{A}^{n+k}$ in $\mathcal{LF}_{n+k}(A)$ we see that:

$$Z \in \mathcal{LF}_n(A) \text{ is stably free} \iff \sigma_{n,k}([Z]) = * \quad \text{for some } k \geq 0. \qquad (3.37)$$

Put $Z_n(A) = \{\zeta \in \overline{GL_n}(A) : s_{n,k}(\zeta) = * \text{ for some } k \geq 1\}$ and put

$$Z(A) = \coprod_{n \geq 1} Z_n(A)$$

$Z(A)$ is the *singular set*. From (3.37) we see that:

$$\nu_n : Z_n(A) \longrightarrow SF_n(\widehat{A}) \cap \mathcal{LF}_n(A) \text{ is bijective.} \qquad (3.38)$$

We say that A is *locally n-free* when $SF_n(A_-) = SF_n(A_+) = \{*\}$; then every stably free $\widehat{A}$-module of rank n is locally free over A so that $SF_n(\widehat{A}) = SF_n(\widehat{A}) \cap \mathcal{LF}_n(\widehat{A})$ and (3.38) becomes:

$$\nu_n : Z_n(A) \xrightarrow{\simeq} SF_n(\widehat{A}) \text{ is bijective when } A \text{ is locally } n\text{-free.} \qquad (3.39)$$

The problem is now to describe the sets $Z_n(A)$ at least to the point of saying whether or not $Z_n(A)$ is trivial. The most obvious way of forcing $Z_n(A)$ to be trivial is to require all $s_{n,k}$ to be injective. Formally:

Suppose that $s_{n,k}$ is injective for all $k \geq 1$; then $\mathcal{Z}_n(\mathcal{A}) = \{*\}$.   (3.40)

If $\mathcal{A}$ is locally n-free and each $s_{n,k} : \overline{GL_n}(\mathcal{A}) \longrightarrow \overline{GL_{n+k}}(\mathcal{A})$ is injective

then $\mathcal{SF}_n(\widehat{A}) = \{*\}$.   (3.41)

One can extend this argument; observe that for $1 \leq k < m$, $s_{n,m} = s_{n+k,m-k} \circ s_{n,k}$ so that if both $s_{n,m}$ and $s_{n,k}$ are bijective then $s_{n+k,m-k}$ is also bijective and hence injective; thus if $\mathcal{A}$ is also locally $(n+k)$-free then $\mathcal{SF}_{n+k}(\widehat{A}) = \{*\}$; that is:

Corollary 3.42 *Suppose that, for each $k \geq 0$, $\mathcal{A}$ is locally $(n+k)$-free and that each $s_{n,k}$ is bijective; then $\widehat{A}$ has no nontrivial stably free module of rank $\geq n$.*

To apply this we seek restrictions which ensure that each $s_{n,k}$ is bijective. We first observe a criterion for surjectivity:

Proposition 3.43 *Let $\mathcal{A}$ be a $\overline{E}_{n+1}$-trivial corner in which A_0 is weakly n-Euclidean; then the stabilization map $s_{n,k} : \overline{GL_n}(\mathcal{A}) \to \overline{GL_{n+k}}(\mathcal{A})$ is surjective for $k \geq 1$.*

Proof We have a commutative diagram

$$GE_{n,k}(A_-)\backslash GE_{n,k}(A_0)/GE_{n,k}(A_+) \xrightarrow{\;\natural_1\;} GE_{n,k}(A_-)\backslash GL_{n+k}(A_0)/GE_{n,k}(A_+)$$

$$\Big\uparrow {\scriptstyle s'_{n,k}} \qquad\qquad\qquad\qquad\qquad\qquad\qquad\qquad \Big\downarrow {\scriptstyle \natural_2}$$

$$GL_n(A_-)\backslash GL_n(A_0)/GL_n(A_+) \xrightarrow{\;s_{n,k}\;} GL_{n+k}(A_-)\backslash GL_{n+k}(A_0)/GL_{n+k}(A_+)$$

where $\natural_1, \natural_2$ are the canonical mappings and $s'_{n,k}$ is also a stabilization mapping. As in (3.23), the hypothesis that $\mathcal{A}$ is $\overline{E}_{n+1}$-trivial implies that $s'_{n,k}$ is bijective. Moreover, as A_0 is assumed to be weakly n-Euclidean then $\natural_1$ is the identity mapping. The conclusion now follows from the surjectivity of the canonical mapping $\natural_2$.   □

In the above we note that if A_+, A_- are also weakly n-Euclidean then the natural map $\natural_2$ is also a bijection. Thus repeating the above proof with this stronger hypothesis we obtain:

Proposition 3.44 *Let $\mathcal{A}$ be an $\overline{E}_{n+1}$-trivial corner in which A_+, A_- and A_0 are all weakly n-Euclidean; then the stabilization map $s_{n,k} : \overline{GL_n}(\mathcal{A}) \to \overline{GL_{n+k}}(\mathcal{A})$ is bijective for $k \geq 1$.*

It now follows from Corollary 3.42 and Proposition 3.44 that:

Corollary 3.45 *Let $\mathcal{A}$ be an $\overline{E}_{n+1}$-trivial corner in which in which A_+, A_- and A_0 are all weakly n-Euclidean and have no nontrivial stably free modules of rank $\geq n$; then $\widehat{A}$ has no nontrivial stably free module of rank $\geq n$.*

There are many variations one can make on the above. Without attempting to be exhaustive we concentrate on those that occur most frequently, namely Karoubi squares and Milnor squares. We observed in Corollary 3.31 that Karoubi squares are $\overline{E}_m$-trivial for $m \geq 3$; thus:

Corollary 3.46 *Let $\widehat{A}$ be a Karoubi square in which A_+, A_- and A_0 are all weakly 2-Euclidean; if A_+, A_- have no nontrivial stably free modules of rank ≥ 2 then $\widehat{A}$ has no nontrivial stably free module of rank ≥ 2.*

Likewise by Proposition 3.24 Milnor squares are $\overline{E}_m$-trivial for $m \geq 2$; hence:

Corollary 3.47 *Let $\widehat{A}$ be a Milnor square in which A_+, A_- and A_0 are all weakly Euclidean and where A_+, A_- have property SFC; then $\widehat{A}$ has property SFC.*

Some examples have slightly different though equally favourable hypotheses. We say the corner $\mathcal{A}$ is *pointlike in dimension one* when $\overline{GL}_1(\mathcal{A}) = A_-^* \backslash A_0^* / A_+^*$ is a singleton. Likewise $\mathcal{A}$ is said to be *of locally free type* when it is locally n-free for all n. Again using the $\overline{E}_2$-triviality of Milnor squares, the following consequence of Proposition 3.44, though crude, is nevertheless useful:

Corollary 3.48 *Let $\mathcal{A}$ be a corner of locally free type satisfying the Milnor patching condition; suppose also that A_0 is weakly Euclidean and that $\mathcal{A}$ is pointlike in dimension one; then $\widehat{A}$ has property SFC.*

Proof As $A_-^* \backslash A_0^* / A_+^*$ consists of a single point then each $s_{1,k}$ is injective. Now each $s_{1,k}$ is surjective by Proposition 3.43. The result now follows from Corollary 3.42. $\qquad\qquad\qquad\qquad\qquad\qquad\qquad\qquad\qquad\qquad\qquad\qquad\qquad\qquad\square$

However, a more refined analysis is possible when the constituent rings are commutative.

3.10 Corners of Determinantal Type

Suppose that in the corner $\mathcal{A}$ the bottom ring A_0 is fully determinantal. There are then additional set valued functors which make if easier to analyze $\overline{GL}_n$. First, however, we establish some basic properties of homomorphisms involving fully determinantal rings. Thus suppose given a ring homomorphism $\varphi : B \rightarrow A$ in which A is fully determinantal with determinant $d_n : GL_n(A) \rightarrow (A^*)^{ab}$ $(1 \leq n)$. Then φ induces homomorphisms $\varphi_* : GL_n(B) \rightarrow GL_n(A)$ and, by composition, $d_n\varphi_* : GL_n(B) \rightarrow (A^*)^{ab}$. We define

$$\mathrm{Im}_n(\varphi) = \mathrm{Im}(d_n\varphi_*) \subset (A^*)^{ab}.$$

Thus we can define a pointed set $\overline{D_n}(A)$ by

$$\overline{D_n}(A) = \mathrm{Im}_n(\varphi_-) \backslash (A_0^*)^{ab} / \mathrm{Im}_n(\varphi_+),$$

where the distinguished point $*$ is the class of 1. By stabilization it is evident that:

If A is fully determinantal then $\mathrm{Im}_n(\varphi) \subset \mathrm{Im}_{n+k}(\varphi)$ for $1 \leq k$.   (3.49)

The inclusions $\mathrm{Im}_n(\varphi_\sigma) \subset \mathrm{Im}_{n+k}(\varphi_\sigma)$ for $1 \leq k$ induce pointed surjections

$$\pi_{n,k} : \overline{D_n}(A) \to \overline{D_{n+k}}(A)$$

and it easy to see that $\pi_{n,m} = \pi_{n+k,m-k} \circ \pi_{n,k}$ for $1 \leq k < m$. The determinant $d_n : GL_n(A_0) \to (A_0^*)^{ab}$ is surjective and induces a surjection, denoted by the same symbol,

$$d_n : \overline{GL_n}(A) \to \overline{D_n}(A).$$

Moreover the following diagram commutes for $1 \leq k$;

$$\begin{array}{ccc}
\overline{GL_n}(A) & \xrightarrow{s_{n,k}} & \overline{GL_{n+k}}(A) \\
\downarrow d_n & & \downarrow d_{n+k} \\
\overline{D_n}(A) & \xrightarrow{\pi_{n,k}} & \overline{D_{n+k}}(A)
\end{array}$$

Now suppose that A is a fully determinantal ring with determinant $\{d_n^A\}_{1 \leq n}$. We say that a ring homomorphism $\varphi : B \to A$ is *compatibly determinantal with respect to* $\{d_n^A\}$ when B admits a full determinant $\{\delta_n\}_{2 \leq n}$ making the following commute for each n:

$$\begin{array}{ccc}
GL_n(B) & \xrightarrow{\varphi_*} & GL_n(A) \\
\downarrow \delta_n & & \downarrow d_n \\
(B^*)^{ab} & \xrightarrow{\varphi} & (A^*)^{ab}
\end{array}$$

Then given $u \in \mathrm{Im}_n(\varphi)$ there exists $X \in GL_n(B)$ such that $d_n\varphi_*(X) = u$. Choose $\lambda \in B^*$ such that $[\lambda] = \delta_n(X) \in (B^*)^{ab}$ and let

$$\Delta = \Delta(\lambda, 1) = \begin{pmatrix} \lambda & & & \\ & 1 & & \\ & & \ddots & \\ & & & 1 \end{pmatrix} \in GL_n(B)$$

so that $\delta_n(\Delta) = [\lambda] = \delta_n(X)$. However, $\varphi\delta_n(X) = d_n\varphi_*(X) = u$. As $\Delta = s_{1,n-1}(\lambda)$ it follows that $\varphi\delta_n s_{1,n-1}(\lambda) = u$. By the stabilization property of determinants $\delta_n s_{1,n-1} = \delta_1$ and so $\varphi\delta_1(\lambda) = u$. Thus $u \in \mathrm{Im}_1(\varphi)$ and hence $\mathrm{Im}_n(\varphi) \subset \mathrm{Im}_1(\varphi)$. Together with (3.49) it follows that:

If $\varphi : B \to A$ is determinantal with respect to $\{d_n^A\}$

then $\mathrm{Im}_n(\varphi) = \mathrm{Im}_1(\varphi)$ for all $n \geq 1$.   (3.50)

A corner $\mathcal{A}$ will be said to be *compatibly determinantal* when A_0 is commutative and the homomorphisms $\varphi_\sigma : A_\sigma \to A_0$ are compatibly determinantal with respect to the canonical determinant of A_0. The fact of A_0 being commutative immediately implies that:

$$\overline{D_1}(\mathcal{A}) \equiv \overline{GL_1}(\mathcal{A}) \text{ when } \mathcal{A} \text{ is compatibly determinantal.} \tag{3.51}$$

Moreover, when $\mathcal{A}$ is compatibly determinantal it follows from (3.50) that $\text{Im}_n(\varphi_\sigma) = \text{Im}_1(\varphi_\sigma)$ for $n \geq 1$. Hence:

If $\mathcal{A}$ is compatibly determinantal then $\overline{D_n}(\mathcal{A}) \equiv \overline{GL_1}(\mathcal{A})$ and each $\pi_{n,k} = \text{Id}$.
$$\tag{3.52}$$

We obtain an improvement on Corollary 3.45.

Theorem 3.53 *Let $\mathcal{A}$ be a compatibly determinantal corner of locally free type in which A_0 is weakly Euclidean; if $\mathcal{A}$ satisfies the Milnor patching condition then $\widehat{A}$ has property SFC.*

Proof We saw in Proposition 3.44 that each $s_{1,k} : \overline{GL_1}(\mathcal{A}) \to \overline{GL_{k+1}}(\mathcal{A})$ is surjective. However, commutativity of the diagram below shows that d_{k+1} is left inverse to $s_{1,k}$.

$$\begin{array}{ccc} \overline{GL_1}(\mathcal{A}) & \xrightarrow{s_{1,k}} & \overline{GL_{k+1}}(\mathcal{A}) \\ \| & & \downarrow d_{k+1} \\ \overline{GL_1}(\mathcal{A}) & \xrightarrow{\text{Id}} & \overline{GL_1}(\mathcal{A}) \end{array}$$

Hence $s_{1,k}$ is bijective and the result follows from (3.40). $\qquad\square$

As any homomorphism between commutative rings is compatibly determinantal we obtain:

Corollary 3.54 *Let $\mathcal{A}$ be a corner of locally free type in which all rings are commutative and A_0 is weakly Euclidean; if $\mathcal{A}$ satisfies the Milnor patching condition then $\widehat{A}$ has property SFC.*

3.11 A Bound for the Singular Set

Let $\mathcal{A}$ be a compatibly determinantal corner; then clearly $d_k : \overline{GL_k}(\mathcal{A}) \to \overline{D_k}(\mathcal{A})$ is surjective. Moreover, $(A_0^*)^{ab} = A_0^*$. We define the *exceptional fibre* $\mathcal{E}_k(\mathcal{A})$ by

$$\mathcal{E}_k(\mathcal{A}) = d_k^{-1}(*).$$

Evidently $\mathcal{E}_k(\mathcal{A}) \subset \overline{GL_k}(\mathcal{A})$ and $s_{n,k}(\mathcal{E}_n(\mathcal{A})) \subset \mathcal{E}_{n+k}(\mathcal{A})$. Moreover, $\mathcal{E}_n(\mathcal{A})$ contains the distinguished point of $\overline{GL_n}(\mathcal{A})$. Thus $\mathcal{A} \mapsto \mathcal{E}_n(\mathcal{A})$ defines a functor

$$\mathcal{E}_n : \{\text{Compatibly determinantal corners}\} \to \textbf{SETS}_*$$

and we have a commutative diagram:

$$
\begin{array}{ccc}
\mathcal{E}_n(\mathcal{A}) & \xrightarrow{s_{n,k}} & \mathcal{E}_{n+k}(\mathcal{A}) \\
\cap & & \cap \\
\overline{GL_n}(\mathcal{A}) & \xrightarrow{s_{n,k}} & \overline{GL_{n+k}}(\mathcal{A}) \\
\downarrow d_n & & \downarrow d_{n+k} \\
\overline{D_n}(\mathcal{A}) & \xrightarrow{\pi_{n,k}} & \overline{D_{n+k}}(\mathcal{A})
\end{array}
$$

Suppose $\mathcal{A}$ is compatibly determinantal and that $\zeta \in \mathcal{Z}_n(\mathcal{A})$; then $\zeta \in \overline{GL_n}(\mathcal{A})$ and $s_{n,k}(\zeta) = *$ for some $k \geq 1$. From the commutativity of the diagram

$$
\begin{array}{ccc}
\overline{GL_n}(\mathcal{A}) & \xrightarrow{s_{n,k}} & \overline{GL_{n+k}}(\mathcal{A}) \\
\downarrow d_n & & \downarrow d_{n+k} \\
\overline{GL_1}(\mathcal{A}) & \xrightarrow{\mathrm{Id}} & \overline{GL_1}(\mathcal{A})
\end{array}
$$

it follows that $d_n(\zeta) = *$ so that $\zeta \in \mathcal{E}_n(\mathcal{A})$. Hence:

Let $\mathcal{A}$ be compatibly determinantal; then $\mathcal{Z}_n(\mathcal{A}) \subset \mathcal{E}_n(\mathcal{A})$ for $n \geq 1$.   (3.55)

If B is a fully determinantal ring with preferred determinant $\{d_n^B\}$ we define

$$
SL_n(B) = \{X \in GL_n(B) \mid d_n^1(X) = [1] \in (B^*)^{ab}\}.
$$

Proposition 3.56 *Let $\varphi : B \to A$ be a compatibly determinantal ring homomorphism; if $Y \in GL_n(B)$ then there exist $\Delta \in GL_n(A)$ and $Y', Y'' \in SL_n(B)$ such that*

$$
\varphi_*(Y) = \varphi_*(Y')\Delta = \Delta\varphi_*(Y'').
$$

Proof Let $(\delta_n)_{1 \leq n}$ denote the determinant of B which renders φ compatible. Choose $\gamma \in B^*$ such that $[\gamma] = \delta_n(Y) \in (B^*)^{ab}$ and put

$$
\Gamma = \Delta(\gamma, 1) = \begin{pmatrix} \gamma & & & \\ & 1 & & \\ & & \ddots & \\ & & & 1 \end{pmatrix} \in GL_n(B).
$$

Now put $\Delta = \varphi_*(\Gamma)$, $Y' = Y\Gamma^{-1}$ and $Y'' = \Gamma^{-1}Y$. Then $\varphi_*(Y) = \varphi_*(Y')\Delta = \Delta\varphi_*(Y'')$. Moreover

$$
\delta_n(Y') = \delta_n(Y)[\gamma^{-1}] = [\gamma][\gamma^{-1}] = [1]
$$

so that $Y' \in SL_n(B)$. Similarly, $Y'' \in SL_n(B)$. $\square$

When $\lambda \in A_0^*$ we denote by $\langle\lambda\rangle_n$ its class in $\mathrm{Im}_n(\varphi_-)\backslash A_0^*/\mathrm{Im}_n(\varphi_+)$.

Lemma 3.57 *Suppose that $\mathcal{A}$ is compatibly determinantal and that $X \in GL_n(A_0)$ satisfies $\langle d_n(X) \rangle_n = \langle 1 \rangle_n$; then there exists $X_\sigma \in SL_n(A_\sigma)$ ($\sigma \in \{-, 0, +\}$) such that*

$$X = \varphi_-(X_-) X_0 \varphi_+(X_+).$$

Proof The hypothesis that $X \in GL_n(A_0)$ satisfies $\langle d_n(X) \rangle_n = \langle 1 \rangle_n$ means that there exist $Z_- \in GL_n(A_-)$ and $Z_+ \in GL_n(A_+)$ such that

$$d_n(X) = \varphi_-(d_n(Z_-)) \varphi_+(d_n(Z_+)) \in A_0^*.$$

Put $Y_- = Z_-^{-1}$ $Y_+ = Z_+^{-1}$. Then $\varphi_-(d_n(Y_-)) d_n(X) \varphi_+(d_n(Y_+)) = 1 \in A_0^*$. Thus

$$d_n[\varphi_-(Y_-) X \varphi_+(Y_+)] = 1 \in A_0^*. \tag{$*$}$$

By (3.55) we may write $\varphi_-(Y_-) = \varphi_-(X_-) \Delta_-$ and $\varphi_+(Y_+) = \Delta_+ \varphi_+(X_+)$ where $\Delta_-, \Delta_+ \in GL_n(A_0)$, $X_- \in SL_n(A_-)$ and $X_+ \in SL_n(A_+)$. Putting $X_0 = \Delta_- X \Delta_+$ then

$$X = \varphi_-(X_-) X_0 \varphi_+(X_+).$$

It suffices to show that $X_0 \in SL_n(A_0)$. However, $X_- \in SL_n(A_-)$ and $X_+ \in SL_n(A_+)$. Thus we have $\varphi_-(d_n(X_-)) = \varphi_+(d_n(X_+)) = [1] \in A_0^*$. Hence

$$
\begin{aligned}
d_n(X_0) &= \varphi_-(d_n(X_-)) d_n(X_0) \varphi_+(d_n(X_+)) \\
&= d_n \varphi_-(X_-) d_n(X_0) d_n \varphi_+(X_+) \\
&= d_n[\varphi_-(X_-) X_0 \varphi_+(X_+)] \\
&= d_n[\varphi_-(X_-) \Delta_- X \Delta_+ \varphi_+(X_+)] \\
&= d_n[\varphi_-(Y_-) X \varphi_+(Y_+)] \\
&= 1
\end{aligned}
$$

so that $X_0 \in SL_n(A_0)$ as claimed. $\qquad\square$

We denote by $\langle X \rangle$, denote the class of $X \in SL_n(A_0)$ in $\overline{SL}_n(\mathcal{A})$ and by $[Y]$ the class of $Y \in GL_n(A_0)$ in $\overline{GL}_n(\mathcal{A})$. We define a mapping $\natural : \overline{SL}_n(\mathcal{A}) \to \overline{GL}_n(\mathcal{A})$ by

$$\natural : \langle X \rangle = [X].$$

It is clear that $\mathrm{Im}(\natural) \subset \mathcal{E}_n(\mathcal{A})$ where $\mathcal{E}_n(\mathcal{A})$ is the exceptional fibre. It follows immediately from (3.55) that:

$$\overline{SL}_n(\mathcal{A}) \overset{\natural}{\to} \mathcal{E}_n(\mathcal{A}) \text{ is surjective for any compatibly determinantal corner } \mathcal{A}. \tag{3.58}$$

There is a corresponding mapping $\widetilde{\natural} : \overline{E}_n(\mathcal{A}) \to \mathcal{E}_n(\mathcal{A})$ by $\widetilde{\natural} : \langle X \rangle = [X]$. Moreover if A_0 is weakly Euclidean then $SL_n(A_0) = E_n(A_0))$ so that the assignment $X \mapsto \langle X \rangle \in \overline{SL}_n(\mathcal{A})$ induces a surjection $\nu : \overline{E}_n(\mathcal{A}) \to \overline{SL}_n(\mathcal{A})$ making the following commute.

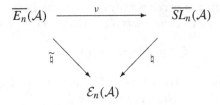

It follows immediately that:

Let $\mathcal{A}$ be a compatibly determinantal corner in which A_0 is weakly Euclidean;

then $\widetilde{\mathfrak{h}} : \overline{E_n}(\mathcal{A}) \to \mathcal{E}_n(\mathcal{A})$ is surjective.   (3.59)

We obtain an analogue of Theorem 3.53.

Theorem 3.60 *Let $\mathcal{A}$ be a compatibly determinantal locally free corner in which A_0 is weakly Euclidean; if $\widehat{A}$ is Karoubi then every nontrivial stably free $\widehat{A}$-module has rank 2.*

Proof When $\widehat{A}$ is Karoubi we saw in Corollary 3.31 that $\overline{E_n}(\mathcal{A}) = \{*\}$ for $3 \leq n$. Thus by (3.59) the only possibilities for $\mathcal{E}_n(\mathcal{A})$ and hence $\mathcal{Z}_n(\mathcal{A})$ to be nontrivial are $n = 1, 2$. However, together with Proposition 3.43, the argument of Theorem 3.53 shows that $s_{1,k}$ is bijective for $k \geq 2$ so that $\mathcal{Z}_1(\mathcal{A}) = \{*\}$. Thus the only possibility for $\mathcal{Z}_n(\mathcal{A})$ to contain a nontrivial elements is $n = 2$. The result now follows from (3.39). $\square$

Chapter 4
Extensions of Modules

In this chapter we develop the classical theory of module extensions and, in particular, its interpretation in terms of cohomology. As explained in the Introduction, to achieve our intended application it is necessary to avoid injective modules. In consequence a feature of our treatment is the development of the exact sequences of Ext using only projective modules.

4.1 The Category of Extensions

We denote by $\mathbf{Ext}^1_\Lambda$ the collection of exact sequences of Λ-modules and homomorphisms

$$\mathbf{E} = (0 \to E_+ \to E_0 \to E_- \to 0)$$

in which the modules E_+, E_0 and E_- are all in $\mathcal{M}od_\Lambda$. $\mathbf{Ext}^1_\Lambda$ can be regarded as a category by taking morphisms to be commutative diagrams of Λ-homomorphisms thus:

$$
\begin{matrix}
\mathbf{E} \\
\downarrow h = \\
\mathbf{F}
\end{matrix}
\begin{pmatrix}
0 \to E_+ \to E_0 \to E_- \to 0 \\
\quad \downarrow h_+ \quad \downarrow h_0 \quad \downarrow h_- \\
0 \to F_+ \to F_0 \to F_- \to 0
\end{pmatrix}.
$$

For $A, B \in \mathcal{M}od_\Lambda$, $\mathbf{Ext}^1_\Lambda(A, B)$ will denote the full subcategory of $\mathbf{Ext}^1_\Lambda$ whose objects $\mathbf{E}$ satisfy $E_- = A$ and $E_+ = B$. If $\mathbf{E}, \mathbf{F} \in \mathbf{Ext}^1_\Lambda(A, B)$, a morphism $\varphi : \mathbf{E} \to \mathbf{F}$ is said to be a *congruence* when it induces the identity at both ends thus:

$$
\begin{matrix}
\mathbf{E} \\
\downarrow \varphi = \\
\mathbf{F}
\end{matrix}
\begin{pmatrix}
0 \to B \to E_0 \to A \to 0 \\
\quad \downarrow \mathrm{Id} \quad \downarrow \varphi_0 \quad \downarrow \mathrm{Id} \\
0 \to B \to F_0 \to A \to 0
\end{pmatrix}.
$$

F.E.A. Johnson, *Syzygies and Homotopy Theory*, Algebra and Applications 17, DOI 10.1007/978-1-4471-2294-4_4, © Springer-Verlag London Limited 2012

We write '$\mathbf{E} \equiv \mathbf{F}$' when $\mathbf{E}$ $\mathbf{F}$ are congruent. By the Five Lemma, congruence is an equivalence relation on $\mathbf{Ext}^1_\Lambda(A, B)$. We denote by $\mathrm{Ext}^1_\Lambda(A, B)$ the collection of equivalence classes in $\mathbf{Ext}^1_\Lambda(A, B)$ under '$\equiv$'. Elementary considerations show that $\mathbf{Ext}^1(A, B)$ is equivalent to a small category, so that $\mathrm{Ext}^1(A, B)$ is actually a set. When the ring Λ is understood we omit the suffix and write $\mathbf{Ext}^1_\Lambda(A, B) = \mathbf{Ext}^1(A, B)$, $\mathrm{Ext}^1_\Lambda(A, B) = \mathrm{Ext}^1(A, B)$.

For any Λ-modules A, B there is a distinguished extension, the *trivial extension*

$$\mathcal{T} = (0 \to B \overset{i_B}{\to} B \oplus A \overset{\pi_A}{\to} A \to 0),$$

where $i_B(b) = (b, 0)$ and $\pi_A(b, a) = a$. An extension

$$\mathcal{F} = (0 \to B \overset{j}{\to} X \overset{p}{\to} A \to 0)$$

is said to *split* when it is congruent to the trivial extension; that is, when there exists an isomorphism $\varphi : X \to B \oplus A$ making the following diagram commute:

$$
\begin{array}{ccccccccc}
0 \to & B & \overset{j}{\to} & X & \overset{p}{\to} & A & \to 0 \\
 & \downarrow \mathrm{Id}_B & & \downarrow \varphi & & \downarrow \mathrm{Id}_A \\
0 \to & B & \overset{i}{\to} & B \oplus A & \overset{\pi}{\to} & A & \to 0
\end{array}
$$

$\mathcal{F}$ is said to *split on the right* when there exists a Λ-homomorphism $s : A \to X$ such that $p \circ s = \mathrm{Id}_A$. $\mathcal{F}$ is said to *split on the left* when there exists a Λ-homomorphism $r : X \to B$ such that $r \circ j = \mathrm{Id}_B$.

Suppose that $\mathcal{F} = (0 \to B \overset{j}{\to} X \overset{p}{\to} A \to 0)$ is a short exact sequence of Λ-modules; if $\varphi : X \to B \oplus A$ is a splitting there is a right splitting s given by $s = \varphi^{-1} \circ i_A$ where $i_A : A \to B \oplus A$ is the standard inclusion $i_A(a) = (0, a)$. There is also a left splitting r given by $r = \pi_B \circ \varphi$ where $\pi_B : B \oplus A \to A$ is the projection $\pi_B(b, a) = b$. Conversely, if $s : A \to X$ is a right splitting, there is a splitting φ whose the inverse, φ^{-1}, takes the form

$$\varphi^{-1}(b, a) = i_B(b) + s(a),$$

whilst if $r : X \to B$ is a left splitting, there is a corresponding splitting φ given by

$$\varphi(x) = (r(x), p(x)).$$

To summarize, we have shown the well known:

If $\mathcal{F} = (0 \to B \overset{j}{\to} X \overset{p}{\to} A \to 0)$ is a short exact sequence of Λ-modules; then

$$\mathcal{F} \text{ splits} \iff \mathcal{F} \text{ splits on the right} \iff \mathcal{F} \text{ splits on the left.} \tag{4.1}$$

In Sect. 4.2 we shall see that $\mathrm{Ext}^1(A, B)$ possesses a natural group structure in which the class of the trivial extension acts as the identity. In order to describe the group multiplication we first recall some natural constructions on $\mathbf{Ext}^1(A, B)$.

Pushout   Let A, B_1, B_2 be Λ-modules; if $f : B_1 \to B_2$ is a Λ-homomorphism and $\mathbf{E} = (0 \to B_1 \overset{i}{\to} E_0 \overset{\eta}{\to} A \to 0) \in \mathbf{Ext}^1(A, B_1)$ we put

$$f_*(\mathbf{E}) = \left(0 \to B_2 \overset{j}{\to} \varinjlim(f, i) \overset{\epsilon}{\to} A \to 0\right),$$

where $\varinjlim(f, i) = (B_2 \oplus E_0)/\mathrm{Im}(f \times -i)$ denotes the colimit and j is the injection $j : B_2 \to \varinjlim(f, i)$; $j(x) = [x, 0]$. The correspondence $\mathbf{E} \mapsto f_*(\mathbf{E})$ determines the 'pushout' functor $f_* : \mathbf{Ext}^1(A, B_1) \to \mathbf{Ext}^1(A, B_2)$. If in addition $g : B_2 \to B_3$, it is straightforward to see that

$$(g \circ f)_*(\mathbf{E}) = g_* f_*(\mathbf{E}).$$

Furthermore, there is a natural transformation $\nu_f : \mathrm{Id} \to f_*$ obtained as follows:

$$
\begin{array}{cc}
\mathbf{E} \\
\downarrow \nu_f \quad = \\
f_*(\mathbf{E})
\end{array}
\left(
\begin{array}{ccc}
0 \to B_1 \overset{i}{\to} & E_0 & \to A \to 0 \\
\downarrow f \quad \downarrow \nu & \downarrow \mathrm{Id} \\
0 \to B_2 \to \varinjlim(f, i) & \to A \to 0
\end{array}
\right),
$$

where $\nu : E_0 \to \varinjlim(f, i)$ is the mapping $\nu(x) = [0, x]$.

Pullback   Let A_1, A_2, B be Λ-modules; if $\mathbf{E} = (0 \to B \to E_0 \overset{\eta}{\to} A_2 \to 0) \in \mathbf{Ext}^1(A_2, B)$ and $f : A_1 \to A_2$ is a Λ-homomorphism we put

$$f^*(\mathbf{E}) = (0 \to B \to \varprojlim(\eta, f) \overset{\epsilon}{\to} A_1 \to 0),$$

where $\varprojlim(\eta, f) = E_0 \underset{\eta, f}{\times} A_1 = \{(x, y) : \eta(x) = f(y)\}$ is the fibre product and $\epsilon : F_0 \to A_1$ is the projection $\epsilon(x, y) = y$. If $g : A_2 \to A_3$ is a Λ-homomorphism, it is straightforward to see that $(g \circ f)^*(\mathbf{E}) = f^* \circ g^*(\mathbf{E})$. The correspondence $\mathbf{E} \mapsto f^*(\mathbf{E})$ thus defines the 'pullback functor' $f^* : \mathbf{Ext}^1(A_2, B) \to \mathbf{Ext}^1(A_1, B)$. There is a natural transformation $\mu_f : f^* \to \mathrm{Id}$ defined by:

$$
\begin{array}{cc}
f^*(\mathbf{E}) \\
\downarrow \mu_f \quad = \\
\mathbf{E}
\end{array}
\left(
\begin{array}{ccc}
0 \to B \to F_0 \to A_1 \to 0 \\
\downarrow \mathrm{Id} \quad \downarrow \mu_0 \quad \downarrow f \\
0 \to B \to E_0 \to A_2 \to 0
\end{array}
\right),
$$

where $\mu_0 : F_0 \to E_0$ is the projection $\mu_0(x, y) = x$.

Direct Product   Let A_1, A_2, B_1, B_2 be Λ-modules and for $r = 1, 2$ let

$$\mathbf{E}(r) = \left(0 \to B_r \to E(r)_0 \to A_r \to 0\right) \in \mathbf{Ext}^n(A_r, B_r).$$

Then $\mathbf{E}(1) \times \mathbf{E}(2) = (0 \to B_1 \times B_2 \to E(1)_0 \times E(2)_0 \to A_1 \times A_2 \to 0)$ is exact, and we get a functorial pairing $\times : \mathbf{Ext}^1(A_1, B_1) \times \mathbf{Ext}^1(A_2, B_2) \to \mathbf{Ext}^1(A_1 \oplus A_2, B_1 \oplus B_2)$. Generalizing the statement of (4.1) for right splittings we have a criterion for a pushout extension to be trivial.

Proposition 4.2 *Let $\mathcal{E} = (0 \to A \xrightarrow{i} B \xrightarrow{p} C \to 0)$ be an exact sequence of Λ-modules and let $\alpha : A \to N$ be a Λ-homomorphism; then the following two statements are equivalent:*

(i) *$\alpha_*(\mathcal{E})$ splits;*
(ii) *there exists a homomorphism $\hat{\alpha} : B \to N$ making the following diagram commute:*

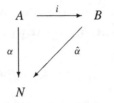

Proof Suppose that $\alpha_*(\mathcal{E})$ splits and let $\varphi : \varinjlim(\alpha, i) \to N \oplus C$ be a splitting. In particular, the following diagram commutes:

$$
\begin{array}{ccccccccc}
0 & \longrightarrow & A & \xrightarrow{i} & B & \xrightarrow{p} & C & \longrightarrow & 0 \\
 & & \downarrow{\alpha} & & \downarrow{\nu} & & \downarrow{\mathrm{Id}_C} & & \\
0 & \longrightarrow & N & \xrightarrow{j} & \varinjlim(\alpha, i) & \xrightarrow{\pi} & C & \longrightarrow & 0 \\
 & & \downarrow{\mathrm{Id}_N} & & \downarrow{\varphi} & & \downarrow{\mathrm{Id}_C} & & \\
0 & \longrightarrow & N & \xrightarrow{i_N} & N \oplus C & \xrightarrow{\pi_C} & C & \longrightarrow & 0
\end{array}
$$

On putting $\hat{\alpha} = \pi_N \circ \varphi \circ \nu$ it is easy to check that the diagram

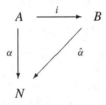

commutes as required. This proves (i) $\Longrightarrow$ (ii).

Conversely, if there is a homomorphism $\hat{\alpha} : B \to N$ such that $\hat{\alpha} \circ i = \alpha$ then the mapping $r : \varinjlim(\alpha, i) \to A$ given by $r[x, y] = x + \hat{\alpha}(y)$ is a left splitting for $\alpha_*(\mathcal{E})$. Thus (ii) $\Longrightarrow$ (i) and this completes the proof. $\qquad\square$

Corresponding to Proposition 4.2 is the dual statement for a pullback extension to be trivial:

Proposition 4.3 *Let $\mathcal{E} = (0 \to A \xrightarrow{i} B \xrightarrow{p} C \to 0)$ be an exact sequence of Λ-modules and let $\gamma : M \to C$ be a Λ-homomorphism; then the following two statements are equivalent:*

(i) *$\gamma^*(\mathcal{E})$ splits;*
(ii) *there exists a homomorphism $\tilde{\gamma} : M \to C$ making the following diagram commute:*

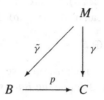

Proof Suppose that $\gamma_*(\mathcal{E})$ splits and let $\varphi : \varprojlim(p, \gamma) \to A \oplus M$ be a splitting. In particular, the following diagram commutes:

$$
\begin{array}{ccccccccc}
0 & \longrightarrow & A & \xrightarrow{\ i_A\ } & A \oplus M & \xrightarrow{\ \pi_M\ } & M & \longrightarrow & 0 \\
 & & \downarrow{\scriptstyle \mathrm{Id}_A} & & \downarrow{\scriptstyle \varphi^{-1}} & & \downarrow{\scriptstyle \mathrm{Id}_M} & & \\
0 & \longrightarrow & A & \xrightarrow{\ j\ } & \varprojlim(\alpha, i) & \xrightarrow{\ \pi\ } & C & \longrightarrow & 0 \\
 & & \downarrow{\scriptstyle \mathrm{Id}_A} & & \downarrow{\scriptstyle \nu} & & \downarrow{\scriptstyle \gamma} & & \\
0 & \longrightarrow & A & \xrightarrow{\ i\ } & B & \xrightarrow{\ p\ } & C & \longrightarrow & 0
\end{array}
$$

On putting $\hat{\gamma} = \nu \circ \varphi^{-1} \circ i_M$ where $i_M : M \to A \oplus M$ is the standard inclusion, it is easy to check that the diagram

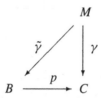

commutes as required, and this proves (i) $\Longrightarrow$ (ii).

Conversely, if there is a homomorphism $\hat{\gamma} : M \to B$ such that $p \circ \hat{\gamma} = \gamma$ then the mapping $s : M \to \varprojlim(p, \gamma)$ given by

$$s(x) = (\hat{\gamma}(x), x)$$

is a right splitting for $\gamma^*(\mathcal{E})$. Thus (ii) $\Longrightarrow$ (i) and this completes the proof.   □

4.2 The Group Structure on Ext^1

There is a natural group structure on $\mathrm{Ext}^1(A, B)$ which we proceed to describe; first note that direct product gives a functorial pairing

$$\times : \mathbf{Ext}^1(A_1, B_1) \times \mathbf{Ext}^1(A_2, B_2) \to \mathbf{Ext}^1(A_1 \oplus A_2, B_1 \oplus B_2).$$

For Λ-modules A, B_1, B_2 there is a functorial pairing, *external sum*,

$$\oplus : \mathbf{Ext}^1(A, B_1) \times \mathbf{Ext}^1(A, B_2) \to \mathbf{Ext}^1(A, B_1 \oplus B_2)$$

given by $\mathcal{E}_1 \oplus \mathcal{E}_2 = \Delta^*(\mathcal{E}_1 \times \mathcal{E}_2)$. Where $\Delta : A \to A \times A$ is the diagonal. The addition map $+ : B \times B \to B$ can also be regarded as a Λ-homomorphism

$$\alpha : B \oplus B \to B; \qquad \alpha(b_1, b_2) = b_1 + b_2.$$

Combining external sum with pushout, we obtain the 'Baer sum' [5]; let $\mathcal{E}_r \in$ **Ext**$^1(A, B)$ for $r = 1, 2$, and define the *Baer sum* $\mathcal{E}_1 + \mathcal{E}_2$ by

$$\mathcal{E}_1 + \mathcal{E}_2 = \alpha_*(\mathcal{E}_1 \oplus \mathcal{E}_2) \quad (= \alpha_* \Delta^*(\mathcal{E}_1 \times \mathcal{E}_2)).$$

This gives a functorial pairing

$$+ : \mathbf{Ext}^1(A, B) \times \mathbf{Ext}^1(A, B) \to \mathbf{Ext}^1(A, B).$$

It is straightforward to see that congruence in **Ext**^{1} is compatible with Baer sum.

From the identity on diagonal maps $(\Delta \times \mathrm{Id}) \circ \Delta = (\mathrm{Id} \times \Delta) \circ \Delta$ we see easily that for $\mathcal{E}_1, \mathcal{E}_2, \mathcal{E}_3 \in \mathbf{Ext}^1(A, B)$ then $\mathcal{E}_1 \oplus (\mathcal{E}_2 \oplus \mathcal{E}_3) \equiv (\mathcal{E}_1 \oplus \mathcal{E}_2) \oplus \mathcal{E}_3$. Furthermore by associativity of addition (in B) there is a commutative diagram of morphisms in **Ext**^{1}

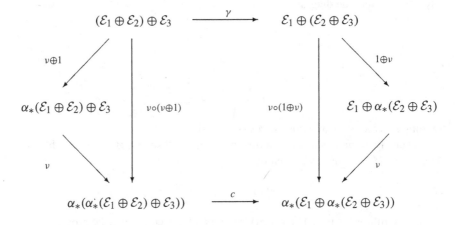

in which each ν is an instance of the natural transformation $\nu_\alpha : \mathrm{Id} \to \alpha_*$ and in which γ and c are congruences. It follows immediately that:

Proposition 4.4 *If* $\mathcal{E}_1, \mathcal{E}_2, \mathcal{E}_3 \in \mathbf{Ext}^1(A, B)$ *then* $\mathcal{E}_1 + (\mathcal{E}_2 + \mathcal{E}_3) \equiv (\mathcal{E}_1 + \mathcal{E}_2) + \mathcal{E}_3$.

Rather more easily, commutativity of addition and the obvious congruence $\mathcal{E} \oplus \mathcal{F} \equiv \mathcal{F} \oplus \mathcal{E}$ show that:

Proposition 4.5 *If* $\mathcal{E}, \mathcal{F} \in \mathbf{Ext}^1(A, B)$ *then* $\mathcal{E} + \mathcal{F} \equiv \mathcal{F} + \mathcal{E}$.

We denote $\mathcal{E}$ the extension $\mathcal{E} = (0 \to B \xrightarrow{i} X \xrightarrow{p} A \to 0)$ and by $\mathcal{T}$ the trivial extension $\mathcal{T} = (0 \to B \to B \oplus A \to A \to 0)$.

Proposition 4.6 *If $\mathcal{E} \in$ **Ext**$^1(A, B)$ then $\mathcal{E} + \mathcal{T} \equiv \mathcal{E}$.*

Proof We may write

$$\mathcal{E} \oplus \mathcal{T} = (0 \to B \times B \xrightarrow{j \times i} X \underset{A}{\times} (B \oplus A) \xrightarrow{\pi} A \to 0),$$

where $X \underset{A}{\times} (B \oplus A) = \{(x, b, a) \in X \times B \times A : p(x) = a\}$, $j \times i(b_1, b_2) = (j(b_1), b_2, 0)$ and $\pi(x, b, a) = p(x)$. There is a morphism of extensions

$$
\begin{array}{c}
\mathcal{E} \oplus \mathcal{T} \\
\downarrow \tilde{\alpha} \\
\mathcal{E}
\end{array}
=
\left(
\begin{array}{ccccc}
0 \to & B \times B & \xrightarrow{j \times i} & X \underset{A}{\times} (B \oplus A) & \xrightarrow{\pi} A \to 0 \\
 & \downarrow \alpha & & \downarrow \tilde{\alpha} & \downarrow \text{Id} \\
0 \to & B & \xrightarrow{j} & X & \to A \to 0
\end{array}
\right)
$$

where, as above, $\alpha(b_1, b_2) = b_1 + b_2$ and $\tilde{\alpha}(x, b, a) = x + j(b)$. Clearly $\tilde{\alpha}$ admits the canonical factorization

$$\mathcal{E} \oplus \mathcal{T} \xrightarrow{\nu} \alpha_*(\mathcal{E} \oplus \mathcal{T}) \xrightarrow{c} \mathcal{E},$$

where $c : \alpha_*(\mathcal{E} \oplus \mathcal{T}) \to \mathcal{E}$ is a congruence. The result follows since $\alpha_*(\mathcal{E} \oplus \mathcal{T}) = \mathcal{E} + \mathcal{T}$.   □

We denote by $-\mathcal{E}$ the extension $-\mathcal{E} = (0 \to B \xrightarrow{i} X \xrightarrow{-p} A \to 0)$.

Proposition 4.7 *If $\mathcal{E} \in$ **Ext**$^1(A, B)$ then $\mathcal{E} + (-\mathcal{E}) \equiv \mathcal{T}$.*

Proof First observe that $\Delta^*(\mathcal{E} \times -\mathcal{E})$ may be identified, by an obvious natural congruence, with the extension

$$(0 \to B \times B \xrightarrow{j \times j} X \underset{p, -p}{\times} X \xrightarrow{p_1} A \to 0),$$

where $p_1(x_1, x_2) = p(x_1)$. There is a homomorphism $\tilde{\alpha} : X \underset{p, -p}{\times} X \to \text{Ker}(p)$ defined by

$$\tilde{\alpha}(x_1, x_2) = x_1 + x_2.$$

On putting $\hat{\alpha}(x_1, x_2) = j^{-1}(\tilde{\alpha}(x_1, x_2) \, (= j^{-1}(x_1 + x_2))$ we see that the following diagram commutes:

$$B \times B \xrightarrow{\ j \times j\ } \underset{p,-p}{X \times X}$$

$$\alpha \downarrow \qquad \nearrow \widehat{\alpha}$$

$$B$$

It follows from Proposition 4.2 that $\mathcal{E} + (-\mathcal{E}) \equiv \mathcal{T}$. $\square$

Observe that we have a congruence

$$0 \to B \xrightarrow{-j} X \xrightarrow{p} A \to 0$$
$$\downarrow \mathrm{Id} \quad \downarrow -\mathrm{Id} \quad \downarrow \mathrm{Id}$$
$$0 \to B \xrightarrow{j} X \xrightarrow{-p} A \to 0$$

so that the additive inverse of $\mathcal{E}$ is equally well represented by the extension

$$(0 \to B \xrightarrow{-j} X \xrightarrow{p} A \to 0).$$

Corollary 4.8 $\mathrm{Ext}^1(A, B)$ *is an abelian group with respect to Baer sum.*

Observe that if Q is projective then any exact sequence $(0 \to N \to X \to Q \to 0)$ splits. However $0 \in \mathrm{Ext}^1(Q, N)$ is defined by the split sequence so that:

If Q is a projective Λ-module then $\mathrm{Ext}^1(Q, N) = 0$ for any N.   (4.9)

If $f : A_1 \to A_2$ is a Λ-homomorphism the correspondence $\mathcal{E} \mapsto f^*(\mathcal{E})$ gives a functor $f^* : \mathbf{Ext}^1(A_2, N) \to \mathbf{Ext}^1(A_1, N)$; it is straightforward to see that $f^*(\mathcal{E}_1 + \mathcal{E}_2) \equiv f^*(\mathcal{E}_1) + f^*(\mathcal{E}_2)$ so that f induces a homomorphism of abelian groups $f^* : \mathrm{Ext}^1(A_2, B) \to \mathrm{Ext}^1(A_1, B)$.

Similarly, if $g : B_1 \to B_2$ is a Λ-homomorphism the correspondence $\mathcal{E} \mapsto g_*(\mathcal{E})$ gives a functor $g_* : \mathbf{Ext}^1(A, B_1) \to \mathbf{Ext}^1(A, B_2)$ and

$$g_*(\mathcal{E}_1 + \mathcal{E}_2) \equiv g_*(\mathcal{E}_1) + g_*(\mathcal{E}_2).$$

Thus g induces a homomorphism of abelian groups $g_* : \mathrm{Ext}^1(A, B_1) \to \mathrm{Ext}^1(A, B_2)$.

4.3 The Exact Sequences of Ext^1

Given an exact sequence of Λ-modules $\mathcal{E} = (0 \to A \xrightarrow{i} B \xrightarrow{p} C \to 0)$, there is a mapping $\delta : \text{Hom}_\Lambda(A, N) \to \text{Ext}^1(C, N)$, the *connecting mapping*, given by

$$\delta(\alpha) = [\alpha_*(\mathcal{E})].$$

It is straightforward to check that:

$\delta : \text{Hom}_\Lambda(A, N) \to \text{Ext}^1(C, N)$ is a homomorphism of abelian groups.   (4.10)

We again omit the suffix when the ring Λ is understood and write $\text{Hom}_\Lambda^1(A, B) = \text{Hom}^1(A, N)$. From (4.10) we obtain a sequence of abelian groups

$$0 \to \text{Hom}(C, N) \xrightarrow{p^*} \text{Hom}(B, N) \xrightarrow{i^*} \text{Hom}(A, N) \xrightarrow{\delta} \text{Ext}^1(C, N)$$

$$\xrightarrow{p^*} \text{Ext}^1(B, N) \xrightarrow{i^*} \text{Ext}^1(A, N). \tag{$*$}$$

We will show that:

Theorem 4.11 *The sequence* $(*)$ *is exact for any* Λ-*module* N.

Proof Exactness of the segment $0 \to \text{Hom}(C, N) \xrightarrow{p^*} \text{Hom}(B, N) \xrightarrow{i^*} \text{Hom}(A, N)$ is straightforward, so it suffices to verify exactness in the following:

$$\text{Hom}(B, N) \xrightarrow{i^*} \text{Hom}(A, N) \xrightarrow{\delta} \text{Ext}^1(C, N), \tag{4.12}$$

$$\text{Hom}(A, N) \xrightarrow{\delta} \text{Ext}^1(C, N) \xrightarrow{p^*} \text{Ext}^1(B, N), \tag{4.13}$$

$$\text{Ext}^1(C, N) \xrightarrow{p^*} \text{Ext}^1(B, N) \xrightarrow{i^*} \text{Ext}^1(A, N). \tag{4.14}$$

To prove exactness of (4.12), let $\beta \in \text{Hom}(B, N)$ and consider the natural transformation

$$\begin{array}{ccc} \mathcal{E} \\ \downarrow \nu \\ (\beta \circ i)_*(\mathcal{E}) \end{array} = \left(\begin{array}{ccccccc} 0 \to & A & \xrightarrow{i} & B & \xrightarrow{p} C & \to 0 \\ & \downarrow \beta \circ i & & \downarrow \nu & \downarrow \text{Id} \\ 0 \to & N & \to & \varinjlim(\beta \circ i, i) & \xrightarrow{\pi} C & \to 0 \end{array} \right).$$

Since the following diagram commutes

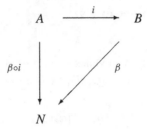

it follows from Proposition 4.2 that the extension $(\beta \circ i)_*(\mathcal{E})$ splits. However $\delta \circ i_*(\beta) = (\beta \circ i)_*(\mathcal{E})$ so that $\delta \circ i_* = 0$ and $\mathrm{Im}(i_*) \subset \mathrm{Ker}(\delta)$.

To show that $\mathrm{Ker}(\delta) \subset \mathrm{Im}(i_*)$ suppose that $\alpha \in \mathrm{Hom}(A, N)$ satisfies $\delta(\alpha) = 0$. Then $\alpha_*(\mathcal{E})$ splits, so that by Proposition 4.2 there exists a homomorphism making the diagram

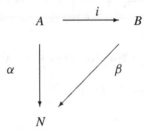

commute. Thus $\alpha = i^*(\beta)$ and $\mathrm{Ker}(\delta) \subset \mathrm{Im}(i_*)$ as required.

To prove exactness of (4.13), let $\alpha \in \mathrm{Hom}(A, N)$; then there a natural transformation $\nu_\alpha : \mathcal{E} \to \alpha_*(\mathcal{E})$ given by

$$
\begin{array}{c}
\mathcal{E} \\
\downarrow \nu_\alpha = \\
\alpha_*(\mathcal{E})
\end{array}
\left(
\begin{array}{ccccccc}
0 \to & A & \xrightarrow{i} & B & \xrightarrow{p} & C \to 0 \\
 & \downarrow \alpha & & \downarrow \nu_\alpha & & \downarrow \mathrm{Id} \\
0 \to & N \to & \varinjlim(\alpha, i) & \xrightarrow{\pi} & C \to 0
\end{array}
\right).
$$

Since the following diagram commutes

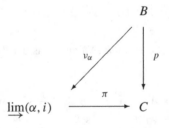

it follows by Proposition 4.3 that the extension $p^*\alpha_*(\mathcal{E})$ splits. However, $p^*\delta[\mathcal{E}] = [p^*\alpha_*(\mathcal{E})]$ so that $p^*\delta = 0$ and $\mathrm{Im}(\delta) \subset \mathrm{Ker}(p^*)$.

Suppose that $\mathcal{F} = (0 \to N \xrightarrow{j} X \xrightarrow{q} C \to 0) \in \mathbf{Ext}^1(C, N)$ satisfies $p^*[\mathcal{F}] = 0$. To show $\mathrm{Ker}(p^*) \subset \mathrm{Im}(\delta)$ we must show that $[\mathcal{F}] = \delta(\alpha)$ $(= [\alpha_*(\mathcal{E})])$ for some $\alpha \in \mathrm{Hom}(A, N)$. Since $p^*\mathcal{F}$ splits then, by Proposition 4.3, there exists a homomorphism $\tilde{p} : B \to X$ making the following diagram commute:

Since $q \circ \tilde{p} \circ i = p \circ i = 0$ it follows that $\mathrm{Im}(\tilde{p} \circ i) \subset \mathrm{Ker}(p) = \mathrm{Im}(j) \cong N$. In particular, there is a unique homomorphism $\alpha : A \to N$ such that $j \circ \alpha = \tilde{p} \circ i$. The mapping $v : N \oplus B \to X$ given by

$$v(x, y) = j(x) + \tilde{p}(y)$$

vanishes on $\mathrm{Im}(\alpha \times -i)$ so induces a homomorphism $\hat{v} : \varinjlim(\alpha, i) \to X$. Hence it also induces a congruence $\hat{v} : \alpha_*(\mathcal{E}) \to \mathcal{F}$ as follows:

$$
\begin{array}{ccccccccc}
0 & \longrightarrow & N & \longrightarrow & \varinjlim(\alpha, i) & \xrightarrow{\pi_C} & C & \longrightarrow & 0 \\
 & & \downarrow \mathrm{Id}_N & & \downarrow \hat{v} & & \downarrow \mathrm{Id}_C & & \\
0 & \longrightarrow & N & \xrightarrow{j} & X & \xrightarrow{q} & C & \longrightarrow & 0
\end{array}
$$

that is, $[\mathcal{F}] = [\alpha_*(\mathcal{E})] = \delta(\alpha)$ as required.

To prove exactness of (4.14), we first show that $\mathrm{Im}(p^*) \subset \mathrm{Ker}(i^*)$. Let $\zeta : A \to C$ be the zero homomorphism and $\mathcal{F} = (0 \to N \xrightarrow{j} X \xrightarrow{\eta} C \to 0) \in \mathbf{Ext}^1(C, N)$. Then

$$\varprojlim(\eta, \zeta) = \mathrm{Ker}(\eta) \oplus A = \mathrm{Im}(j) \oplus A \cong N \oplus A$$

showing that the extension $\zeta^*(\mathcal{F})$ splits. As $p \circ i = \zeta$ then $i^* \circ p^*([\mathcal{F}]) = (p \circ i)^*[\mathcal{F}] = 0$ so that $\mathrm{Im}(p^*) \subset \mathrm{Ker}(i^*)$. To show that $\mathrm{Ker}(i^*) \subset \mathrm{Im}(p^*)$ let $\mathcal{F} = (0 \to N \xrightarrow{j} X \xrightarrow{\eta} B \to 0)$ be such that $i^*(\mathcal{F})$ is trivial. Then by Proposition 4.3, there exists a homomorphism $\hat{i} : A \to X$ making the following commute:

Define $Y = X/\mathrm{Im}(\hat{i})$, and let $\natural : X \to Y$ be the canonical mapping, $\natural(x) = x + \mathrm{Im}(\hat{i})$.

Since $p \circ \eta \circ \hat{i} = 0$, there is a homomorphism $q : Y \to C$ given by

$$q(x + \mathrm{Im}(\hat{i})) = p \circ \eta(x).$$

It is straightforward to check that the following sequence is exact

$$\mathcal{G} = (0 \longrightarrow N \xrightarrow{\natural \circ j} Y \xrightarrow{q} C \longrightarrow 0)$$

and also that the square below commutes.

$$\begin{array}{ccc} X & \xrightarrow{\eta} & B \\ \downarrow{\natural} & & \downarrow{p} \\ Y & \xrightarrow{q} & C \end{array}$$

We may therefore construct a congruence of extensions $\hat{h} : \mathcal{F} \to p^*(\mathcal{G})$ as follows:

$$\begin{array}{ccccccccc} 0 & \longrightarrow & N & \xrightarrow{j} & X & \xrightarrow{\eta} & B & \longrightarrow & 0 \\ & & \downarrow{\mathrm{Id}_N} & & \downarrow{\nu} & & \downarrow{\mathrm{Id}_B} & & \\ 0 & \longrightarrow & N & \longrightarrow & \varprojlim(q, p) & \xrightarrow{\pi_B} & B & \longrightarrow & 0 \end{array}$$

where $\nu(x) = (\natural(x), \eta(x))$; that is, $[\mathcal{F}] = p^*[\mathcal{G}]$.

The above results have natural duals. We record the main statements. Given an exact sequence $\mathcal{E} = (0 \to A \xrightarrow{i} B \xrightarrow{p} C \to 0)$ there is a mapping $\partial : \mathrm{Hom}(M, C) \to \mathrm{Ext}^1(M, A)$ given by $\partial(\gamma) = [\gamma^*(\mathcal{E})]$. One sees easily that:

The connecting map $\partial : \mathrm{Hom}(M, C) \to \mathrm{Ext}^1(M, A)$ is additive.   (4.15)

From this we obtain a sequence of homomorphisms of abelian groups:

$$0 \to \mathrm{Hom}(M, A) \xrightarrow{i_*} \mathrm{Hom}(M, B) \xrightarrow{p_*} \mathrm{Hom}(M, C) \xrightarrow{\partial} \mathrm{Ext}^1(M, A)$$

$$\xrightarrow{i_*} \mathrm{Ext}^1(M, B) \xrightarrow{p_*} \mathrm{Ext}^1(M, C).$$ (**)

$\square$

Theorem 4.16 *The sequence* (**) *is exact for any* Λ-*module* M.

The proof of Theorem 4.16 is step-by-step dual to that of Theorem 4.11. We leave the details to the reader.

We refer to (*) as the *direct exact sequence* and to (**) as the *reverse exact sequence*. They are functorial on $\mathcal{M}od_\Lambda$ in a way which, for the sake of precision, we formalize. Thus let $\mathcal{E}xact(n)$ denote the category whose objects are exact sequences of abelian groups of length n

$$\mathbf{E} = (E_1 \to E_2 \to \cdots \to E_{n-1} \to E_n)$$

and whose morphisms are commutative diagrams

$$\mathbf{E} \atop \downarrow = \atop \mathbf{F} \left(\begin{array}{ccccc} E_1 & \to & E_2 & \to & \cdots & \to & E_{n-1} & \to & E_n \\ \downarrow & & \downarrow & & & & \downarrow & & \downarrow \\ F_1 & \to & F_2 & \to & \cdots & \to & F_{n-1} & \to & F_n \end{array} \right).$$

There is a full subcategory $\mathcal{E}\mathrm{xact}_0(n)$ of $\mathcal{E}\mathrm{xact}(n)$ consisting of objects

$$\mathbf{E} = (E_1 \to E_2 \to \cdots \to E_{n-1} \to E_n)$$

in which the first arrow $E_1 \to E_2$ is injective. We may thus portray the objects of $\mathcal{E}\mathrm{xact}_0(n)$ as exact sequences $(0 \to E_1 \to E_2 \to \cdots \to E_{n-1} \to E_n)$. Given an exact sequence of Λ-modules $\mathcal{E} = (0 \to A \to B \xrightarrow{p} B \to 0)$ then $(*)$ gives a *covariant functor* $\mathrm{Hom}(\mathcal{E}, -) : \mathcal{M}\mathrm{od}_\Lambda \to \mathcal{E}\mathrm{xact}_0(6)$

$$\mathrm{Hom}(\mathcal{E}, -) = \left(0 \to \mathrm{Hom}(C, -) \xrightarrow{p^*} \mathrm{Hom}(B, -) \xrightarrow{i^*} \cdots \xrightarrow{p^*} \mathrm{Ext}^1(B, -) \right.$$
$$\left. \xrightarrow{i^*} \mathrm{Ext}^1(A, -) \right)$$

whilst $(**)$ gives a *contravariant functor* $\mathrm{Hom}(-, \mathcal{E}) : \mathcal{M}\mathrm{od}_\Lambda \to \mathcal{E}\mathrm{xact}_0(6)$

$$0 \to \mathrm{Hom}(-, A) \xrightarrow{i_*} \mathrm{Hom}(-, B) \xrightarrow{p_*} \mathrm{Hom}(-, C) \xrightarrow{\partial} \mathrm{Ext}^1(-, A)$$
$$\xrightarrow{i_*} \mathrm{Ext}^1(-, B) \xrightarrow{p_*} \mathrm{Ext}^1(-, C).$$

4.4 The Standard Cohomology Theory of Modules

We begin by recalling briefly, without proofs, the basics of the Eilenberg-Maclane cohomology theory [13, 68]. Let M be a Λ-module; a *resolution* of M is an exact sequence of Λ-homomorphisms

$$\mathbf{A} = (\cdots \xrightarrow{\partial_{n+1}^{\mathcal{A}}} A_n \xrightarrow{\partial_n^{\mathcal{A}}} A_{n-1} \xrightarrow{\partial_{n-1}^{\mathcal{A}}} \cdots \xrightarrow{\partial_1^{\mathcal{A}}} A_0 \xrightarrow{\epsilon} M \to 0),$$

abbreviated to $\mathbf{A} = (A_* \to M)$. We say that a resolution $\mathbf{A}$ is *projective* (resp. *free*) when each A_r is a projective (resp. free) Λ-module. Projective resolutions will be denoted thus $\mathbf{P} = (P_* \to M)$ and free resolutions thus $\mathbf{F} = (F_* \to M)$:

Every Λ-module has a free (and hence a projective) resolution. (4.17)

If $\mathbf{A} = (A_* \to M_1)$, $\mathbf{B} = (B_* \to M_2)$ are resolutions and $f : M_1 \to M_2$ is a module homomorphism then by a morphism of resolutions, $\varphi : \mathbf{A} \to \mathbf{B}$ over f we mean a collection (φ_r) of Λ-homomorphisms completing a commutative diagram:

$$\mathbf{A} \atop \downarrow \varphi = \atop \mathbf{B} \left(\begin{array}{ccccccc} \cdots & \to & A_n & \to & A_{n-1} & \to & \cdots & \to & A_0 & \to & M_1 & \to & 0 \\ & & \downarrow \varphi_n & & \downarrow \varphi_{n-1} & & & & \downarrow \varphi_0 & & \downarrow \varphi_- \\ \cdots & \to & B_n & \to & B_{n-1} & \to & \cdots & \to & B_0 & \to & M_2 & \to & 0 \end{array} \right).$$

Such a morphism over f is also called a *lifting* of f. Given a Λ-homomorphism $f : M_1 \to M_2$, morphisms $\varphi, \psi : \mathbf{A} \to \mathbf{B}$ over f are said to be *homotopic over f* (written $\varphi \simeq_f \psi$) when there exists a collection $\eta = (\eta_r)_{r \geq 0}$ of Λ-homomorphisms $\eta_r : A_r \to B_{r+1}$ such that

(i) $\varphi_0 - \psi_0 = \partial_1 \eta_0$ and   (ii) $\varphi_r - \psi_r = \partial_{r+1} \eta_r + \eta_{r-1} \partial_r$ for all $r \geq 1$.

Let $f : M_1 \to M_2$ be a Λ-homomorphism; if $\mathbf{B} = (B_* \to M_2)$ is a resolution and $\mathbf{P} = (P_* \to M_1)$ is a projective resolution, then:

Proposition 4.18 *There exists a lifting $\tilde{f} : \mathbf{P} \to \mathbf{B}$ of f; moreover any two liftings of $f : M_1 \to M_2$ are homotopic over f.*

If M, N are Λ-modules then choosing a projective resolution $\mathbf{P} = (P_* \to M)$ of M we construct a cochain complex (P_r^N, ∂_r^N) thus:

$$P_r^N = \begin{cases} \text{Hom}_\Lambda(P_r, N) & \text{for } r \geq 0, \\ 0 & \text{for } r < 0, \end{cases}$$

where for $r \geq 0$, $\partial_r^N : P_r^N \to P_{r+1}^N$ is the induced map $\partial_r^N(\alpha) = \alpha \circ \partial_r$. We denote by $H_{\mathbf{P}}^*(M, N)$ the cohomology of this cochain complex; that is

$$H_{\mathbf{P}}^k(M, N) \cong \frac{\text{Ker}(\text{Hom}(P_k, N) \overset{\partial_{k+1}^*}{\to} \text{Hom}(P_{k+1}, N))}{\text{Im}(\text{Hom}(P_{k-1}, N) \overset{\partial_k^*}{\to} \text{Hom}(P_k, N))} \quad \text{for } k \geq 1.$$

Here $H_{\mathbf{P}}^0(M, N) = \text{Ker}(\text{Hom}(P_0, N) \overset{\partial_1^*}{\to} \text{Hom}(P_1, N))$ and $H_{\mathbf{P}}^k(M, N) = 0$ for $k < 0$. If $\mathbf{Q} = (Q_* \to L)$ is a projective resolution of L, $f : M \to L$ is a homomorphism and $\tilde{f} : \mathbf{P} \to \mathbf{Q}$ is a lifting over f then the induced maps $f_r^N : \text{Hom}(Q_r, N) \to \text{Hom}(P_r, N)$ give rise to homomorphisms in cohomology $f_{\mathbf{PQ}} : H_{\mathbf{Q}}^n(L, N) \to H_{\mathbf{P}}^n(M, N)$ which are independent of the particular lifting of f. Given a projective resolution $\mathbf{R} \to K$ and a homomorphism $g : L \to K$ then $(g \circ f)_{\mathbf{PR}} : H_{\mathbf{R}}^n(K, N) \to H_{\mathbf{P}}^n(M, N)$ satisfies the transitivity property:

$$(g \circ f)_{\mathbf{PR}}^* = f_{\mathbf{PQ}}^* g_{\mathbf{QR}}^*.$$

In particular if $\mathbf{P}, \mathbf{Q}$ are projective resolutions of M and $H_{\mathbf{P}}^*(M, N)$ (resp. $H_{\mathbf{Q}}^*(M, N)$) is the cohomology computed using $\mathbf{P}$ (resp. $\mathbf{Q}$) then there is a *transition homomorphism* $t_{\mathbf{PQ}} = \text{Id}_{\mathbf{PQ}}^* : H_{\mathbf{Q}}^*(M, N) \to H_{\mathbf{P}}^*(M, N)$. If $\mathbf{R}$ is also a projective resolution of M then

$$t_{\mathbf{PR}} = t_{\mathbf{PQ}} \circ t_{\mathbf{QR}}$$

whilst in the special case where $\mathbf{Q} = \mathbf{P}$ then $t_{\mathbf{PP}} = \text{Id} : H_{\mathbf{P}}^*(M, N) \to H_{\mathbf{P}}^*(M, N)$. It follows that each transition homomorphism $t_{\mathbf{PQ}}$ is an isomorphism and $t_{\mathbf{PQ}}^{-1} = t_{\mathbf{QP}}$.

In particular:

$$\text{The isomorphism type of } H_{\mathbf{P}}^n(M, N) \text{ is independent of } \mathbf{P}. \qquad (4.19)$$

Moreover:

There is a natural equivalence of functors $\natural_{\mathbf{P}} : \text{Hom}(-, N) \xrightarrow{\simeq} H_{\mathbf{P}}^0(-, N)$. (4.20)

We may eliminate the subscript $\mathbf{P}$ from $H_{\mathbf{P}}^n$ by making, for each module M, a specific choice of projective resolution $\mathbf{P}(M)$ for M. We then write

$$H^*(M, N) = H_{\mathbf{P}(M)}^*(M, N).$$

Moreover, if $f : L \to M$ is a homomorphism we write

$$f^* = f_{\mathbf{P}(M)\mathbf{P}(L)}^* : H^n(M, N) \to H^n(L, N).$$

In the special case of a projective module P, as a projective resolution of P we may take $(\cdots \to 0 \to 0 \to P \xrightarrow{\epsilon} P \to 0)$ where $\epsilon = \text{Id}_P$. It follows that:

If P is projective, then for any $n \geq 1$, $H^n(P, C) = 0$ for any module C. (4.21)

4.5 The Cohomological Interpretation of Ext[1]

We first give a model for the cohomology group $H^1(M, N)$ which confers certain advantages at the cost of a slight degree of unconventionality. By a *projective 0-complex* over Λ we mean a short exact sequence $\mathcal{P} = (0 \to K \xrightarrow{i} P \xrightarrow{p} M \to 0)$ where P is projective. Given such a projective 0-complex we define

$$H_{\mathcal{P}}^1(M, N) = \frac{\text{Hom}(K, N)}{\text{Im}(\text{Hom}(P, N) \xrightarrow{i^*} (\text{Hom}(K, N)))}.$$

We first show that this construction is functorial on morphisms of projective 0-complexes. Specifically, given a morphism of projective 0-complexes

$$\begin{array}{c} \mathcal{P} \\ \downarrow \varphi = \\ \mathcal{Q} \end{array} \begin{pmatrix} 0 \to K \xrightarrow{i} P \xrightarrow{\epsilon} M \to 0 \\ \downarrow \varphi_+ \quad \downarrow \varphi_0 \quad \downarrow \varphi_- \\ 0 \to K' \xrightarrow{j} Q \xrightarrow{\eta} M' \to 0 \end{pmatrix}$$

we construct an induced morphism $\varphi^* : H_{\mathcal{Q}}^1(M', N) \to H_{\mathcal{P}}^1(M, N)$. To do this observe that the induced map

$$(\varphi_+)^* : \text{Hom}(K', N) \to \text{Hom}(K, N)$$

has the property that $\mathrm{Im}(\mathrm{Hom}(Q, N) \xrightarrow{j^*} (\mathrm{Hom}(K', N)) \subset \mathrm{Im}(\mathrm{Hom}(P, N) \xrightarrow{i^*}$
$(\mathrm{Hom}(K, N))$ since if $\alpha = j^*(\beta)$ for $\beta \in \mathrm{Hom}(Q, N)$ then $\varphi_+^*(\alpha) = \varphi_+^* j^*(\beta) = i^*\varphi_0^*(\beta)$. We define

$$\varphi^* : H_Q^1(M', N) \to H_P^1(M, N)$$

to be the homomorphism induced from $\varphi_+^* : \mathrm{Hom}(K', N) \to \mathrm{Hom}(K, N)$. It is straightforward to check that this construction is functorial on morphisms of projective 0-complexes; that is, if $\varphi : \mathcal{P} \to \mathcal{Q}$, $\psi : \mathcal{Q} \to \mathcal{R}$ are morphisms as follows:

$$
\begin{array}{ccc}
\mathcal{P} & & \begin{pmatrix} 0 \to K \to & P \to M \to & 0 \\ & \downarrow \varphi_+ & \downarrow \varphi_0 \ \downarrow \varphi_- & \\ \end{pmatrix} \\
\downarrow \varphi & & \\
\mathcal{Q} & = & \begin{pmatrix} 0 \to K' \to & Q \to M' \to & 0 \\ & \downarrow \psi_+ & \downarrow \psi_0 \ \downarrow \psi_- & \\ \end{pmatrix} \\
\downarrow \psi & & \\
\mathcal{R} & & \begin{pmatrix} 0 \to K'' \to & R \to M'' \to & 0 \end{pmatrix}
\end{array}
$$

then $(\psi\varphi)^* = \varphi^*\psi^*$. Next we show that φ^* depends only upon φ_-.

Proposition 4.22 *If $\varphi : \mathcal{P} \to \mathcal{Q}$ has the property that $\varphi_0 = 0$ then $\varphi^* = 0$.*

Proof We have a commutative diagram

$$
\begin{array}{ccccccc}
0 \to & K & \xrightarrow{i} & P & \xrightarrow{\epsilon} & M & \to 0 \\
& \downarrow \varphi_+ & & \downarrow \varphi_0 & & \downarrow 0 & \\
0 \to & K' & \xrightarrow{j} & Q & \xrightarrow{\eta} & M' & \to 0
\end{array}
$$

from which it is clear that $\mathrm{Im}(\varphi_0) \subset \mathrm{Ker}(\eta) = \mathrm{Im}(j)$. Put $\tilde{\varphi} = j^{-1}\varphi_0 : P \to K'$ so that the following diagram commutes:

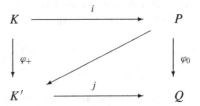

As $\varphi_+^*(\beta) = i^*(\beta\tilde{\varphi}) \in \mathrm{Im}\,(\mathrm{Hom}(P, N) \to \mathrm{Hom}(K, N))$ for $\beta \in \mathrm{Hom}(K', N)$ it follows that $\varphi^* = 0$. $\square$

Corollary 4.23 *If $\varphi_1, \varphi_2 : \mathcal{P} \to \mathcal{Q}$ satisfy $(\varphi_1)_- = (\varphi_2)_-$ then $(\varphi_1)^* = (\varphi_2)^*$.*

Proof For then $(\varphi_1 - \varphi_2)_- = 0$ so that $(\varphi_1 - \varphi_2)^* = 0$ and so $\varphi_1^* = \varphi_2^*$. $\square$

We can now liberate this construction from global dependence upon the category of projective 0-complexes, at least in part; suppose that $\mathcal{P}$, $\mathcal{Q}$ are projective

0-complexes

$$\mathcal{P} = (0 \to K \xrightarrow{i} P \xrightarrow{\epsilon} M \to 0),$$
$$\mathcal{Q} = (0 \to K' \xrightarrow{j} Q \xrightarrow{\eta} M' \to 0)$$

and that $f : M \to M'$ is a Λ-homomorphism. By the universal property of projective we may construct a lifting $\hat{f}$ of f thus:

$$
\begin{array}{c}
\mathcal{P} \\
\downarrow \hat{f} = \\
\mathcal{Q}
\end{array}
\begin{pmatrix}
0 \to K \xrightarrow{i} P \xrightarrow{\epsilon} M \to 0 \\
\downarrow f_+ \quad \downarrow f_0 \quad \downarrow f \\
0 \to K' \xrightarrow{j} Q \xrightarrow{\eta} M' \to 0
\end{pmatrix}.
$$

We may then define $f^*_{\mathcal{P}\mathcal{Q}} : H^1_{\mathcal{Q}}(M', N) \to H^1_{\mathcal{P}}(M, N)$ by $f^*_{\mathcal{P}\mathcal{Q}} = (\hat{f})^*$. This is meaningful since if $\tilde{f}$ is another lifting of f then $(\hat{f} - \tilde{f})_+ = 0$ and so $(\tilde{f})^* = (\hat{f})^*$.

The construction $f \mapsto f_{\mathcal{P}\mathcal{Q}}$ is functorial in the following sense; suppose that $\mathcal{P}$, $\mathcal{Q}$, $\mathcal{R}$ are projective 0-complexes

$$\mathcal{P} = (0 \to K \to P \to M \to 0),$$
$$\mathcal{Q} = (0 \to K' \to Q \to M' \to 0),$$
$$\mathcal{R} = (0 \to K'' \to R \to M'' \to 0)$$

and $g : M \to M'$, $f : M' \to M''$ are Λ-homomorphisms ; then

$$(f \circ g)^*_{\mathcal{P}\mathcal{R}} = g^*_{\mathcal{P}\mathcal{Q}} f^*_{\mathcal{Q}\mathcal{R}}.$$

We note that the suffices $\mathcal{P}$, $\mathcal{Q}$ are analogous to the roles of bases in the change of basis formula in linear algebra. There is a particular special case one should consider, namely the effect of lifting the identity; in the case of projective 0-complexes over the same module M thus

$$\mathcal{P} = (0 \to K \to P \to M \to 0),$$
$$\mathcal{P}' = (0 \to K' \to P' \to M \to 0)$$

we will write $\tau_{\mathcal{P}'\mathcal{P}} = (\mathrm{Id}_M)^*_{\mathcal{P}'\mathcal{P}}$. Although this notation consciously suppresses the module M it should cause no confusion. We obtain the general transformation rule.

$$f^*_{\mathcal{P}'\mathcal{Q}'} = \tau_{\mathcal{P}'\mathcal{P}} f^*_{\mathcal{P}\mathcal{Q}} \tau_{\mathcal{Q}\mathcal{Q}'}.$$

We proceed by showing that the model we have just given for H^1 is isomorphic to the standard model. Specifically, we will show:

Proposition 4.24 *If $\mathcal{P} = (0 \to K \to P \to M \to 0)$ is a projective 0-complex and $\mathbf{Q} = (\cdots \to Q_n \to Q_{n-1} \to \cdots \to Q_0 \to M \to 0)$ is a projective resolution then there exists an isomorphism $\natural_{\mathcal{Q}\mathcal{P}} : H^1_{\mathcal{P}}(M, N) \to H^1_{\mathbf{Q}}(M, N)$ which is natural in the sense that the following diagrams commute:*

$$H_Q^1(M, N) \xrightarrow{\; \natural_{QQ} \;} H_Q^1(M, N)$$

$$\downarrow {\scriptstyle \tau_{PQ}} \qquad\qquad\qquad \downarrow {\scriptstyle \iota_{PQ}}$$

$$H_P^1(M, N) \xrightarrow{\; \natural_{PP} \;} H_P^1(M, N)$$

Proof Let $\mathcal{P} = (0 \to K \xrightarrow{i} P \xrightarrow{\epsilon} M \to 0)$ be a projective 0-complex. Then we may extend $\mathcal{P}$ to a projective resolution

$$\mathbf{P} = (\cdots \to P_n \xrightarrow{\partial_n} P_{n-1} \xrightarrow{\partial_{n-1}} \cdots \xrightarrow{\partial_3} P_2 \xrightarrow{\partial_2} P_1 \xrightarrow{\partial_1} P_0 \xrightarrow{\epsilon} M \to 0)$$

in which $P_0 = P$ and, on factorising $\partial_{r+1} : P_{r+1} \to P_r$ as

$$P_{r+1} \xrightarrow{\;\;\partial_{r+1}\;\;} P_r$$
$$\pi_{r+1} \searrow \qquad \nearrow i_{r+1}$$
$$K_{r+1}$$

with π_{r+1} surjective and i_{r+1} injective, in which $K_1 = K$ and $i_1 = i$. In particular, we have a pair of short exact sequences

$$0 \to K_1 \xrightarrow{i_1} P_0 \xrightarrow{\epsilon} M \to 0,$$

$$0 \to K_2 \xrightarrow{i_2} P_1 \xrightarrow{\pi_1} K_1 \to 0.$$

From the exact sequence $0 \to \mathrm{Hom}(K_n, N) \xrightarrow{\pi_n^N} \mathrm{Hom}(P_n, N) \xrightarrow{i_n^N} \cdots$ we see that π_r^N is injective for $r = 1, 2$. Now we have a diagram in which the row is exact and the triangle commutes:

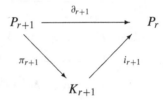

$$0 \longrightarrow \mathrm{Hom}(K_1, N) \xrightarrow{\;\pi_1^N\;} \mathrm{Hom}(P_1, N) \xrightarrow{\;i_2^N\;} \mathrm{Hom}(K_2, N)$$

Since π_2^N is injective and $\partial_2^N = \pi_2^N \circ i_2^N$ then $\mathrm{Ker}(\partial_2^N) = \mathrm{Ker}(i_2^N)$ so that, by exactness $\pi_1^N : \mathrm{Hom}(K_1, N) \to \mathrm{Ker}(\partial_2^N)$ is an isomorphism. Since $\pi_1^N : \mathrm{Hom}(K_1, N) \to$

$\text{Hom}(P_1, N)$ is injective it follows that π_1^N induces an isomorphism

$$\pi_1^N : \cfrac{\text{Hom}(K_1, N)}{\text{Im}(\text{Hom}(P_0, N) \overset{i_1^N}{\to} \text{Hom}(K, N))}$$

$$\longrightarrow \cfrac{\text{Im}(\text{Hom}(K_1, N) \overset{\pi_1^N}{\to} \text{Hom}(P_1, N))}{\text{Im}(\text{Hom}(P_0, N) \overset{\pi_1^N i_1^N}{\to} \text{Hom}(P_1, N))}.$$

As above $\text{Ker}(\partial_2^N) = \text{Im}(\text{Hom}(K_1, N) \overset{\pi_1^N}{\to} \text{Hom}(P_1, N))$ and $\partial_1^N = \pi_1^N i_1^N$ so that π_1^N induces an isomorphism, denoted by $\natural_{\mathcal{PP}}$, thus

$$\natural_{\mathbf{PP}} = \pi_1^N : \cfrac{\text{Hom}(K_1, N)}{\text{Im}(\text{Hom}(P_0, N) \overset{i_1^N}{\to} \text{Hom}(K, N))} \longrightarrow \cfrac{\text{Ker}(\partial_2^N)}{\text{Im}(\partial_1^N)}.$$

However

$$H_{\mathcal{P}}^1(M, N) = \cfrac{\text{Hom}(K_1, N)}{\text{Im}(\text{Hom}(P_0, N) \overset{i_1^N}{\to} \text{Hom}(K, N))} \quad \text{and} \quad H_{\mathbf{P}}^1(M, N) = \cfrac{\text{Ker}(\partial_2^N)}{\text{Im}(\partial_1^N)}$$

so that $\natural_{\mathbf{PP}}$ gives an isomorphism $\natural_{\mathbf{PP}} : H_{\mathcal{P}}^1(M, N) \longrightarrow H_{\mathbf{P}}^1(M, N)$. In general if $\mathbf{Q}$ is an arbitrary projective resolution of M we define

$$\natural_{\mathbf{QP}} = t_{\mathbf{QP}} \natural_{\mathbf{PP}} : H_{\mathcal{P}}^1(M, N) \to H_{\mathbf{Q}}^1(M, N),$$

where $\mathbf{P}$ is a projective resolution of M extending $\mathcal{P}$. That this is independent of $\mathbf{P}$ can be seen easily from the fact that if $\mathbf{P}'$ is also a projective resolution of M extending $\mathcal{P}$ then the following diagram commutes:

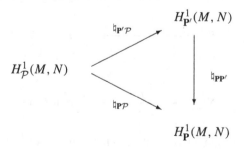

This completes the proof.   □

Let $\mathcal{P} = (0 \to K \to P \to M \to 0)$ be a projective 0-complex. Given a Λ-module N and a homomorphism $f : K \to N$ we may form the pushout sequence thus:

$$\begin{matrix} \mathcal{P} \\ \downarrow v_f \\ f_*(\mathcal{P}) \end{matrix} = \begin{pmatrix} 0 \to K \overset{i}{\to} & P & \overset{\epsilon}{\to} M \to 0 \\ \downarrow f & \downarrow \natural & \downarrow \mathrm{Id} \\ 0 \to N \to & \varinjlim(f, i) & \to M \to 0 \end{pmatrix}.$$

It follows that we have a group homomorphism $v_\mathcal{P} : \mathrm{Hom}(K, N) \to \mathrm{Ext}^1(M, N)$ given by $v_\mathcal{P}(f) = f_*(\mathcal{P})$. If $\alpha \in \mathrm{Ext}^1(M, N)$ is represented by the extension

$$\alpha = (0 \to N \to A \to M \to 0)$$

then from the universal property of projectives there is a morphism of extensions lifting the identity of M thus:

$$\begin{matrix} \mathcal{P} \\ \downarrow g \\ \alpha \end{matrix} = \begin{pmatrix} 0 \to K \overset{i}{\to} P \overset{\epsilon}{\to} M \to 0 \\ \downarrow g_+ \quad \downarrow g_0 \quad \downarrow \mathrm{Id} \\ 0 \to N \to A \to M \to 0 \end{pmatrix}.$$

It follows by general nonsense that $(g_+)_*(\mathcal{P}) \equiv \alpha$ so that:

Proposition 4.25 $v_\mathcal{P} : \mathrm{Hom}(K, N) \to \mathrm{Ext}^1(M, N)$ *is surjective.*

Proposition 4.26 $\mathrm{Ker}(v_\mathcal{P}) = \mathrm{Im}(\mathrm{Hom}(P, N) \overset{i^*}{\to} \mathrm{Hom}(K, N))$.

Proof Given a pushout morphism

$$\begin{matrix} \mathcal{P} \\ \downarrow \\ f_*(\mathcal{P}) \end{matrix} = \begin{pmatrix} 0 \to K \overset{i}{\to} & P & \overset{\epsilon}{\to} M \to 0 \\ \downarrow f & \downarrow \natural & \downarrow \mathrm{Id} \\ 0 \to N \overset{j}{\to} & \varinjlim(f, i) & \overset{\eta}{\to} M \to 0 \end{pmatrix}$$

we see that $f_*(\mathcal{P}) \equiv 0$ if and only if $f_*(\mathcal{P})$ splits. By (4.1), Proposition 4.2 there exists a homomorphism $\hat{f} : P \to N$ making the following diagram commute;

equivalently, $f = i^*(\hat{f})$ or alternatively $f \in \mathrm{Im}(\mathrm{Hom}(P, N) \overset{i^*}{\to} \mathrm{Hom}(K, N))$.   □

As

$$H_\mathcal{P}^1(M, N) = \frac{\mathrm{Hom}(K, N)}{\mathrm{Im}(\mathrm{Hom}(P, N) \overset{i^*}{\to} \mathrm{Hom}(K, N))}$$

then denoting the map induced by v_P by the same symbol we see that:

Proposition 4.27 v_P *induces an isomorphism* $v_P : H_P^1(M, N) \xrightarrow{\simeq} \mathrm{Ext}^1(M, N)$.

This construction is compatible with the transition isomorphisms constructed above in the sense that the following diagram commutes:

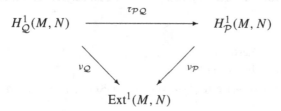

4.6 The Exact Sequences in Cohomology

Let $\mathcal{E} = (0 \to A \xrightarrow{i} B \xrightarrow{p} C \to 0)$ be an exact sequence of Λ-modules. If P is a projective module it is straightforward to see that the induced sequence

$$0 \to \mathrm{Hom}(P, A) \xrightarrow{i_*} \mathrm{Hom}(P, B) \xrightarrow{p_*} \mathrm{Hom}(P, C) \to 0$$

is exact. More generally,

If $P_* \to M$ is a projective resolution of M then the sequence of cochain complexes

$$0 \to \mathrm{Hom}(P_*, A) \xrightarrow{i_*} \mathrm{Hom}(P_*, B) \xrightarrow{p_*} \mathrm{Hom}(P_*, C) \to 0 \text{ is exact.} \qquad (4.28)$$

Applying the cohomology functor to (4.28) gives a long exact sequence in cohomology

$$\cdots \to H^{n-1}(M, C) \xrightarrow{\partial} H^n(M, A) \xrightarrow{i_*} H^n(M, B) \xrightarrow{p_*} H^n(M, C)$$
$$\xrightarrow{\partial} H^{n+1}(M, A) \to \cdots$$

in which the terms $H^r(-, N)$ are zero for $r < 0$. In view of the natural equivalence $H^0(M, -) \cong \mathrm{Hom}(M, -)$ the initial portion of the sequence can be written:

$$0 \to \mathrm{Hom}(M, A) \xrightarrow{i_*} \mathrm{Hom}(M, B) \xrightarrow{p_*} \mathrm{Hom}(M, C) \xrightarrow{\partial} H^1(M, A)$$
$$\xrightarrow{i_*} H^1(M, B) \xrightarrow{p_*} H^1(M, C) \to \cdots$$

The above is the *direct exact sequence*; in addition, there is a *reverse exact sequence* as we now proceed to establish. Thus suppose given a commutative diagram of

Λ-modules and homomorphisms

$$0 \to X \xrightarrow{j} Y \xrightarrow{\pi} Z \to 0$$
$$\downarrow \epsilon_A \quad \downarrow \epsilon_B \quad \downarrow \epsilon_C \qquad\qquad (4.29)$$
$$0 \to A \xrightarrow{i} B \xrightarrow{p} C \to 0$$

in which both rows are exact and ϵ_A, ϵ_B, ϵ_C are all surjective. It then follows that:

Lemma 4.30

(i) $j(\mathrm{Ker}(\epsilon_A)) \subset \mathrm{Ker}(\epsilon_B)$;

(ii) $\pi(\mathrm{Ker}(\epsilon_B)) \subset \mathrm{Ker}(\epsilon_C)$ and

(iii) the sequence $0 \to \mathrm{Ker}(\epsilon_A) \xrightarrow{j} \mathrm{Ker}(\epsilon_B) \xrightarrow{\pi} \mathrm{Ker}(\epsilon_C) \to 0$ is exact.

Proof (i) and (ii) are clear. Moreover, in (iii) $j : \mathrm{Ker}(\epsilon_A) \to \mathrm{Ker}(\epsilon_B)$ is injective since it is the restriction of the injective mapping $j : X \to Y$, and since $\pi \circ j = 0$ then evidently

$$\mathrm{Im}(j : \mathrm{Ker}(\epsilon_A) \to \mathrm{Ker}(\epsilon_B)) \subset \mathrm{Ker}(\pi : \mathrm{Ker}(\epsilon_B) \to \mathrm{Ker}(\epsilon_C)).$$

It suffices to prove:

(a) if $y \in \mathrm{Ker}(\epsilon_B)$ satisfies $\pi(y) = 0$ then there exists $x \in \mathrm{Ker}(\epsilon_A)$ such that $j(x) = y$.

(b) if $z \in \mathrm{Ker}(\epsilon_C)$ then there exists $y \in \mathrm{Ker}(\epsilon_B)$ such that $\pi(y) = z$.

(a) Let $y \in \mathrm{Ker}(\epsilon_B)$ satisfy $\pi(y) = 0$. Choose $x \in X$ such that $j(x) = y$. Then $i\epsilon_A(x) = \epsilon_B j(x) = \epsilon_B(y) = 0$. However i is injective so that $\epsilon_A(x) = 0$. As required there exists $x \in \mathrm{Ker}(\epsilon_A)$ such that $j(x) = y$.

(b) Let $z \in \mathrm{Ker}(\epsilon_C)$ and choose $y_1 \in Y$ such that $\pi(y_1) = z$. Then

$$p\epsilon_B(y_1) = \epsilon_C \pi(y_1) = \epsilon_C(z) = 0,$$

so that $\epsilon_B(y_1) \in \mathrm{Ker}(p) = \mathrm{Im}(i)$. Now choose $a \in A$ such that $i(a) = \epsilon_B(y_1)$, and choose $x \in X$ such that $\epsilon_A(x) = a$. Then $\epsilon_B j(x) = i\epsilon_A(x) = \epsilon_B(y_1)$. Put $y = y_1 - j(x)$. Then $y \in \mathrm{Ker}(\epsilon_B)$ and $\pi(y) = \pi(y_1) - \pi j(x) = \pi(y_1) = z$. $\square$

Lemma 4.31 *Let* $0 \to A \xrightarrow{i} B \xrightarrow{p} C \to 0$ *be an exact sequence of Λ-modules and homomorphisms, and let* $\epsilon_A : P \to A$ *and* $\epsilon_C : Q \to C$ *be surjective homomorphisms in which* P, Q *are projective. Then there exists a surjective homomorphism* $\epsilon_B : P \oplus Q \to B$ *making the following diagram commute*

$$0 \to P \xrightarrow{j} P \oplus Q \xrightarrow{\pi} Q \to 0$$
$$\downarrow \epsilon_A \qquad \downarrow \epsilon_B \qquad \downarrow \epsilon_C$$
$$0 \to A \xrightarrow{i} \quad B \quad \xrightarrow{p} C \to 0$$

where $j(x) = (x, 0)$ *and* $\pi(x, y) = y$.

Proof Since $p : B \to C$ is surjective and Q is projective then there exists a homomorphism $\widetilde{\epsilon}_C : Q \to B$ making the following diagram commute.

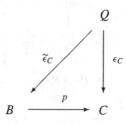

Put $\epsilon_B = (i\epsilon_A, \widetilde{\epsilon}_C) : P \oplus Q \to B$. It is easy to check that the diagram

$$0 \to P \xrightarrow{j} P \oplus Q \xrightarrow{\pi} Q \to 0$$
$$\downarrow \epsilon_A \qquad \downarrow \epsilon_B \qquad \downarrow \epsilon_C$$
$$0 \to A \xrightarrow{i} \quad B \quad \xrightarrow{p} C \to 0$$

commutes. To see that ϵ_B is surjective, let $b \in B$ and choose $z \in Q$ such that $\epsilon_C(z) = p(b)$. Then $b - \widetilde{\epsilon}_C(z) \in \mathrm{Ker}(p) = \mathrm{Im}(i)$. Choose $a \in A$ such that $i(a) = b - \widetilde{\epsilon}_C(z)$ and choose $x \in P$ such that $\epsilon_A(x) = a$. Then $\epsilon_B(x, z) = b$ and ϵ_B is surjective. $\qquad\square$

Theorem 4.32 *Let* $\mathcal{E} = (0 \to A \xrightarrow{i} B \xrightarrow{p} C \to 0)$ *be an exact sequence of* Λ-*modules and let* $P_* \to A$ *and* $R_* \to C$ *be projective resolutions. Then there is a projective resolution* $Q_* \to B$ *and exact sequence of chain complexes* $0 \to P_* \to Q_* \to R_* \to 0$ *covering* $\mathcal{E}$.

Proof Let $(P_* \to A)$ and $(R_* \to C)$ be projective resolutions. For each n write

$$D_n^A = \mathrm{Im}(\partial_n^A)(= \mathrm{Ker}(\partial_{n-1}^A) \subset P_{n-1})$$

and consider the canonical factorization of ∂_n^A through its image

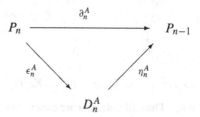

where η_n^A is inclusion and $\epsilon_n^A(x) = \partial_n^A(x)$. Likewise there is a canonical factorization of ∂_n^C. When convenient we will suppress the symbols η_n^A, η_n^C denoting inclusions, and for inductive purposes it is convenient to write $D_0^A = A$ and $\epsilon_{-1}^A = \epsilon^A$, and likewise $D_0^C = C$ and $\epsilon_{-1}^C = \epsilon^C$. We shall consider commutative diagrams $\mathcal{D}_r$ of the form

$$
\mathcal{D}_r = \begin{pmatrix}
 & 0 & & 0 & & 0 & \\
 & \downarrow & & \downarrow & & \downarrow & \\
0 \longrightarrow & D_r^A & \overset{i_r}{\longrightarrow} & D_r^B & \overset{p_r}{\longrightarrow} & D_r^C & \longrightarrow 0 \\
 & \downarrow \eta_r^A & & \downarrow \eta_r^B & & \downarrow \eta_r^C & \\
0 \longrightarrow & P_r & \overset{j_r}{\longrightarrow} & P_r \oplus R_r & \overset{\pi_r}{\longrightarrow} & R_r & \longrightarrow 0 \\
 & \downarrow \epsilon_{r-1}^A & & \downarrow \epsilon_{r-1}^B & & \downarrow \epsilon_{r-1}^C & \\
0 \longrightarrow & D_{r-1}^A & \overset{i_{r-1}}{\longrightarrow} & D_{r-1}^B & \overset{p_{r-1}}{\longrightarrow} & D_{r-1}^C & \longrightarrow 0 \\
 & \downarrow & & \downarrow & & \downarrow & \\
 & 0 & & 0 & & 0 &
\end{pmatrix}
$$

in which rows and columns are exact, where $j_r(x) = (x, 0)$ and $\pi_r(x, y) = y$, and where $\eta_r^A, \eta_r^B, \eta_r^C$, are set-theoretic inclusions, so that, in particular, i_r is the restriction of j_r and p_r is the restriction of π_r.

We argue by induction; the inductive base will be to construct $\mathcal{D}_1$; the inductive step will be to construct $\mathcal{D}_n$ from $\mathcal{D}_{n-1}$. It is more convenient to treat the inductive step first. Suppose $\mathcal{D}_{n-1}$ is constructed. From it we may extract the top row

$$
\mathcal{E}_{n-1} = (0 \longrightarrow D_{n-1}^A \overset{i_{n-1}}{\longrightarrow} D_{n-1}^B \overset{p_{n-1}}{\longrightarrow} D_{n-1}^C \longrightarrow 0)
$$

and in addition we have surjective homomorphisms $\epsilon_n^A : P_n \to D_{n-1}^A$ and $\epsilon_n^C : R_n \to D_{n-1}^C$ where P_n, R_n are projective. We may therefore apply Lemma 4.31 to obtain a commutative diagram

$$
\mathcal{D}_n = \begin{pmatrix}
 & 0 & & 0 & & 0 & \\
 & \downarrow & & \downarrow & & \downarrow & \\
0 \longrightarrow & D_n^A & \overset{i_n}{\longrightarrow} & D_n^B & \overset{p_n}{\longrightarrow} & D_n^C & \longrightarrow 0 \\
 & \downarrow \eta_n^A & & \downarrow \eta_n^B & & \downarrow \eta_n^C & \\
0 \longrightarrow & P_n & \overset{j_n}{\longrightarrow} & P_n \oplus R_n & \overset{\pi_n}{\longrightarrow} & R_n & \longrightarrow 0 \\
 & \downarrow \epsilon_{n-1}^A & & \downarrow \epsilon_{n-1}^B & & \downarrow \epsilon_{n-1}^C & \\
0 \longrightarrow & D_{n-1}^A & \overset{i_{n-1}}{\longrightarrow} & D_{n-1}^B & \overset{p_{n-1}}{\longrightarrow} & D_{n-1}^C & \longrightarrow 0 \\
 & \downarrow & & \downarrow & & \downarrow & \\
 & 0 & & 0 & & 0 &
\end{pmatrix},
$$

where $\epsilon_n^B : P_n \oplus R_n \to D_{n-1}^B$ is surjective, $D_n^B = \mathrm{Ker}(\epsilon_n^B)$ and where η_n^B is the inclusion of D_n^B in $P_n \oplus R_n$. Thus all columns are exact. Also $0 \longrightarrow D_n^A \overset{i_n}{\longrightarrow} D_n^B \overset{p_n}{\longrightarrow} D_n^C \longrightarrow 0$ is exact by (4.29) so that $\mathcal{D}_n$ is a commutative diagram of the required type.

With only a slight modification the same argument now establishes the induction base, starting with the exact sequence $\mathcal{E}_0 = (0 \longrightarrow A \overset{i}{\longrightarrow} B \overset{p}{\longrightarrow} C \longrightarrow 0)$ (corresponding to $n = 1$) described for purposes of formal agreement as

$$\mathcal{E}_0 = (0 \longrightarrow D_0^A \xrightarrow{i_0} D_0^B \xrightarrow{p_0} D_0^C \longrightarrow 0).$$

We now put $Q_n = P_n \oplus R_n$ and define $\partial_n^B : Q_n \to Q_{n-1}$ by $\partial_n^B = \eta_n^B \circ \epsilon_n^B$. Then it is straightforward to check that

$$Q_* \to B = (\cdots \to Q_n \xrightarrow{\partial_n^B} Q_{n-1} \to \cdots \to Q_1 \xrightarrow{\partial_1^B} Q_0 \xrightarrow{\epsilon^B} A \to 0)$$

is a projective resolution of B. Moreover, the infinite commutative diagram in (4.33) below gives an exact sequence of projective resolutions $0 \to P_* \to Q_* \to R_* \to 0$ covering the initial exact sequence $0 \to A \to B \to C \to 0$.   □

In the construction below at each level $Q_n = P_n \oplus R_n$. However, in general $\partial_n^B \neq \partial_n^A \oplus \partial_N^C$.

$$
\begin{array}{ccccccc}
& \vdots & & \vdots & & \vdots & \\
0 \longrightarrow & P_n & \xrightarrow{j_n} & Q_n & \xrightarrow{\pi_n} & R_n & \longrightarrow 0 \\
& \downarrow \partial_n^A & & \downarrow \partial_n^B & & \downarrow \partial_n^C & \\
0 \longrightarrow & P_{n-1} & \xrightarrow{j_{n-1}} & Q_{n-1} & \xrightarrow{\pi_{n-1}} & R_{n-1} & \longrightarrow 0 \\
& \downarrow \partial_{n-1}^A & & \downarrow \partial_{n-1}^B & & \downarrow \partial_{n-1}^C & \\
& \vdots & & \vdots & & \vdots & \\
& \downarrow \partial_1^A & & \downarrow \partial_1^B & & \downarrow \partial_1^C & \\
0 \longrightarrow & P_0 & \xrightarrow{j_0} & Q_0 & \xrightarrow{\pi_0} & R_0 & \longrightarrow 0 \\
& \downarrow \epsilon^A & & \downarrow \epsilon^B & & \downarrow \epsilon^C & \\
0 \longrightarrow & A & \xrightarrow{i} & B & \xrightarrow{p} & C & \longrightarrow 0 \\
& \downarrow & & \downarrow & & \downarrow & \\
& 0 & & 0 & & 0 &
\end{array}
$$

(4.33)

Applying the cohomology functor we get. as promised, a long *reverse exact sequence*

$$\cdots \xrightarrow{i^*} H^{n-1}(A, N) \xrightarrow{\delta} H^n(C, N) \xrightarrow{p^*} H^n(B, N) \xrightarrow{i^*} H^n(A, N)$$

$$\xrightarrow{\delta} H^{n+1}(C, N) \xrightarrow{p^*} \cdots$$

Here $H^r(-, N) = 0$ for $r < 0$; in view of the natural equivalence $H^0(M, -) \cong \mathrm{Hom}(M, -)$ the initial portion of the sequence can be written:

$$0 \to \mathrm{Hom}(C, N) \xrightarrow{p^*} \mathrm{Hom}(B, N) \xrightarrow{i^*} \mathrm{Hom}(A, N) \xrightarrow{\delta} H^1(C, N)$$

$$\xrightarrow{p^*} H^1(B, N) \xrightarrow{i^*} H^1(A, N) \to \cdots$$

Up to this point we have preserved the notational distinction between cohomology $H^1(-, -)$ and the group of extensions $\text{Ext}^1(-, -)$. In the light of Proposition 4.27 it will frequently seem pedantic to maintain the distinction. Furthermore (and without any explicit appeal to Yoneda's cohomological interpretation of higher extensions [101]) we shall, as and when seems appropriate, adopt the standard convention of writing

$$\text{Ext}^n(M, N) = H^n(M, N).$$

Chapter 5
The Derived Module Category

Linear algebra over a field is rendered tractable by the fact that every module over a field is free; that is, has a spanning set of linearly independent vectors. Over more general rings, when a module M is not free we make a first approximation to its being free by taking a surjective homomorphism $\varphi : F \to M$ where F is free. The kernel $\mathrm{Ker}(\varphi)$ may then be regarded as the 'first derivative' module of M. These considerations may be made precise by working in the 'derived module category', the quotient of the category of Λ-modules obtained by quotienting out by morphisms which factorize through a projective. The invariants of this first derivative lead to a nonstandard definition of module cohomology.

5.1 The Derived Module Category

If $f : M \to N$ is a morphism in $\mathcal{M}od_\Lambda$ we say that f *factors through a projective module*, written '$f \approx 0$', when f can be written as a composite $f = \xi \circ \eta$ thus

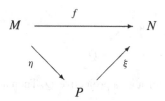

where $P \in \mathcal{M}od_\Lambda$ is a projective module and $\eta : M \to P$ and $\xi : P \to N$ are Λ-homomorphisms. As projective modules are direct summands of free modules the condition '$f \approx 0$' is evidently equivalent to the requirement that $f : M \to N$ factors through a free module. We define

$$\langle M, N \rangle = \{ f \in \mathrm{Hom}_\Lambda(M, N) : f \approx 0 \}.$$

On taking either ξ or η to be zero we see that $0 \in \langle M, N \rangle$. Moreover, if $f, g : M \to N$ are Λ-homomorphisms and $f = \alpha \circ \beta$, $g = \gamma \circ \delta$ are factorizations

F.E.A. Johnson, *Syzygies and Homotopy Theory*, Algebra and Applications 17, DOI 10.1007/978-1-4471-2294-4_5, © Springer-Verlag London Limited 2012

through the projectives P, Q respectively; then

$$f - g = (\alpha, \gamma)\begin{pmatrix} \beta \\ -\delta \end{pmatrix}$$

is a factorization of $f - g$ through the projective $P \oplus Q$. It follows that:

$$\langle M, N \rangle \text{ is an additive subgroup of } \mathrm{Hom}_\Lambda(M, N). \qquad (5.1)$$

We extend $\approx$ to a binary relation on $\mathrm{Hom}_\Lambda(M, N)$ by means of

$$f \approx g \iff f - g \approx 0.$$

So extended, $\approx$ is an equivalence relation compatible with composition; that is given Λ-homomorphisms $f, f' : M_0 \to M_1, g, g' : M_1 \to M_2$ then:

$$f \approx f' \quad \text{and} \quad g \approx g' \implies g \circ f \approx g' \circ f'. \qquad (5.2)$$

We obtain the *derived module category* $\mathcal{D}er = \mathcal{D}er(\Lambda)$, whose objects are right Λ-modules, and in which, for any two objects M, N, the set of morphisms $\mathrm{Hom}_{\mathcal{D}er}(M, N)$ is given by

$$\mathrm{Hom}_{\mathcal{D}er}(M, N) = \mathrm{Hom}_\Lambda(M, N)/\langle M, N \rangle.$$

Since $\langle M, N \rangle$ is a subgroup of $\mathrm{Hom}_\Lambda(M, N)$, it follows that:

$$\mathrm{Hom}_{\mathcal{D}er}(M, N) \text{ has the natural structure of an abelian group.} \qquad (5.3)$$

We extend these considerations to the functor Ext^1; suppose that $f, g : M_1 \to M_2$ are Λ-homomorphisms such that $f \approx g$ and let

$$
\begin{array}{ccc}
M_1 & \xrightarrow{\quad f-g \quad} & M_2 \\
& {\scriptstyle \alpha}\searrow \quad \nearrow{\scriptstyle \beta} & \\
& Q &
\end{array}
$$

be a factorization of $f - g$ through a projective Q. Then application of $\mathrm{Ext}^1(-, N)$ yields a factorization

$$
\begin{array}{ccc}
\mathrm{Ext}^1(M_2, N) & \xrightarrow{\quad f^*-g^* \quad} & \mathrm{Ext}^1(M_1, N) \\
& {\scriptstyle \beta^*}\searrow \quad \nearrow{\scriptstyle \alpha^*} & \\
& \mathrm{Ext}^1(Q, N) &
\end{array}
$$

of $(f - g)^* = f^* - g^*$ through $\mathrm{Ext}^1(Q, N)$. As Q is projective then $\mathrm{Ext}^1(Q, N) = 0$

by (4.9); thus if $f \approx g$ then $f^* = g^* : \mathrm{Ext}^1(M_2, N) \to \mathrm{Ext}^1(M_1, N)$; that is:

For any Λ-module N, the correspondence $M \mapsto \mathrm{Ext}^1(M, N)$ defines

a contravariant functor $\mathrm{Ext}^1(-, N) : \mathcal{D}er \to \mathrm{Ab}$.   (5.4)

We obtain a useful characterization of projective modules.

Proposition 5.5 *For any module Q over Λ the following are equivalent:*

(i) *Q is projective.*
(ii) *$\mathrm{Hom}_{\mathcal{D}er}(Q, N) = 0$ for all Λ-modules N.*
(iii) *$\mathrm{Hom}_{\mathcal{D}er}(N, Q) = 0$ for all Λ-modules N.*
(iv) *$\mathrm{Hom}_{\mathcal{D}er}(Q, Q) = 0$.*
(v) *$\mathrm{Ext}^1(Q, N) = 0$ for all Λ-modules N.*
(vi) *$H^1(Q, N) = 0$ for all Λ-modules N.*

Proof The implications (i) $\Longrightarrow$ (ii) $\Longrightarrow$ (iv) and (i) $\Longrightarrow$ (iii) $\Longrightarrow$ (iv) are clear. To see that (iv) $\Longrightarrow$ (ii), suppose that $\mathrm{Hom}_{\mathcal{D}er}(Q, Q) = 0$; then $\mathrm{Id}_Q \approx 0$. However any Λ-homomorphism $\alpha : Q \to N$ can be written as $\alpha = \alpha \circ \mathrm{Id}_Q$ so that $\alpha \approx 0$, by (5.2). Thus $\mathrm{Hom}_{\mathcal{D}er}(Q, N) = 0$ as required; moreover the proof that (iv) $\Longrightarrow$ (iii) follows similarly, on factorizing $\beta : N \to Q$ in the form $\beta = \mathrm{Id}_Q \circ \beta$.

To show that (ii) $\Longrightarrow$ (i), suppose that $\mathrm{Hom}_{\mathcal{D}er}(Q, N) = 0$ for all Λ-modules N, let $\alpha : Q \to N$ be a Λ-homomorphism and let $p : M \to N$ be a surjective Λ-homomorphism. To show that Q is projective it suffices to show there exists a Λ-homomorphism $\widehat{\alpha} : Q \to M$ making the following commute:

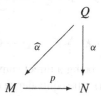

However, since by hypothesis $\mathrm{Hom}_{\mathcal{D}er}(Q, N) = 0$ then α factors through a projective P:

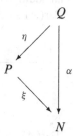

As p is surjective, there exists a homomorphism $\widehat{\xi} : P \to M$ such that $\xi = p \circ \widehat{\xi}$ making the following diagram commute:

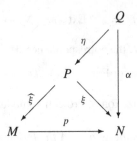

Hence taking $\widehat{\alpha} = \widehat{\xi} \circ \eta$ we see that $p \circ \widehat{\alpha} = \alpha$ as required, and this establishes the equivalences (i) $\Longleftrightarrow$ (ii) $\Longleftrightarrow$ (iii) $\Longleftrightarrow$ (iv). The implication (i) $\Longrightarrow$ (v) simply restates (4.9) so to complete the proof it suffices to show that (v) $\Longrightarrow$ (i). Let Q be a Λ-module and let $p : F \to Q$ be a surjective homomorphism from a free module F. Putting $K = \mathrm{Ker}(p)$, from the exact sequence

$$0 \to K \xrightarrow{j} F \xrightarrow{p} Q \to 0$$

then from Sect. 2.1 we obtain the exact sequence

$$0 \to \mathrm{Hom}_\Lambda(Q, K) \xrightarrow{p^*} \mathrm{Hom}_\Lambda(F, K) \xrightarrow{j^*} \mathrm{Hom}_\Lambda(K, K) \xrightarrow{\delta} \mathrm{Ext}^1(Q, K).$$

Under the hypothesis of (v) for Q, we have $\mathrm{Ext}^1(Q, K) = 0$ so that the above exact sequence reduces to $0 \to \mathrm{Hom}_\Lambda(Q, K) \xrightarrow{p^*} \mathrm{Hom}_\Lambda(F, K) \xrightarrow{j^*} \mathrm{Hom}_\Lambda(K, K) \to 0$. In particular, $j^* : \mathrm{Hom}_\Lambda(F, K) \to \mathrm{Hom}_\Lambda(K, K)$ is surjective. Choosing a homomorphism $r : F \to K$ such that $r \circ j = j^*(r) = \mathrm{Id}_K$, r is then a left splitting of the sequence

$$0 \to K \xrightarrow{j} F \xrightarrow{p} Q \to 0.$$

Thus $F \cong K \oplus Q$, and hence Q, being a direct summand of the free module F, is projective. This proves (v) $\Longrightarrow$ (i). The equivalence of (v) and (vi) follows from the cohomological interpretation of Ext^1 so completing the proof. $\qquad\qquad\square$

Suppose that $\mathrm{E} : \mathcal{Mod}_\Lambda \to \mathrm{Ab}$ is an additive functor. We say that E descends to $\mathcal{Der}(\Lambda)$ when there is a functor $\underline{\mathrm{E}} : \mathcal{Der}(\Lambda) \to \mathrm{Ab}$ making the following diagram commute

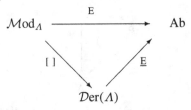

(5.4) may be extended to give:

Let $E : \mathcal{M}od_\Lambda \to Ab$ be an additive functor; then

$$E \text{ descends to } \mathcal{D}er(\Lambda) \iff E(Q) = 0 \text{ for any projective module } Q. \quad (5.6)$$

The functor E may be either covariant or contravariant. By Proposition 5.5 $H^n(Q, N) = 0$ for any projective module Q, so that:

Proposition 5.7 $H^n(-, N)$ descends to $\mathcal{D}er(\Lambda)$ for each $n \geq 1$.

Let M, P be Λ-modules with inclusion $i_M : M \to M \oplus P$ and projection $\pi_M : M \oplus P \to M$; if P is projective then $i_M \circ \pi_M \approx \mathrm{Id}_M$ by means of the factorization:

$$M \oplus P \xrightarrow{\mathrm{Id} - i_M \circ \pi_M} M \oplus P$$

with π_P and i_P through P.

As $\pi_M \circ i_M = \mathrm{Id}_M$ it follows that:

Proposition 5.8 Let M, P be Λ-modules with P projective and let $i_M : M \to M \oplus P$ and $\pi_M : M \oplus P \to M$ denote respectively the canonical inclusions and projections. Then i_M and π_M are mutually inverse isomorphisms in $\mathcal{D}er$.

This gives a criterion for deciding when Λ-modules become isomorphic in $\mathcal{D}er$:

Theorem 5.9 Let M_1, M_2 be modules over Λ; then

$$M_1 \cong_{\mathcal{D}er} M_2 \iff M_1 \oplus P_1 \cong_\Lambda M_2 \oplus P_2 \text{ for some projectives } P_1, P_2.$$

Proof The implication ($\Longleftarrow$) is an easy deduction from Proposition 5.8. To show ($\Longrightarrow$), suppose $f : M_1 \to M_2$ is a Λ homomorphism which defines an isomorphism in $\mathcal{D}er$. Then by Proposition 5.7,

$$f^* : H^k(M_2, N) \xrightarrow{\simeq} H^k(M_1, N)$$

is an isomorphism for all $k \geq 1$. Let $\eta : F \to M_2$ be a surjective Λ-homomorphism where F is a free module over Λ and let $p : M_1 \oplus F \to M_2$ be the Λ-homomorphism $p(x, y) = f(x) + \eta(y)$. Since both i and f are isomorphisms in $\mathcal{D}er$ it follows from the factorization

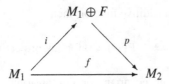

that p is also an isomorphism in $\mathcal{D}er$. Put $K = \mathrm{Ker}(p)$. Applying $H^k(-, K)$ to the exact sequence $0 \to K \xrightarrow{i} M_1 \oplus F \xrightarrow{p} M_2 \to 0$ we get an exact sequence in cohomology

$$0 \to \mathrm{Hom}_\Lambda(M_2, K) \xrightarrow{p^*} \mathrm{Hom}_\Lambda(M_1 \oplus F, K) \xrightarrow{i^*} \mathrm{Hom}_\Lambda(K, K) \xrightarrow{\delta_*} H^1(M_2, K)$$

$$\xrightarrow{p^*} H^1(M_1 \oplus F, K).$$

Since p is an isomorphism in $\mathcal{D}er$, $p^* : H^1(M_2, K) \to H^1(M_1 \oplus F, K)$ is an isomorphism and the above exact sequence reduces to

$$0 \to \mathrm{Hom}_\Lambda(M_2, K) \xrightarrow{p^*} \mathrm{Hom}_\Lambda(M_1 \oplus F, K) \xrightarrow{i^*} \mathrm{Hom}_\Lambda(K, K) \to 0.$$

Choosing $r : M_1 \oplus F \to K$ such that $i^*(r) = \mathrm{Id}_K$ we see that $r \circ i = \mathrm{Id}_M$. In particular, r splits the sequence $0 \to K \xrightarrow{i} M_1 \oplus F \xrightarrow{p} M_2 \to 0$ on the left, so that $M_1 \oplus F \cong_\Lambda M_2 \oplus K$. Now F is projective since it is free. To establish the conclusion as stated it remains to show that K is projective.

Since p is an isomorphism in $\mathcal{D}er$ then $p^* : H^k(M_2, N) \to H^k(M_1 \oplus F, N)$ is an isomorphism for all $k \geq 1$ and any coefficient module N. Appealing to this in the following portion of the long exact sequence

$$H^1(M_2, N) \xrightarrow{p^*} H^1(M_1 \oplus F, N) \xrightarrow{i^*} H^1(K, N) \xrightarrow{\delta^*} H^2(M_2, N) \xrightarrow{p^*} H^2(M_1 \oplus F, N)$$

we see easily that $H^1(K, N) = 0$ for all Λ-modules N. Thus K is projective by Proposition 5.5, and this completes the proof.   □

5.2 Coprojectives and De-stabilization

We first recall.

Proposition 5.10 (Schanuel's Lemma) *For $r = 1, 2$ let $(0 \to D_r \xrightarrow{i_r} P_r \xrightarrow{f_r} M \to 0$ be short exact sequences of Λ-modules in which P_1 and P_2 are projective; then*

$$D_1 \oplus P_2 \cong D_2 \oplus P_1.$$

Proof Form the *fibre product* $Q = P_1 \underset{f_1, f_2}{\times} P_2 = \{(x, y) \in P_1 \times P_2 : f_1(x) = f_2(y)\}$. There is a short exact sequence $0 \to D_2 \to Q \xrightarrow{\pi_1} P_1 \to 0$ where $\pi_1(x, y) = x$.

The sequence splits as P_1 is projective, so that $Q \cong D_2 \oplus P_1$. Likewise, as P_2 is projective, the short exact sequence $0 \to D_1 \to Q \xrightarrow{\pi_2} P_2 \to 0$ with $\pi_2(x, y) = y$ also splits, and $Q \cong D_1 \oplus P_2$. Now $D_1 \oplus P_1 \cong Q \cong D_2 \oplus P_1$ as claimed. $\square$

There is a dual form of Schanuel's Lemma which holds in restricted circumstances. To analyse the restriction condition, we first observe that Baer extension theory [5, 68] may be formulated within an arbitrary Abelian category $\mathcal{D}$ [74]. Then the dual category $\mathcal{D}^*$ obtained by formally reversing the directions of arrows is also an Abelian category; one has the obvious identity

$$\mathrm{Ext}^1_{\mathcal{D}}(N, M) = \mathrm{Ext}^1_{\mathcal{D}^*}(M, N),$$

and every theorem in $\mathcal{D}$ has a corresponding dual in $\mathcal{D}^*$. In the abelian category $\mathcal{M}od_\Lambda$, Schanuel's Lemma may be traced back to the fact that $\mathrm{Ext}^1(\Lambda, N) = 0$. It follows that if one wants the dual form of Schanuel's Lemma to hold in $\mathcal{M}od_\Lambda$ (rather than in the formal dual to $\mathcal{M}od_\Lambda$) then the dual condition $\mathrm{Ext}^1(N, \Lambda) = 0$ becomes significant.

To formulate the dual version of Schanuel's Lemma efficiently, we first consider a pair of exact sequences in $\mathcal{M}od_\Lambda$

$$\mathcal{E} = (0 \to K \xrightarrow{i} P \xrightarrow{p} M \to 0); \qquad \mathcal{F} = (0 \to K \xrightarrow{j} Q \xrightarrow{q} N \to 0)$$

in which P and Q are projective and form the pushout square

$$
\begin{array}{ccc}
K & \xrightarrow{\ \ j\ \ } & Q \\
\downarrow{\scriptstyle i} & & \downarrow{\scriptstyle \eta_Q} \\
P & \xrightarrow[\ \ \eta_P\ \ ]{} & \varinjlim(i, j)
\end{array}
$$

where $\varinjlim(i, j) = (P \oplus Q)/\mathrm{Im}(i \times -j)$. Taking the canonical inclusions and projections

$$i_P : P \to P \oplus Q; \quad i_P(x) = (x, 0);$$

$$i_Q : Q \to P \oplus Q; \quad i_Q(y) = (0, y);$$

$$\pi_P : P \oplus Q \to P; \quad \pi_P(x, y) = x;$$

$$\pi_Q : P \oplus Q \to Q; \quad \pi_Q(x, y) = y$$

then put $\eta_P = \mu \circ i_P; \eta_Q = \mu \circ i_Q$ where $\mu : P \oplus Q \to (P \oplus Q)/\mathrm{Im}(i \times -j)$ is the identification map. Then there are internal direct sum decompositions

$$P \oplus \mathrm{Im}(j) = \mathrm{Im}(i_P) \dotplus \mathrm{Im}(i \times -j), \tag{5.11}$$

$$\mathrm{Im}(i) \oplus Q = \mathrm{Im}(i_Q) \dotplus \mathrm{Im}(i \times -j). \tag{5.12}$$

Note that $p \circ \pi_P : P \oplus Q \to M$ vanishes on $\mathrm{Im}(i \times -j)$ and so induces a homomorphism

$$\xi_P : \varinjlim(i, j) \to M; \quad \xi_P([x, y]) = p(x).$$

Similarly $q \circ \pi_Q : P \oplus Q \to N$ vanishes on $\mathrm{Im}(i \times -j)$ and so induces a homomorphism

$$\xi_Q : \varinjlim(i, j) \to N; \quad \xi_Q([x, y]) = q(y).$$

From the internal direct sum decompositions (5.11), (5.12) we obtain exact sequences

$$0 \to Q \xrightarrow{\eta_Q} \varinjlim(i, j) \xrightarrow{\xi_P} M \to 0, \tag{5.13}$$

$$0 \to P \xrightarrow{\eta_P} \varinjlim(i, j) \xrightarrow{\xi_Q} N \to 0. \tag{5.14}$$

We obtain the following (weak) dual form of Schanuel's Lemma:

Theorem 5.15 Let $\mathcal{E} = (0 \to K \xrightarrow{i} P \xrightarrow{p} M \to 0); \mathcal{F} = (0 \to K \xrightarrow{j} Q \xrightarrow{q} N \to 0)$ be projective 0-complexes in Mod_Λ. If $\mathrm{Ext}^1(M, Q) = \mathrm{Ext}^1(N, P) = 0$ then

$$M \oplus Q \cong N \oplus P.$$

Proof Since $\mathrm{Ext}^1(M, Q) = 0$ then (5.13) above splits giving an isomorphism

$$\varinjlim(i, j) \cong M \oplus Q.$$

Similarly, since $\mathrm{Ext}^1(N, P) = 0$ then (5.14) splits showing also that

$$\varinjlim(i, j) \cong N \oplus P$$

from which the result is obvious. $\square$

We turn the hypotheses of Theorem 5.15 into a definition; we say that a Λ-module M is *coprojective* when $\mathrm{Ext}^1(M, Q) = 0$ for any projective module Q. We first note:

Proposition 5.16 *The following conditions upon a module M are equivalent*:

 (i) *M is coprojective*;
 (ii) $\mathrm{Ext}^1(M, \Lambda) = 0$;
 (iii) $H^1(M, Q) = 0$ *for any projective module Q*;
 (iv) $H^1(M, \Lambda) = 0$.

Proof The equivalences between (i) and (iii) and (ii) and (iv) follow from Proposition 4.27. The implication (i) $\Longrightarrow$ (ii) is trivial. To prove (ii) $\Longrightarrow$ (i) suppose that

$\text{Ext}^1(M, \Lambda) = 0$. If F is a free module, $F \cong \bigoplus_i \Lambda_i$ with $\Lambda_i \cong \Lambda$ for each i, it follows from the additivity properties of Ext^1 that $\text{Ext}^1(M, F) \cong \bigoplus_i \text{Ext}^1(M, \Lambda_i) \cong 0$. By representing Q as a direct summand $Q \oplus Q' \cong F$ we see that $\text{Ext}^1(M, Q) \oplus \text{Ext}^1(M, Q') \cong \text{Ext}^1(M, F) = 0$ and so also $\text{Ext}^1(M, Q) \cong 0$. $\qquad\square$

The following de-stabilization result is essential at a number of points:

Proposition 5.17 *Let* $0 \to J \oplus Q_0 \xrightarrow{j} Q_1 \to M \to 0$ *be an exact sequence of* Λ-modules in which Q_0, Q_1 are projective; if M is coprojective then $Q_1/j(Q_0)$ is projective.

Proof Let $i : J \to J \oplus Q_0$ be the inclusion, $i(x) = (x, 0)$, and let $\pi : J \oplus Q_0 \to J$ be the projection $\pi(x, y) = x$. Put $L = \varinjlim(i \circ \pi, j)$; then we have a commutative diagram

$$
\begin{array}{cc}
\mathcal{E} & \\
\downarrow \nu(\alpha) & = \\
(i \circ \pi)_*(\mathcal{E}) &
\end{array}
\left(
\begin{array}{ccc}
0 \to J \oplus Q_0 \xrightarrow{j} Q_1 \to M \to 0 \\
\quad\quad \downarrow i \circ \pi \quad\quad \downarrow \nu \quad \downarrow \text{Id} \\
0 \to J \oplus Q_0 \to L \to \quad M \to 0
\end{array}
\right),
$$

where $\nu : Q_1 \to L = \varinjlim(i \circ \pi, j)$ is the natural map. As

$$H^1(M, J \oplus Q_0) \cong H^1(M, J) \oplus H^1(M, Q_0)$$

and $H^1(M, Q_0) = 0$ it follows that $\pi_* : H^1(M, J \oplus Q_0) \to H^1(M, J)$ is an isomorphism; likewise $i_* : H^1(M, J) \to H^1(M, J \oplus Q_0)$ is an isomorphism. However,

$$\pi_* \circ i_* = \text{Id} : H^1(M, J) \to H^1(M, J),$$

hence that $i_* \circ \pi_* = (i \circ \pi)_* = \text{Id} : H^1(M, J \oplus Q_0) \to H^1(M, J \oplus Q_0)$.

Let $c = c_{\mathcal{E}} \in H^1(M, J \oplus Q_0)$ be the element classifying the extension $\mathcal{E}$. Then $(i \circ \pi)_*(\mathcal{E})$ is classified by $(i \circ \pi)_*(c) = c$. Thus $(i \circ \pi)_*(\mathcal{E})$ is congruent to $\mathcal{E}$, so that $L \cong Q_1$, and in particular, L is projective. Now put $S = \varinjlim(\pi, j)$. It is straightforward to check that $S = Q_1/j(Q_0)$, thus it suffices to show that S is projective. We have a commutative diagram

$$
\begin{array}{cc}
\pi_*(\mathcal{E}) & \\
\downarrow \nu(\alpha) & = \\
(i \circ \pi)_*(\mathcal{E}) &
\end{array}
\left(
\begin{array}{ccc}
0 \to \quad J \to \quad S \to M \to 0 \\
\quad\quad \downarrow i \quad\quad \downarrow \mu \quad \downarrow \text{Id} \\
0 \to J \oplus Q_0 \to L \to M \to 0
\end{array}
\right),
$$

where $\mu : S \to L$ is the induced map on pushouts. We obtain a commutative diagram for any coefficient module B;

$$
\begin{array}{ccccccc}
H^1(M, B) & \to & H^1(L, B) & \to & H^1(J \oplus Q_0, B) & \to & H^2(M, P) \\
\downarrow \text{Id} & & \downarrow \mu^* & & \downarrow i^* & & \downarrow \text{Id} \\
H^1(M, B) & \to & H^1(S, B) & \to & H^1(J, B) & \to & H^2(M, P)
\end{array}
$$

Clearly Id : $H^k(M, B) \to H^k(M, B)$ is an isomorphism for $k = 1, 2$, and as Q_0 is projective, $i^* : H^1(J \oplus Q_0, B) \to H^1(J, B)$ is an isomorphism. Thus $\mu^* : H^1(L, B) \to H^1(S, B)$ is surjective. However, since L is projective, $H^1(L, B) = 0$. Hence $H^1(S, B) = 0$ for all coefficient modules B, so that, by Proposition 5.5, $S = Q_1/j(Q_0)$ is projective, as desired. $\square$

5.3 Corepresentability of Ext^1

Let $\mathcal{P} = (0 \to K \xrightarrow{i} P \xrightarrow{\epsilon} M \to 0)$ be a projective 0-complex; then for any Λ-module N we define

$$\mathcal{H}_{\mathcal{P}}^1(M, N) = \mathrm{Hom}_{\mathcal{D}\mathrm{er}}(K, N).$$

Functoriality of this construction is similar to that of $H_{\mathcal{P}}^1(M, N)$ introduced in Sect. 4.5. If $\mathcal{P}' = (0 \to K' \xrightarrow{j} P' \xrightarrow{\eta} M' \to 0)$ is a projective 0-complex and $f : M' \to M$ is a Λ-homomorphism then we can lift f to a morphism

$$\begin{matrix} \mathcal{P}' \\ \downarrow \hat{f} \\ \mathcal{P} \end{matrix} = \begin{pmatrix} 0 \to K' \xrightarrow{j} P' \xrightarrow{\eta} M' \to 0 \\ \quad\; \downarrow f_+ \quad\; \downarrow f_0 \quad\; \downarrow f \\ 0 \to K \xrightarrow{i} P \xrightarrow{\epsilon} M \to 0 \end{pmatrix}.$$

Then f_+ induces a homomorphism $f_{\mathcal{P}'\mathcal{P}} : \mathrm{Hom}_{\mathcal{D}\mathrm{er}}(K, N) \to \mathrm{Hom}_{\mathcal{D}\mathrm{er}}(K', N)$ by means of

$$f_{\mathcal{P}'\mathcal{P}}(\alpha) = \alpha \circ f_+.$$

Then since $\mathrm{Hom}_{\mathcal{D}\mathrm{er}}(K, N) = \mathcal{H}_{\mathcal{P}}^1(M, N)$ and $\mathrm{Hom}_{\mathcal{D}\mathrm{er}}(K', N) = \mathcal{H}_{\mathcal{P}'}^1(M', N)$ we have defined

$$f_{\mathcal{P}'\mathcal{P}} : \mathcal{H}_{\mathcal{P}}^1(M, N) \to \mathcal{H}_{\mathcal{P}'}^1(M', N).$$

Likewise if $\mathcal{P}, \mathcal{Q}$ are projective 0-complexes over M we denote by

$$\tau_{\mathcal{P}\mathcal{Q}} : \mathcal{H}_{\mathcal{Q}}^1(M, N) \to \mathcal{H}_{\mathcal{P}}^1(M, N)$$

the transition isomorphism obtained by lifting the identity $\tau_{\mathcal{P}\mathcal{Q}} = (\mathrm{Id}_M)_{\mathcal{P}\mathcal{Q}}^*$. Defining $\langle K, N \rangle = \{\alpha \in \mathrm{Hom}_\Lambda(K, N) : \alpha \approx 0\}$ we see that if

$$\mathcal{P} = (0 \to K \xrightarrow{i} P \xrightarrow{\epsilon} M \to 0)$$

is a projective 0-complex then $\mathrm{Im}(\mathrm{Hom}(P, N) \xrightarrow{i^*} \mathrm{Hom}(K, N)) \subset \langle K, N \rangle$.

Let $\lambda_{\mathcal{P}} : \mathrm{Hom}^1(K, N)/\mathrm{Im}(i^*) \to \mathrm{Hom}^1(K, N)/\langle K, N \rangle$ be the natural surjection. Then $\lambda_{\mathcal{P}}$ defines a homomorphism $\lambda_{\mathcal{P}} : H_{\mathcal{P}}^1(M, N) \to \mathcal{H}_{\mathcal{P}}^1(M, N)$ so that we have an exact sequence

$$0 \to \langle K, N \rangle/\mathrm{Im}(i^*) \to H_{\mathcal{P}}^1(M, N) \xrightarrow{\lambda_{\mathcal{P}}} \mathcal{H}_{\mathcal{P}}^1(M, N) \to 0. \qquad (5.18)$$

The mappings $\lambda_{\mathcal{P}}$ are compatible with transition; that is we have commutative diagrams

$$
\begin{array}{ccc}
H_Q^1(M, N) & \xrightarrow{\ \tau_{\mathcal{P}Q}\ } & H_{\mathcal{P}}^1(M, N) \\
\downarrow{\lambda_Q} & & \downarrow{\lambda_{\mathcal{P}}} \\
\mathcal{H}_Q^1(M, N) & \xrightarrow{\ \tau_{\mathcal{P}Q}\ } & \mathcal{H}_{\mathcal{P}}^1(M, N)
\end{array}
$$

Moreover, with the above notation:

Proposition 5.19 *There is a commutative diamond*:

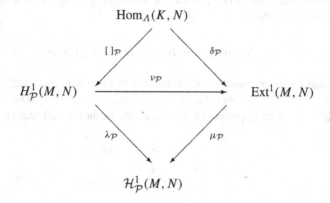

The mappings $[\ ]_{\mathcal{P}}$, $\delta_{\mathcal{P}}$, $\nu_{\mathcal{P}}$, $\lambda_{\mathcal{P}}$, $\mu_{\mathcal{P}}$ *depend only upon the congruence class of* $\mathcal{P}$ *and are described thus*:

$[\ ]_{\mathcal{P}}$ is the canonical surjection from the description

$$
H_{\mathcal{P}}^1(M, N) = \frac{\mathrm{Hom}_\Lambda(K, N)}{\mathrm{Im}(\mathrm{Hom}(P, N) \xrightarrow{i^*} \mathrm{Hom}(K, N))};
$$

$\delta_{\mathcal{P}} : \mathrm{Hom}_\Lambda(K, N) \twoheadrightarrow \mathrm{Ext}^1(M, N)$ is the mapping $\delta_{\mathcal{P}}(f) = f_*(\mathcal{P})$;

$\nu_{\mathcal{P}} : H_{\mathcal{P}}^1(M, N) \to \mathrm{Ext}^1(M, N)$ is the mapping described in Proposition 4.27;

$\lambda_{\mathcal{P}} : H_{\mathcal{P}}^1(M, N) \to \mathcal{H}_{\mathcal{P}}^1(M, N)$ is the mapping described in (5.18);

$\mu_{\mathcal{P}} : \mathrm{Ext}^1(M, N) \to \mathcal{H}_{\mathcal{P}}^1(M, N)$ may be described intrinsically as follows: represent $\alpha \in \mathrm{Ext}^1(M, N)$ by the extension $\alpha = (0 \to N \xrightarrow{j} A \xrightarrow{\pi} M \to 0)$; by the universal property of projectives there exists a morphism $\hat{f} : \mathcal{P} \to \alpha$ over the identity of M thus:

$$
\begin{array}{cc}
\mathcal{P} \\
\downarrow \hat{f} = \\
\alpha
\end{array}
\left(
\begin{array}{ccccccc}
0 \to & K & \xrightarrow{i} & P & \xrightarrow{\epsilon} & M & \to 0 \\
& \downarrow f_+ & & \downarrow f_0 & & \downarrow \mathrm{Id}_M & \\
0 \to & N & \xrightarrow{j} & A & \xrightarrow{\pi} & M & \to 0
\end{array}
\right);
$$

then we define $\mu_\mathcal{P}(\alpha) = [f_+] \in \mathrm{Hom}_{\mathcal{D}\mathrm{er}}(K, N) = \mathcal{H}_\mathcal{P}^1(M, N)$. It is easy to verify that $\mu_\mathcal{P} \circ \nu_\mathcal{P} = \lambda_\mathcal{P}$.

Let $\mathcal{P} = (0 \to K \xrightarrow{i} P \xrightarrow{\epsilon} M \to 0)$ be a fixed projective 0-complex; then the correspondences $N \mapsto H_\mathcal{P}^1(M, N)$ and $N \mapsto \mathcal{H}_\mathcal{P}^1(M, N)$ are (covariant) functors $\mathcal{M}\mathrm{od}_\Lambda \to \mathbf{Ab}$ and the homomorphism $\lambda_\mathcal{P} : H_\mathcal{P}^1(M, -) \to \mathcal{H}_\mathcal{P}^1(M, -)$ is a surjective natural transformation. It is straightforward to check:

The natural transformation $\lambda_\mathcal{P} : H_\mathcal{P}^1(M, -) \to \mathcal{H}_\mathcal{P}^1(M, -)$ is surjective.   (5.20)

It is natural to ask under what conditions $\lambda_\mathcal{P}$ is a natural equivalence. By definition $\mathcal{H}_\mathcal{P}^1(M, \Lambda) = 0$ so a necessary condition is that $H_\mathcal{P}^1(M, \Lambda) = 0$. By Proposition 5.16 this forces M to be coprojective. Conversely, suppose M is coprojective. We have an exact sequence

$$0 \to \langle K, N \rangle / \mathrm{Im}(i^*) \to H_\mathcal{P}^1(M, N) \xrightarrow{\lambda_\mathcal{P}} \mathcal{H}_\mathcal{P}^1(M, N) \to 0;$$

to show that $\lambda_\mathcal{P}$ is an isomorphism it suffices to show that $\langle K, N \rangle \subset \mathrm{Im}(i^*)$. Suppose that $f : K \to N$ factors through a projective, say $f = \xi \circ \eta$, where $\eta : K \to Q$ and $\xi : Q \to N$, with Q projective. Construct the pushout exact sequence

$$\begin{array}{ccccccc} 0 \to & K & \xrightarrow{i} & P & \xrightarrow{\epsilon} & M & \to 0 \\ & \downarrow \eta & & \downarrow \mu & & \downarrow \mathrm{Id} & \\ 0 \to & Q & \to & \varinjlim(\eta, i) & \to & M & \to 0 \end{array}$$

where $\mu : P \to \varinjlim(\eta, i)$ is the mapping $\mu(x) = [0, x]$. As M is coprojective $\mathrm{Ext}^1(M, Q) = 0$ and the exact sequence $0 \to Q \to \varinjlim(\eta, i) \to M \to 0$ splits. If $r : \varinjlim(\eta, i) \to Q$ is a left splitting then $\eta = r \circ \mu \circ i$ and $f = \xi \circ r \circ \mu \circ i = i^*(\xi \circ r \circ \mu)$. In particular, $f \in \mathrm{Im}(i^*)$; we obtain the following which was pointed out by Humphreys in his thesis [44]:

$H_\mathcal{P}^1(M, -) \xrightarrow{\lambda_\mathcal{P}} \mathcal{H}_\mathcal{P}^1(M, -)$ is a natural equivalence $\Leftrightarrow$ M is coprojective.   (5.21)

Alternatively, observing the diamond of Proposition 5.19 and appealing to Proposition 4.27 we have:

$\mathrm{Ext}^1(M, -) \xrightarrow{\mu_\mathcal{P}} \mathcal{H}_\mathcal{P}^1(M, -)$ is a natural equivalence $\Leftrightarrow$ M is coprojective.   (5.22)

5.4 The Exact Sequences in the Derived Module Category

In Chap. 4 with any exact sequence of Λ-modules $\mathcal{E} = (0 \to A \to B \xrightarrow{p} B \to 0)$ we associated two distinct exact sequences, namely the direct exact sequence:

$$0 \to \operatorname{Hom}_\Lambda(M, A) \xrightarrow{i_*} \operatorname{Hom}_\Lambda(M, B) \xrightarrow{p_*} \operatorname{Hom}_\Lambda(M, C)$$
$$\xrightarrow{\partial} \operatorname{Ext}^1(M, A) \xrightarrow{i_*} \operatorname{Ext}^1(M, B) \xrightarrow{p_*} \operatorname{Ext}^1(M, C)$$

and the reverse exact sequence:

$$0 \to \operatorname{Hom}_\Lambda(C, N) \xrightarrow{p^*} \operatorname{Hom}_\Lambda(B, N) \xrightarrow{i^*} \operatorname{Hom}_\Lambda(A, N) \xrightarrow{\delta} \operatorname{Ext}^1(C, N)$$
$$\xrightarrow{p^*} \operatorname{Ext}^1(B, N) \xrightarrow{i^*} \operatorname{Ext}^1(A, N).$$

These sequences are functorial on the category $\mathcal{M}od_\Lambda$; we now consider how to modify these sequences so as to become functorial on the derived category. We deal first with the direct sequence. Here the difficulty lies in the nature of the first few terms

$$0 \to \operatorname{Hom}_\Lambda(M, A) \xrightarrow{i_*} \operatorname{Hom}_\Lambda(M, B) \xrightarrow{p_*} \operatorname{Hom}_\Lambda(M, C)$$

since the correspondence $M \mapsto \operatorname{Hom}_\Lambda(M, N)$ is evidently not fuctorial on $\mathcal{D}er$. The obvious remedy is to replace $\operatorname{Hom}_\Lambda(M, N)$ systematically by $\operatorname{Hom}_{\mathcal{D}er}(M, N)$. There is a cost for doing this; if $i : A \to B$ is injective then, in general, $\operatorname{Hom}_{\mathcal{D}er}(M, A) \xrightarrow{i_*} \operatorname{Hom}_{\mathcal{D}er}(M, B)$ need not be injective; so we do not expect to be able to extend to the left by 0. We state our conclusions as a theorem and elaborate the remaining detail in the proof:

Theorem 5.23 *Let* $\mathcal{E} = (0 \to A \xrightarrow{i} B \xrightarrow{p} C \to 0)$ *be an exact sequence of Λ-modules; then for any Λ-module M there is an exact sequence:*

$$\operatorname{Hom}_{\mathcal{D}er}(M, A) \xrightarrow{i_*} \operatorname{Hom}_{\mathcal{D}er}(M, B) \xrightarrow{p_*} \operatorname{Hom}_{\mathcal{D}er}(M, C) \xrightarrow{\partial_*} \operatorname{Ext}^1(M, A)$$
$$\xrightarrow{i_*} \operatorname{Ext}^1(M, B) \xrightarrow{p_*} \operatorname{Ext}^1(M, C).$$

Proof First note that for any Λ-homomorphism $f : X \to Y$ and any Λ-module M, the induced map $f_* : \operatorname{Hom}_\Lambda(M, X) \to \operatorname{Hom}_\Lambda(M, Y)$ satisfies $f_*\langle M, X\rangle \subset \langle M, Y\rangle$ and so induces a homomorphism $f_* : \operatorname{Hom}_{\mathcal{D}er}(M, X) \to \operatorname{Hom}_{\mathcal{D}er}(M, Y)$.

It follows that the sequence

$$\operatorname{Hom}_{\mathcal{D}er}(M, A) \xrightarrow{i_*} \operatorname{Hom}_{\mathcal{D}er}(M, B) \xrightarrow{p_*} \operatorname{Hom}_{\mathcal{D}er}(M, C),$$

is well defined and satisfies $p_* \circ i_* = 0$. Thus $\operatorname{Im}(i_*) \subset \operatorname{Ker}(p_*)$. To show that $\operatorname{Ker}(p_*) \subset \operatorname{Im}(i_*)$ suppose that $\beta \in \operatorname{Hom}_\Lambda(M, B)$ is such that $p \circ \beta$ factorizes through a projective Q thus:

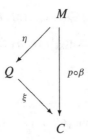

We must find $\alpha \in \mathrm{Hom}_\Lambda(M, A)$ such that $i_*(\alpha) \approx \beta$. Since Q is projective and p is surjective, there exists a homomorphism $\widehat{\xi} : Q \to B$ such that $\xi = p \circ \widehat{\xi}$; in particular the diagram below commutes:

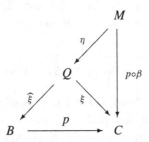

Now $\beta - \widehat{\xi} \circ \eta \in \mathrm{Hom}_\Lambda(M, \mathrm{Ker}(p))$ since $p(\beta - \widehat{\xi} \circ \eta) \equiv 0$. Since $\mathrm{Ker}(p) = \mathrm{Im}(i)$ we may define $\alpha = i^{-1}(\beta - \widehat{\xi} \circ \eta) \in \mathrm{Hom}_\Lambda(M, A)$. Then $\beta - i_*(\alpha) \approx 0$ by virtue of the identity $\beta - i_*(\alpha) = \widehat{\xi} \circ \eta$.

Next we show that the connecting homomorphism $\partial : \mathrm{Hom}_\Lambda(M, C) \to \mathrm{Ext}^1(M, A)$ vanishes on $\langle M, C \rangle$. Recall that ∂ is defined by $\partial(\gamma) = [\gamma^*(\mathcal{E})]$. If $\gamma = \xi \circ \eta$ is a factorization of γ through a projective Q then $\xi^*(\mathcal{E}) \in \mathrm{Ext}^1(Q, A)$ is trivial since Q is projective. Thus $\gamma^*(\mathcal{E}) = \eta^*\xi^*(\mathcal{E})$ is also trivial, so that $\partial(\gamma) = 0$ for $\gamma \in \langle M, C \rangle$. The connecting homomorphism $\partial : \mathrm{Hom}_\Lambda(M, C) \to \mathrm{Ext}^1(M, A)$ induces a homomorphism $\partial_* : \mathrm{Hom}_{\mathcal{D}\mathrm{er}}(M, C) \to \mathrm{Ext}^1(M, A)$. Furthermore, it is clear from the above that

$$\mathrm{Im}(\partial_*) = \mathrm{Im}(\partial) = \mathrm{Ker}(i^* : \mathrm{Ext}^1(M, A) \to \mathrm{Ext}^1(M, B)).$$

From the exactness of the exactness of the original direct exact sequence on $\mathcal{M}\mathrm{od}_\Lambda$ that the segment

$$\mathrm{Hom}_{\mathcal{D}\mathrm{er}}(M, C) \xrightarrow{\partial_*} \mathrm{Ext}^1(M, A) \xrightarrow{i_*} \mathrm{Ext}^1(M, B) \xrightarrow{p_*} \mathrm{Ext}^1(M, C)$$

is well defined and exact.

Finally we must show the segment

$$\mathrm{Hom}_{\mathcal{D}\mathrm{er}}(M, B) \xrightarrow{p_*} \mathrm{Hom}_{\mathcal{D}\mathrm{er}}(M, C) \xrightarrow{\partial_*} \mathrm{Ext}^1(M, A)$$

is exact. It is easy to see that $\partial_* \circ p_* = 0$ so it suffices to show that $\mathrm{Ker}(\partial_*) \subset \mathrm{Im}(p_*)$. Suppose that $\gamma \in \mathrm{Hom}_\Lambda(M, C)$ satisfies $\partial(\gamma) = 0$; that is, $\gamma^*(\mathcal{E})$ splits. Then by

Proposition 4.3 there exists a homomorphism $\hat{\gamma} : M \to B$ making the following diagram commute:

Then $p_*(\tilde{\gamma}) = \gamma$, and $\mathrm{Ker}(\partial_*) \subset \mathrm{Im}(p_*)$ as claimed, and this completes the proof.   □

We stress the functoriality of the above; as in Chap. 4, let $\mathcal{E}\mathrm{xact}(6)$ denote the category whose morphisms are commutative ladders with exact rows thus:

$$\begin{array}{cccccccccccc}
A_1 & \to & A_2 & \to & A_3 & \to & A_4 & \to & A_5 & \to & A_6 \\
\downarrow & & \downarrow & & \downarrow & & \downarrow & & \downarrow & & \downarrow \\
B_1 & \to & B_2 & \to & B_3 & \to & B_4 & \to & B_5 & \to & B_6
\end{array}$$

With $\mathcal{E}$ as above, for any Λ-module M we write $\mathrm{Hom}(M, \mathcal{E})$ for the direct sequence as modified above; that is:

$$\mathrm{Hom}_{\mathcal{D}\mathrm{er}}(M, A) \overset{i_*}{\to} \mathrm{Hom}_{\mathcal{D}\mathrm{er}}(M, B) \overset{p_*}{\to} \cdots \overset{i_*}{\to} \mathrm{Ext}^1(M, B) \overset{p_*}{\to} \mathrm{Ext}^1(M, C);$$

then the correspondence $M \mapsto \mathrm{Hom}(M, \mathcal{E})$ defines a contravariant functor

$$\mathrm{Hom}(-, \mathcal{E}) : \mathcal{D}\mathrm{er} \to \mathcal{E}\mathrm{xact}(6). \tag{5.24}$$

For a Λ-homomorphism $f : X \to Y$ it is clear that $f^* : \mathrm{Hom}_\Lambda(Y, N) \to \mathrm{Hom}_\Lambda(X, N)$ has the property that $f_*(\langle Y, N \rangle) \subset \langle X, N \rangle$ so that there is induced a well defined mapping $f^* : \mathrm{Hom}_{\mathcal{D}\mathrm{er}}(Y, N) \to \mathrm{Hom}_{\mathcal{D}\mathrm{er}}(X, N)$. It follows that for any exact sequence of Λ-modules $\mathcal{E} = (0 \to A \overset{i}{\to} B \overset{p}{\to} C \to 0)$ and any Λ-module N there is a sequence

$$\mathrm{Hom}_{\mathcal{D}\mathrm{er}}(C, N) \overset{p^*}{\to} \mathrm{Hom}_{\mathcal{D}\mathrm{er}}(B, N) \overset{i^*}{\to} \mathrm{Hom}_{\mathcal{D}\mathrm{er}}(A, N) \tag{5.25}$$

which, so far, we do not claim is exact. There is also a sequence

$$\mathrm{Ext}^1(C, N) \overset{p^*}{\to} \mathrm{Ext}^1(B, N) \overset{i^*}{\to} \mathrm{Ext}^1(A, N) \tag{5.26}$$

and the first difficulty arises in attempting to link them up. For the corresponding situation over $\mathcal{M}\mathrm{od}_\Lambda$ there is a connecting homomorphism $\delta : \mathrm{Hom}_\Lambda(A, N) \to \mathrm{Ext}^1(C, N)$ given by $\delta(\alpha) = [\alpha_*(\mathcal{E})]$. Without some extra condition, in general δ *does not factorize through* $\mathrm{Hom}_{\mathcal{D}\mathrm{er}}(A, N)$; the appropriate condition is to require C to be coprojective:

Proposition 5.27 *Let* $\mathcal{E} = (0 \to A \xrightarrow{i} B \xrightarrow{p} C \to 0)$ *be an exact sequence of* Λ*-modules. If* C *is coprojective then the connecting homomorphism* $\delta : \mathrm{Hom}_\Lambda(A, N) \to \mathrm{Ext}^1(C, N)$ *given by* $\delta(\alpha) = [\alpha_*(\mathcal{E})]$ *factors through* $\mathrm{Hom}_{\mathcal{D}\mathrm{er}}(A, N)$ *according to the diagram below; in particular,* $\mathrm{Im}(\delta_*) = \mathrm{Im}(\delta)$.

$$\mathrm{Hom}_\Lambda(A, N) \xrightarrow{\quad\delta\quad} \mathrm{Ext}^1(C, N)$$

$$\natural \searrow \qquad \nearrow \delta_*$$

$$\mathrm{Hom}_{\mathcal{D}\mathrm{er}}(A, N)$$

Proof Suppose that $\alpha \in \mathrm{Hom}_\Lambda(A, N)$ factors through a projective Q:

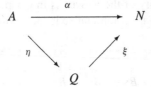

Then $\alpha_*(\mathcal{E}) = \xi_* \eta_*(\mathcal{E})$. As C is coprojective then $[\eta_*(\mathcal{E})] \in \mathrm{Ext}^1(C, Q) = 0$, so that $\delta(\alpha) = [\alpha_*(\mathcal{E})] = [\xi_* \eta_*(\mathcal{E})] = \xi_*[\eta_*(\mathcal{E})] = \xi_*(0) = 0$. In particular, δ vanishes on $\langle A, N \rangle$ so that $\delta : \mathrm{Hom}_\Lambda(A, N) \to \mathrm{Ext}^1(C, N)$ factors through $\mathrm{Hom}_{\mathcal{D}\mathrm{er}}(A, N)$ thus:

$$\mathrm{Hom}_\Lambda(A, N) \xrightarrow{\quad\delta\quad} \mathrm{Ext}^1(C, N)$$

$$\natural \searrow \qquad \nearrow \delta_*$$

$$\mathrm{Hom}_{\mathcal{D}\mathrm{er}}(A, N)$$

hence $\mathrm{Im}(\delta_*) = \mathrm{Im}(\delta)$, as claimed.   □

In consequence, we obtain:

Proposition 5.28 *Let* $\mathcal{E} = (0 \to A \xrightarrow{i} B \xrightarrow{p} C \to 0)$ *be an exact sequence of* Λ*-modules in which* C *is coprojective; then for any* Λ*-module* N *we have an exact sequence*:

$$\mathrm{Hom}_{\mathcal{D}\mathrm{er}}(C, N) \xrightarrow{p^*} \mathrm{Hom}_{\mathcal{D}\mathrm{er}}(B, N) \xrightarrow{i^*} \mathrm{Hom}_{\mathcal{D}\mathrm{er}}(A, N) \xrightarrow{\delta_*} \mathrm{Ext}^1(C, N)$$

$$\xrightarrow{p^*} \mathrm{Ext}^1(B, N) \xrightarrow{i^*} \mathrm{Ext}^1(A, N).$$

Proof Exactness of the segment

$$\mathrm{Hom}_{\mathcal{D}\mathrm{er}}(A, N) \xrightarrow{\delta_*} \mathrm{Ext}^1(C, N) \xrightarrow{p^*} \mathrm{Ext}^1(B, N) \xrightarrow{i^*} \mathrm{Ext}^1(A, N)$$

follows from Theorem 4.11 using the fact that $\mathrm{Im}(\delta_*) = \mathrm{Im}(\delta)$. It suffices to show that the sequence

$$\mathrm{Hom}_{\mathcal{D}\mathrm{er}}(C, N) \xrightarrow{p^*} \mathrm{Hom}_{\mathcal{D}\mathrm{er}}(B, N) \xrightarrow{i^*} \mathrm{Hom}_{\mathcal{D}\mathrm{er}}(A, N) \xrightarrow{\delta_*} \mathrm{Ext}^1(C, N)$$

is exact. However, it follows directly from (5.25) that both $\delta_* \circ i^* = 0$ and $i^* \circ p^* = 0$ so that $\mathrm{Im}(i^*) \subset \mathrm{Ker}(\delta_*)$ and $\mathrm{Im}(p^*) \subset \mathrm{Ker}(i^*)$. It suffices to show that

(a) $\mathrm{Ker}(\delta_*) \subset \mathrm{Im}(i^*)$ and
(b) $\mathrm{Ker}(i^*) \subset \mathrm{Im}(p^*)$.

For (a), suppose that $[\alpha] \in \mathrm{Hom}_{\mathcal{D}\mathrm{er}}(A, N)$ is the class of $\alpha \in \mathrm{Hom}_\Lambda(A, N)$ and satisfies $\delta_*([\alpha]) = 0$. Then $\delta(\alpha) = 0$ and so there exists $\beta \in \mathrm{Hom}_\Lambda(B, N)$ such that $\imath^*(\beta) = \alpha$. If $[\beta]$ denotes the class of β in $\mathrm{Hom}_{\mathcal{D}\mathrm{er}}(B, N)$ then $i^*([\beta]) = [\alpha]$ and so $\mathrm{Ker}(\delta_*) \subset \mathrm{Im}(i^*)$ as required.

For (b), let $\beta \in \mathrm{Hom}_\Lambda(B, N)$ and suppose that $\beta \circ i$ factors through a projective Q.

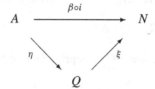

We must find $\gamma \in \mathrm{Hom}_\Lambda(C, N)$ such that $p^*(\gamma) \approx \beta$. First note that we have a commutative diagram of exact sequences

$$\begin{pmatrix} \mathcal{E} \\ \downarrow \nu \\ \eta_*(\mathcal{E}) \end{pmatrix} = \begin{pmatrix} 0 \longrightarrow A \xrightarrow{\ i\ } B \xrightarrow{\ p\ } C \longrightarrow 0 \\ \ \ \ \ \ \downarrow \eta \ \ \ \ \ \downarrow \nu \ \ \ \ \ \downarrow \mathrm{Id}_C \\ 0 \longrightarrow Q \xrightarrow{\ j\ } \varprojlim(\eta, i) \xrightarrow{\ \pi\ } C \longrightarrow 0 \end{pmatrix}.$$

As $\mathrm{Ext}^1(C, Q) = 0$ the extension $\eta_*(\mathcal{E})$ splits, so that, by Proposition 4.2, there exists a homomorphism $\hat{\eta} : B \to N$ such that $\hat{\eta} \circ i = \eta$. Then $(\beta - \xi\hat{\eta}) \circ i \equiv 0$, so that $\beta - \xi\hat{\eta}$ induces a homomorphism $B/\mathrm{Im}(i) \to N$. Identifying $B/\mathrm{Im}(i)$ with C, there is a unique homomorphism $\gamma : C \to N$ defined by the condition that

$$\gamma(y) = (\beta - \xi \circ \hat{\eta})(x) \quad \text{when } p(x) = y \ (x \in B, \ y \in C).$$

Tautologically $\beta - p^*(\gamma) = \xi \circ \hat{\eta}$; however $\xi \circ \hat{\eta}$ factors through the projective Q so that $p^*(\gamma) \approx \beta$, and $[p^*(\gamma)] = [\beta] \in \mathrm{Hom}_{\mathcal{D}\mathrm{er}}(B, N)$ as required. $\square$

Whilst, for fixed N, the correspondence $X \mapsto \mathrm{Ext}^1(X, N)$ is functorial on $\mathcal{D}\mathrm{er}$, in general, for fixed M, the correspondence $Y \mapsto \mathrm{Ext}^1(M, Y)$ *is not functorial on $\mathcal{D}\mathrm{er}$*. Similar to the proof of (5.21) we have:

$$\mathrm{Ext}^1(M, -) \text{ is functorial on } \mathcal{D}\mathrm{er} \iff M \text{ is coprojective.} \qquad (5.29)$$

From the point of view of functoriality the modified reverse sequence Proposition 5.28 has a hybrid nature. Whilst the initial stage

$$N \mapsto \left(\mathrm{Hom}_{\mathcal{D}\mathrm{er}}(C, N) \overset{p^*}{\to} \mathrm{Hom}_{\mathcal{D}\mathrm{er}}(B, N) \overset{i^*}{\to} \mathrm{Hom}_{\mathcal{D}\mathrm{er}}(A, N)\right)$$

is functorial on $\mathcal{D}\mathrm{er}$ the final stage $N \mapsto (\mathrm{Ext}^1(C, N) \overset{i^*}{\to} \mathrm{Ext}^1(B, N) \overset{i^*}{\to} \mathrm{Ext}^1(A, N))$ is not, in general, functorial on $\mathcal{D}\mathrm{er}$. However, denote by $\mathrm{Hom}(\mathcal{E}, N)$ the sequence

$$\mathrm{Hom}_{\mathcal{D}\mathrm{er}}(C, N) \overset{p^*}{\to} \mathrm{Hom}_{\mathcal{D}\mathrm{er}}(B, N) \overset{i^*}{\to} \cdots \overset{p^*}{\to} \mathrm{Ext}^1(B, N) \overset{i^*}{\to} \mathrm{Ext}^1(A, N)$$

it transpires that if A, B, C are coprojective then the correspondence $N \mapsto \mathrm{Hom}(\mathcal{E}, N)$ defines a covariant functor $\mathrm{Hom}(\mathcal{E}, -) : \mathcal{D}\mathrm{er} \to \mathcal{E}\mathrm{xact}(6)$. Rather more useful is the observation that if only B, C are coprojective then the correspondence $N \mapsto \mathrm{Hom}(\mathcal{E}, N)$ is functorial on $\mathcal{D}\mathrm{er}$ as far as

$$\mathrm{Hom}_{\mathcal{D}\mathrm{er}}(C, N) \overset{p^*}{\to} \mathrm{Hom}_{\mathcal{D}\mathrm{er}}(B, N) \overset{i^*}{\to} \mathrm{Hom}_{\mathcal{D}\mathrm{er}}(A, N) \overset{\delta}{\to} \mathrm{Ext}^1(C, N)$$

$$\overset{p^*}{\to} \mathrm{Ext}^1(B, N).$$

This is true, for example, when C is coprojective and B is projective (cf. Theorem 5.41 below.)

5.5   Generalized Syzygies

A module D is said to be a *generalized first syzygy* of a M when there is an exact sequence of Λ-homomorphisms of the form $(0 \to D \to P \to M \to 0)$ where P is projective. The above sequence may be transformed by the addition of a projective module Q thus:

$$0 \to D \oplus Q \to P \oplus Q \to M \to 0$$

showing that if D is a generalised first syzygy then so also is $D \oplus Q$. In particular, a module M may have many generalised first syzygies, which are distinct as Λ-modules. Nevertheless by Schanuel's Lemma and the criterion of Theorem 5.9 they become isomorphic in $\mathcal{D}\mathrm{er}$; that is:

Corollary 5.30 *Let D_1, D_2 both be generalised first syzygies of the Λ-module M; then*

$$D_1 \cong_{\mathcal{D}\mathrm{er}} D_2.$$

Thus to any Λ-module M we associate a *unique isomorphism class* $\mathbf{D}_1(M)$ in $\mathcal{D}\mathrm{er}$ represented by any generalised first syzygy of M; formally, we have

$D \in \mathbf{D}_1(M)$ if and only if there exists an exact sequence $0 \to D' \to Q \to M \to 0$ such that for some projective modules P, P':

$$D \oplus P \cong D' \oplus P'.$$

We refer to $\mathbf{D}_1(M)$ as the *first derived module* of M. It is natural to imagine that each $D \in \mathbf{D}_1(M)$ is a generalised syzygy; that is, occurs in an exact sequence

$$0 \to D \to Q \to M \to 0,$$

where Q is projective. We caution the reader against taking this attractively simple view as, in general, it is false.[1] It is, however, correct under the additional hypothesis that M is coprojective, as we now prove:

Proposition 5.31 *Let Λ be a ring and let M be a Λ-module such that $H^1(M, \Lambda)=0$. Then each $D \in D_1(M)$ occurs in an exact sequence $0 \to D \to S \to M \to 0$ where S is projective.*

Proof We may choose an exact sequence $\mathcal{S} = (0 \to K \xrightarrow{i} F_X \xrightarrow{p} M \to 0)$ where X is a set of generators for M. Since $D \in D_1(M)$ then, by definition, there are projective modules P, Q such that $D \oplus P \cong K \oplus Q$. We may now modify $\mathcal{S}$ to get an exact sequence

$$0 \to K \oplus Q \xrightarrow{i'} F_X \oplus Q \xrightarrow{p'} M \to 0,$$

where $i' = i \oplus \mathrm{Id} : K \oplus Q \xrightarrow{i'} F_X \oplus Q$ and where p' is the obvious composite of p with the projection $F_X \oplus Q \to F_X$. Now let $\varphi : D \oplus P \cong K \oplus Q$ be an isomorphism. Putting $j = i' \circ \varphi$ we get an exact sequence

$$0 \to D \oplus P \xrightarrow{j} F_X \oplus Q \xrightarrow{p'} M \to 0.$$

Evidently j induces an imbedding $j : P \to F_X \oplus Q$. Putting $S = (F_X \oplus Q)/j(P)$ and identifying D with $(D \oplus P)/P$ we get an exact sequence

$$0 \to D \xrightarrow{j_*} S \xrightarrow{\pi} M \to 0,$$

where j_*, π are induced by j and p' respectively. Now $H^1(M, \Lambda) = 0$, by hypothesis, so that S is projective by Proposition 5.17.   □

We may modify an exact sequence of Λ-modules $0 \to D \to P \to M \to 0$ so as to have the form $0 \to D \to P \oplus Q \to M \oplus Q \to 0$. If P, Q are both projective then

---

[1] For example, if C_∞ denotes the infinite cyclic group and $\mathbf{Z}$ is the trivial module over $\Lambda = \mathbf{Z}[C_\infty]$ then $\mathbf{Z}$ occurs in an exact sequence $0 \to \Lambda \to \Lambda \to \mathbf{Z} \to 0$. Thus $\Lambda \in \mathbf{D}_1(\mathbf{Z})$. As $\Lambda \sim 0$ then $0 \in \mathbf{D}_1(\mathbf{Z})$. However, in any exact sequence $0 \to 0 \to X \to \mathbf{Z} \to 0$ the module X is isomorphic to $\mathbf{Z}$ and so is not projective.

so also is $P \oplus Q$ and we obtain:

$$\mathbf{D}_1(M \oplus Q) = \mathbf{D}_1(M) \quad \text{if } Q \text{ is projective.} \tag{5.32}$$

Thus $\mathbf{D}_1(M)$ depends only upon the isomorphism class of M in $\mathcal{D}er$ rather its isomorphism class as a Λ-module. The construction may be iterated; $\mathbf{D}_1(M)$ itself has a generalised first syzygy $\mathbf{D}_1(\mathbf{D}_1(M))$ and we may write recursively $\mathbf{D}_{n+1}(M) = \mathbf{D}_1(\mathbf{D}_n(M))$. We refer to $\mathbf{D}_n(M)$ as the nth *derived module* of M. Evidently we have:

$$\mathbf{D}_{m+n}(M) = \mathbf{D}_m(\mathbf{D}_n(M)). \tag{5.33}$$

Generalised syzygies arise as the intermediate steps in projective resolutions; let

$$\mathbf{P} = (\cdots \rightarrow P_n \overset{\partial_n}{\rightarrow} P_{n-1} \rightarrow \cdots \rightarrow P_1 \overset{\partial_1}{\rightarrow} P_0 \overset{\epsilon}{\rightarrow} M \rightarrow 0)$$

be a projective resolution of M and consider the canonical factorization of ∂_n

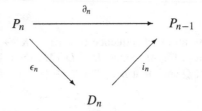

where $D_n = \mathrm{Im}(\partial_n)$, $i_n : D_n \rightarrow P_{n-1}$ is the inclusion and where $\epsilon_n(x) = \partial_n(x)$. The resulting decomposition breaks $(P_* \rightarrow M)$ into a collection of connected short exact sequences

$$0 \rightarrow D_1 \overset{i_1}{\rightarrow} P_0 \overset{\epsilon}{\rightarrow} M \rightarrow 0$$
$$0 \rightarrow D_n \overset{i_n}{\rightarrow} P_{n-1} \overset{\epsilon_{n-1}}{\rightarrow} D_{n-1} \rightarrow 0$$

and shows, recursively, that $D_n \in \mathbf{D}_n(M)$. In this connection we have the phenomenon of *dimension shifting* in cohomology. Thus suppose that

$$0 \rightarrow D \overset{i}{\rightarrow} P \overset{\epsilon}{\rightarrow} M \rightarrow 0 \tag{$*$}$$

is an exact sequence of Λ-modules with P projective, and let

$$(\cdots \rightarrow Q_n \overset{\delta_n}{\rightarrow} Q_{n-1} \rightarrow \cdots \rightarrow Q_1 \overset{\delta_1}{\rightarrow} Q_0 \overset{\eta}{\rightarrow} D \rightarrow 0) \tag{$**$}$$

be a projective resolution of D. Splicing $(*)$ and $(**)$ together we get a projective resolution of M

$$(\cdots \rightarrow Q_n \overset{\delta_n}{\rightarrow} Q_{n-1} \rightarrow \cdots \rightarrow Q_1 \overset{\delta_1}{\rightarrow} Q_0 \overset{i\eta}{\rightarrow} P \overset{\epsilon}{\rightarrow} M \rightarrow 0)$$

which, on re-indexing thus

$$P_0 = P; \qquad \partial_1 = i \circ \eta;$$

$$P_{n+1} = Q_n; \qquad \partial_{n+1} = \delta_n;$$

we see in particular that, for $n \geq 1$, $H^n(D, N) \cong H^{n+1}(M, N)$. Expressed in terms of derived modules, we have more generally, that for any Λ-modules M, N:

Proposition 5.34 $H^n(\mathbf{D}_m(M), N) \cong H^{n+m}(M, N)$ *for* $m, n \geq 1$.

5.6 Corepresentability of Cohomology

In Sect. 4.5, for a projective 0-complex $\mathcal{P} = (0 \to K \xrightarrow{i} P \xrightarrow{\epsilon} M \to 0)$, we introduced the notation

$$\mathcal{H}^1_{\mathcal{P}}(M, N) = \mathrm{Hom}_{\mathrm{Der}}(K, N)$$

and constructed a natural transformation $\lambda_{\mathcal{P}} : H^1_{\mathcal{P}}(M, N) \to \mathcal{H}^1_{\mathcal{P}}(M, N)$ where $H^1_{\mathcal{P}}(M, N)$ is a (nonstandard) model for $H^1(M, N)$. Here we generalize this; we now take $\mathcal{P}$ to be a projective $(n-1)$-complex

$$\mathcal{P} = (0 \to K_n \xrightarrow{i_n} P_{n-1} \xrightarrow{\partial_{n-1}} P_{n-2} \xrightarrow{\partial_{n-2}} \cdots \xrightarrow{\partial_2} P_1 \xrightarrow{\partial_1} P_0 \xrightarrow{\epsilon} M \to 0)$$

and construct a natural transformation $\lambda_{\mathcal{P}} : H^n_{\mathcal{P}}(M, N) \to \mathcal{H}^n_{\mathcal{P}}(M, N)$ where

$$\mathcal{H}^n_{\mathcal{P}}(M, N) = \mathrm{Hom}_{\mathrm{Der}}(K_n, N)$$

and where $H^n_{\mathcal{P}}(M, N)$ is a corresponding (nonstandard) model for $H^n(M, N)$. We begin with the latter and to elucidate its features we mimic the discussion of Sect. 5.3 in greater generality but less detail. Let

$$\mathcal{P} = (0 \to K_n \xrightarrow{i_n} P_{n-1} \xrightarrow{\partial_{n-1}} P_{n-2} \xrightarrow{\partial_{n-2}} \cdots \xrightarrow{\partial_2} P_1 \xrightarrow{\partial_1} P_0 \xrightarrow{\epsilon} M \to 0)$$

be a projective $(n-1)$-complex and let

$$\mathbf{P} = (\cdots \xrightarrow{\partial_{k+1}} P_k \xrightarrow{\partial_k} P_{k-1} \xrightarrow{\partial_{k-1}} \cdots \xrightarrow{\partial_2} P_1 \xrightarrow{\partial_1} P_0 \xrightarrow{\epsilon} M \to 0)$$

be a projective resolution extending $\mathcal{P}$. Then we define

$$H^n_{\mathcal{P}}(M, N) = \frac{\mathrm{Hom}(K_n, N)}{\mathrm{Im}(\mathrm{Hom}(P_{n-1}, N) \xrightarrow{i_n^*} \mathrm{Hom}(K_n, N))}.$$

To consider the effect of morphisms suppose given a morphism of projective n-complexes

$$\begin{matrix} \mathcal{P} \\ \downarrow \varphi = \\ \mathcal{Q} \end{matrix} \begin{pmatrix} 0 \to K_n \xrightarrow{i_n} P_{n-1} \to & \cdots \to & P_0 \to & M \to 0 \\ \downarrow \varphi_+ \quad \downarrow \varphi_{n-1} & & \downarrow \varphi_0 & \downarrow \varphi_- \\ 0 \to K'_n \xrightarrow{j_n} Q_{n-1} \to & \cdots \to & Q_0 \to & M' \to 0 \end{pmatrix}.$$

We construct an induced morphism $\varphi^* : H_Q^n(M', N) \to H_P^n(M, N)$. To do this observe that for the induced map

$$(\varphi_+)^* : \mathrm{Hom}(K_n', N) \to \mathrm{Hom}(K_n, N)$$

we have

$$\mathrm{Im}(\mathrm{Hom}(Q_{n-1}, N) \xrightarrow{j_n^*} \mathrm{Hom}(K_n', N)) \subset \mathrm{Im}(\mathrm{Hom}(P_{n-1}, N) \xrightarrow{i_n^*} \mathrm{Hom}(K_n, N)).$$

We define

$$\varphi^* : H_Q^n(M', N) \to H_P^n(M, N)$$

to be the homomorphism induced from $\varphi_+^* : \mathrm{Hom}(K_n', N) \to \mathrm{Hom}(K_n, N)$. As in Corollary 4.23, φ^* depends only upon φ_- that is:

Proposition 5.35 *If $\varphi_1, \varphi_2 : P \to Q$ satisfy $(\varphi_1)_- = (\varphi_2)_-$ then $(\varphi_1)^* = (\varphi_2)^*$.*

It is straightforward to check that this construction is functorial on morphisms of projective n-complexes; that is, if $\varphi : P \to Q$, $\psi : Q \to R$ are morphisms as follows:

$$
\begin{array}{ccc}
P & & \left(\begin{array}{cccccccc} 0 \to & K_n & \to & P_{n-1} & \to \cdots \to & P_0 & \to & M \to 0 \\ & \downarrow \varphi_+ & & \downarrow \varphi_{n-1} & & \downarrow \varphi_0 & & \downarrow \varphi_- \\ 0 \to & K_n' & \to & Q_{n-1} & \to \cdots \to & Q_0 & \to & M' \to 0 \\ & \downarrow \psi_+ & & \downarrow \psi_{n-1} & & \downarrow \psi_0 & & \downarrow \psi_- \\ 0 \to & K_n'' & \to & R_{n-1} & \to \cdots \to & R_0 & \to & M'' \to 0 \end{array}\right) \\
\downarrow \varphi & = & \\
Q & & \\
\downarrow \psi & & \\
R & &
\end{array}
$$

then $(\psi\varphi)^* = \varphi^*\psi^*$. Suppose that P, Q are projective n-complexes as below.

$$P = (0 \to K_n \xrightarrow{i_n} P_{n-1} \to \cdots \to P_0 \to M \to 0),$$

$$Q = (0 \to K_n' \xrightarrow{j_n} Q_{n-1} \to \cdots \to Q_0 \to M' \to 0)$$

and suppose that $f : M \to M'$ is a Λ-homomorphism. By the universal property of projectives we may construct a lifting $\hat{f}$ of f thus:

$$
\begin{array}{cc}
P & \left(\begin{array}{cccccccc} 0 \to & K_n & \xrightarrow{i_n} & P_{n-1} & \to \cdots \to & P_0 & \to & M \to 0 \\ & \downarrow f_+ & & \downarrow f_{n-1} & & \downarrow f_0 & & \downarrow f \\ 0 \to & K_n' & \xrightarrow{j_n} & Q_{n-1} & \to \cdots \to & Q_0 & \to & M' \to 0 \end{array}\right) \\
\downarrow \hat{f} = & \\
Q &
\end{array}
$$

and define $f_{PQ}^* : H_Q^n(M', N) \to H_P^n(M, N)$ by $f_{PQ}^* = (\hat{f})^*$. The construction $f \mapsto f_{PQ}$ is again functorial; if in addition R is a projective n-complex over M'' and $g : M \to M'$, $f : M' \to M''$ are Λ-homomorphisms; then

$$(f \circ g)_{PR}^* = g_{PQ}^* f_{QR}^*.$$

We show that this model for H^n is isomorphic to the standard model; specifically that:

Proposition 5.36 *If $\mathcal{P}$ is a projective n-complex over M, and $\mathbf{Q}$ is a projective res-olution over M then there exists an isomorphism $\natural_{\mathbf{Q}\mathcal{P}} : H_{\mathcal{P}}^n(M, N) \to H_{\mathbf{Q}}^n(M, N)$ which is natural in the sense that the following diagrams commute:*

$$
\begin{array}{ccc}
H_{\mathbf{Q}}^n(M, N) & \xrightarrow{\;\;\natural_{\mathbf{Q}\mathbf{Q}}\;\;} & H_{\mathbf{Q}}^n(M, N) \\
\Big\downarrow{\scriptstyle \tau_{\mathcal{P}\mathbf{Q}}} & & \Big\downarrow{\scriptstyle \iota_{\mathbf{P}\mathbf{Q}}} \\
H_{\mathcal{P}}^n(M, N) & \xrightarrow{\;\;\natural_{\mathbf{P}\mathcal{P}}\;\;} & H_{\mathbf{P}}^n(M, N)
\end{array}
$$

Proof Given a projective n-complex $\mathcal{P} = (0 \to K_n \xrightarrow{i_n} P_{n-1} \to \cdots \to P_0 \to M \to 0)$ extend $\mathcal{P}$ to a projective resolution $\mathbf{P} = (\cdots \xrightarrow{\partial_{k+1}} P_k \xrightarrow{\partial_k} \cdots \xrightarrow{\partial_2} P_1 \xrightarrow{\partial_1} P_0 \xrightarrow{\epsilon} M \to 0)$ where

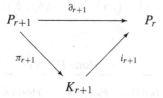

is the canonical factorization of $\partial_{r+1} : P_{r+1} \to P_r$ in which π_{r+1} is surjective and i_{r+1} injective. In particular, we have short exact sequences

$$0 \to K_1 \xrightarrow{i_1} P_0 \xrightarrow{\epsilon} M \to 0$$
$$0 \to K_{r+1} \xrightarrow{i_{r+1}} P_r \xrightarrow{\pi_r} K_r \to 0$$

From the exact sequence $0 \to \mathrm{Hom}_\Lambda(K_r, N) \xrightarrow{\pi_r^N} \mathrm{Hom}_\Lambda(P_r, N) \xrightarrow{i_r^N} \cdots$ we see that each π_r^N is injective. Now we have a diagram in which the row is exact and the triangle commutes:

$$
\begin{array}{ccccc}
0 \longrightarrow & \mathrm{Hom}(K_n, N) & \xrightarrow{\;\pi_n^N\;} & \mathrm{Hom}(P_n, N) & \xrightarrow{\;i_{n+1}^N\;} & \mathrm{Hom}(K_{n+1}, N) \\
 & & & & \searrow{\scriptstyle \partial_{n+1}^N} \quad & \Big\downarrow{\scriptstyle \pi_{n+1}^N} \\
 & & & & & \mathrm{Hom}(P_{n+1}, N)
\end{array}
$$

As π_{n+1}^N is injective and $\partial_{n+1}^N = \pi_{n+1}^N \circ i_{n+1}^N$ then $\mathrm{Ker}(\partial_{n+1}^N) = \mathrm{Ker}(i_{n+1}^N)$; by exactness $\pi_n^N : \mathrm{Hom}(K_n, N) \to \mathrm{Ker}(\partial_{n+1}^N)$ is an isomorphism. Since $\pi_n^N : \mathrm{Hom}(K_n, N) \to \mathrm{Hom}(P_n, N)$ is injective it follows that π_n^N induces an isomorphism

$$\pi_n^N : \cfrac{\operatorname{Hom}(K_n, N)}{\operatorname{Im}(\operatorname{Hom}(P_{n-1}, N) \overset{i_n^N}{\to} \operatorname{Hom}(K_n, N))}$$

$$\longrightarrow \cfrac{\operatorname{Im}(\operatorname{Hom}(K_n, N) \overset{\pi_n^N}{\to} \operatorname{Hom}(P_n, N))}{\operatorname{Im}(\operatorname{Hom}(P_{n-1}, N) \overset{\pi_n^N i_n^N}{\to} \operatorname{Hom}(P_n, N))}.$$

As above $\operatorname{Ker}(\partial_{n+1}^N) = \operatorname{Im}(\operatorname{Hom}(K_n, N) \overset{\pi_n^N}{\to} \operatorname{Hom}(P_n, N))$ and $\partial_n^N = \pi_n^N i_n^N$ so that π_n^N induces an isomorphism, denoted by $\natural_{PP}$, thus

$$\natural_{PP} = \pi_n^N : \cfrac{\operatorname{Hom}(K_n, N)}{\operatorname{Im}(\operatorname{Hom}(P_{n-1}, N) \overset{i_n^N}{\to} \operatorname{Hom}(K_n, N))} \longrightarrow \cfrac{\operatorname{Ker}(\partial_{n+1}^N)}{\operatorname{Im}(\partial_n^N)}.$$

However

$$H_{\mathcal{P}}^n(M, N) = \cfrac{\operatorname{Hom}(K_n, N)}{\operatorname{Im}(\operatorname{Hom}(P_{n-1}, N) \overset{i_n^N}{\to} \operatorname{Hom}(K_n, N))} \quad \text{and}$$

$$H_{\mathbf{P}}^n(M, N) = \cfrac{\operatorname{Ker}(\partial_{n+1}^N)}{\operatorname{Im}(\partial_n^N)}$$

so that $\natural_{PP}$ gives an isomorphism $\natural_{PP} : H_{\mathcal{P}}^n(M, N) \longrightarrow H_{\mathbf{P}}^n(M, N)$. In general if $\mathbf{Q}$ is an arbitrary projective resolution of M we define

$$\natural_{\mathbf{Q}\mathcal{P}} = t_{\mathbf{QP}}\natural_{\mathbf{P}\mathcal{P}} : H_{\mathcal{P}}^n(M, N) \to H_{\mathbf{Q}}^n(M, N),$$

where $\mathbf{P}$ is a projective resolution of M extending $\mathcal{P}$. If $\mathbf{P}'$ is also a projective resolution of M extending $\mathcal{P}$ then the diagram below commutes:

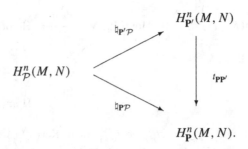

It follows that $\natural_{\mathbf{Q}\mathcal{P}}$ is independent of the choice of $\mathbf{P}$ so completing the proof.   □

Evidently $\mathrm{Im}(i_n^N) \subset \langle K_n, N \rangle \subset \mathrm{Hom}(K_n, N)$ so that we obtain an exact sequence

$$0 \to \langle K_n, N \rangle / \mathrm{Im}(i_n^N) \to \mathrm{Hom}(K_n, N) / \mathrm{Im}(i_n^N) \to \mathrm{Hom}(K_n, N) / \langle K_n, N \rangle \to 0.$$

Denoting by $\lambda_{\mathcal{P}} \colon \mathrm{Hom}(K_n, N) / \mathrm{Im}(i_n^N) \to \mathrm{Hom}(K_n, N) / \langle K_n, N \rangle$ the canonical surjection we now have a short exact sequence

$$0 \to \langle K_n, N \rangle / \mathrm{Im}(i_n^N) \to H_{\mathcal{P}}^n(M, N) \overset{\lambda_{\mathcal{P}}}{\to} \mathcal{H}_{\mathcal{P}}^n(M, N) \to 0.$$

We can extend the definition of $\mathcal{H}_{\mathcal{P}}^n(M, N)$ to $n = 0$ by writing

$$\mathcal{H}_{\mathcal{P}}^n(M, N) = \mathrm{Hom}_{\mathcal{D}\mathrm{er}}(M, N) = \mathrm{Hom}_\Lambda(M, N) / \langle M, N \rangle.$$

We arrive at the Corepresentation Theorem for cohomology.

Theorem 5.37 *Let M be a Λ-module such that $H^n(M, \Lambda) = 0$ where $n \geq 0$ and let $\mathcal{P}$ be a projective $(n-1)$-complex over M; then $\lambda_{\mathcal{P}} \colon H_{\mathcal{P}}^n(M, N) \to \mathcal{H}_{\mathcal{P}}^n(M, N)$ is an isomorphism for any Λ-module N.*

Proof We first consider dimension shifting in the nonstandard model. Suppose that $\mathcal{P} = (0 \to K_{\nu+1} \to P_\nu \overset{\partial_\nu}{\to} \cdots \to P_1 \overset{\partial_1}{\to} P_0 \to M \to 0)$ is a projective ν-complex and that $0 < m < \nu$. Let

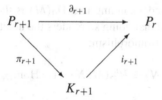

be the canonical factorization of $\partial_{r+1} \colon P_{r+1} \to P_r$ in which π_{r+1} is surjective and i_{r+1} injective. By the m-coskeleton $\mathcal{P}\langle m \rangle$ of $\mathcal{P}$ we mean the exact sequence

$$\mathcal{P} = (0 \to K_{\nu+1} \to P_\nu \overset{\partial_\nu}{\to} \cdots \to P_{m+1} \overset{\partial_{m+1}}{\to} P_m \to K_m \to 0).$$

Observe that $\mathcal{P}\langle m \rangle$ is a projective $(\nu - m)$-complex In the particular case where $\nu = n - 1$ we see that

$$H_{\mathcal{P}\langle m \rangle}^{n-m}(K_m, N) \equiv H_{\mathcal{P}}^n(M, N)$$

Moreover $\lambda_{\mathcal{P}} \equiv \lambda_{\mathcal{P}\langle m \rangle}$; that is:

$$H_{\mathcal{P}\langle m\rangle}^{n-m}(K_m, N) \quad \equiv \quad H_{\mathcal{P}}^n(M, N)$$

$$\downarrow \lambda_{\mathcal{P}\langle m\rangle} \qquad\qquad\qquad \downarrow \lambda_{\mathcal{P}}$$

$$\mathcal{H}_{\mathcal{P}\langle m\rangle}^{n-m}(K_m, N) \quad \equiv \quad \mathcal{H}_{\mathcal{P}}^n(M, N)$$

By hypothesis $H^n(M, \Lambda) = 0$ so that $H^1(K_{n-1}, \Lambda) = 0$; hence by (5.21)

$$\lambda_{\mathcal{P}\langle n-1\rangle} : H_{\mathcal{P}\langle n-1\rangle}^1(K_{n-1}, N) \to \mathcal{H}_{\mathcal{P}\langle n-1\rangle}^1(K_{n-1}, N)$$

is an isomorphism. By above, $\lambda_{\mathcal{P}} : H_{\mathcal{P}}^n(M, N) \to \mathcal{H}_{\mathcal{P}}^n(M, N)$ is an isomorphism. $\square$

The above Corepresentation Theorem is proved in the thesis of Humphreys [44]. We note that the case $n = 0$ in Theorem 5.37 distinguishes between finite and infinite groups. Thus let $\Lambda = \mathbf{Z}[G]$ be an integral group ring acting trivially on $\mathbf{Z}$.

If G is finite then $H^0(\mathbf{Z}, \Lambda) \cong \mathbf{Z}$ and $\mathcal{H}^0(\mathbf{Z}, \mathbf{Z}) \cong \mathbf{Z}/|G|$. $\qquad$ (5.38)

If G is infinite then $H^0(\mathbf{Z}, \Lambda) = 0$ and $\mathcal{H}^0(\mathbf{Z}, \mathbf{Z}) \cong \mathbf{Z}$. $\qquad$ (5.39)

Note also that in the case $n = 0$ the definition is entirely independent of $\mathcal{P}$ so retaining complete definiteness of description we may omit the subscript and write

$$\mathcal{H}^0(M, N) = \mathrm{Hom}_{\mathcal{D}\mathrm{er}}(M, N).$$

In general, we may omit the suffix $\mathcal{P}$ and write $\mathcal{H}^n(M, N) = \mathrm{Hom}_{\mathcal{D}\mathrm{er}}(\mathbf{D}_n(M), N)$ for all $n \geq 0$, with the understanding that $\mathbf{D}_0(M)$ is the class of M in the derived module category $\mathcal{D}\mathrm{er}$ and that doing so renders the notation imprecise in that it only defines $\mathcal{H}^n(M, N)$ up to isomorphism;

$$\mathcal{H}^n(M, N) \cong \mathcal{H}_{\mathcal{P}}^n(M, N) \quad (= \mathrm{Hom}_{\mathcal{D}\mathrm{er}}(K_n, N)).$$

The groups $\mathcal{H}^n(M, N)$ are the *generalized Tate cohomology groups*. As we shall see, they have some (and in some restricted contexts *all*) of the properties one might expect of cohomology groups defined on the derived category $\mathcal{D}\mathrm{er}$ (as opposed to the category $\mathcal{M}\mathrm{od}_\Lambda$ of Λ-modules).

5.7 Swan's Projectivity Criterion

In this section we generalize the criterion for projectivity given by Swan [91]. Let

$$\mathcal{P} = (0 \to D \xrightarrow{i} P \xrightarrow{p} M \to 0)$$

be a projective 0-complex. If $\alpha : D \to D$ is a Λ-homomorphism we have a morphism of exact sequences;

$$
\begin{pmatrix} \mathcal{P} \\ \downarrow \nu \\ \alpha_*(\mathcal{P}) \end{pmatrix} = \begin{pmatrix} 0 \longrightarrow & D & \overset{i}{\longrightarrow} & P & \overset{p}{\longrightarrow} & M & \longrightarrow 0 \\ & \downarrow \alpha & & \downarrow \nu & & \downarrow \mathrm{Id} & \\ 0 \longrightarrow & D & \overset{j}{\longrightarrow} & \varinjlim(\alpha, i) & \overset{\pi}{\longrightarrow} & M & \longrightarrow 0 \end{pmatrix}.
$$

Putting $S(\alpha) = \varinjlim(\alpha, i)$ we obtain:

Proposition 5.40 *If α is an isomorphism in $\mathcal{D}\mathrm{er}(\Lambda)$ then $S(\alpha)$ is a projective Λ-module.*

Proof By Theorem 5.23, (5.24) any $N \in \mathcal{M}\mathrm{od}_\Lambda$ gives a commutative ladder with exact rows:

$$
\begin{array}{ccccccc}
\mathcal{H}^0(N, D) & \overset{i_*}{\to} & \mathcal{H}^0(N, P) & \overset{p_*}{\to} & \mathcal{H}^0(N, M) & \overset{\partial}{\to} & \mathrm{Ext}^1(N, D) \\
\downarrow \alpha_* & & \downarrow \nu_* & & \downarrow \mathrm{Id} & & \downarrow \alpha_* \\
\mathcal{H}^0(N, D) & \overset{j_*}{\to} & \mathcal{H}^0(N, S(\alpha)) & \overset{\pi_*}{\to} & \mathcal{H}^0(N, M) & \overset{\partial}{\to} & \mathrm{Ext}^1(N, D)
\end{array}
$$

Evidently Id is an isomorphism on $\mathcal{H}^0(N, M)$. Moreover, since α is an isomorphism in $\mathcal{D}\mathrm{er}$ then α_* is an isomorphism on both $\mathcal{H}^0(N, D)$ and $\mathrm{Ext}^1(N, D)$. Since P is projective then $\mathcal{H}^0(N, P) = \mathrm{Hom}_{\mathcal{D}\mathrm{er}}(N, P) = 0$. It now follows easily that $\mathcal{H}^0(N, S(\alpha)) = 0$. Thus for any Λ-module N it follows that $\mathrm{Hom}_{\mathcal{D}\mathrm{er}}(N, S(\alpha)) = 0$. Thus $S(\alpha)$ is projective by Proposition 5.5, and this completes the proof. $\square$

The converse is true provided that M is coprojective.

Theorem 5.41 *If $\mathrm{Ext}^1(M, \Lambda) = 0$ then $S(\alpha)$ is a projective Λ-module if and only if α is an isomorphism in $\mathcal{D}\mathrm{er}$.*

Proof By Proposition 5.40 it suffices to prove ($\Longrightarrow$). Assume that $S(\alpha)$ is projective; as P is also projective then $\mathcal{H}^0(P, N) = \mathcal{H}^0(S(\alpha), N) = 0$ and also $\mathrm{Ext}^1(P, N) = \mathrm{Ext}^1(S(\alpha), N) = 0$ for any Λ-module N. Taking $N = D$ then from Proposition 5.28 (which is where we require the hypothesis that $\mathrm{Ext}^1(M, \Lambda) = 0$) we obtain a commutative diagram with exact rows:

$$
\begin{array}{ccccc}
0 \to & \mathcal{H}^0(D, D) & \overset{j^*}{\to} & \mathrm{Ext}^1(M, D) & \to 0 \\
& \downarrow \alpha^* & & \downarrow \mathrm{Id} & \\
0 \to & \mathcal{H}^0(D, D) & \overset{j^*}{\to} & \mathrm{Ext}^1(M, D) & \to 0
\end{array}
$$

In particular, $\alpha^* : \mathcal{H}^0(D, D) \to \mathcal{H}^0(D, D)$ is bijective. We may choose a Λ-homomorphism $\beta : D \to D$ such that $\alpha^*(\beta) = \beta \circ \alpha \approx \mathrm{Id}$. Then

$$\alpha^*(\alpha \circ \beta) = (\alpha \circ \beta) \circ \alpha$$
$$= \alpha \circ (\beta \circ \alpha)$$
$$\approx \alpha \circ \text{Id}$$
$$\approx \text{Id} \circ \alpha$$
$$= \alpha^*(\text{Id}).$$

By injectivity of $\alpha^* : \mathcal{H}^0(D, D) \to \mathcal{H}^0(D, D)$ it follows that $\alpha \circ \beta \approx \text{Id}$, and $\alpha : D \to D$ is an isomorphism in $\mathcal{D}\text{er}$. This proves ($\Longrightarrow$) and completes the proof.   □

Chapter 6
Finiteness Conditions

As no restrictions were placed on the modules involved, the derived module category $\mathcal{D}er$ of Chap. 5 has an absolute character. This is unrealistic, and in practice it is necessary to limit the size of modules. Although our primary interest is in modules which are finitely generated, it is inconvenient to impose this restriction from the outset. We assume that the ring Λ is *weakly coherent* and initially restrict attention to modules which are countably generated. The appropriate derived module category is denoted by $\mathcal{D}er_\infty$. Recalling that any countably generated module M has a hyperstabilization $\widehat{M} = M \oplus \Lambda^\infty$, the objects of $\mathcal{D}er_\infty$ have a convenient representation as hyperstable modules. Thereafter we consider the difficulties involved in imposing the more stringent restriction of finite generation; compare [84].

6.1 Hyperstable Modules and the Category $\mathcal{D}er_\infty$

Λ will denote a weakly coherent associative ring with unity. and $\mathcal{M}od_\infty$ ($= \mathcal{M}od_\infty(\Lambda)$) will denote the category of countably generated right Λ-modules. If $f : M \to N$ is a Λ-homomorphism between countably generated Λ modules M, N we write '$f \approx_\infty 0$', when f can be written as a composite $f = \xi \circ \eta$, for some *countably generated* projective module $P \in \mathcal{M}od_\infty$, and Λ-homomorphisms $\eta : M \to P$ and $\xi : P \to N$ thus:

$$
\begin{array}{ccc}
M & \xrightarrow{\;\;f\;\;} & N \\
& \eta \searrow \quad \nearrow \xi & \\
& P &
\end{array}
\tag{6.1}
$$

For $M, N \in \mathcal{M}od_\infty$ we define $\langle M, N \rangle_\infty = \{ f \in \mathrm{Hom}_\Lambda(M, N) : f \approx_\infty 0 \}$.

Proposition 6.2 *If $M, N \in \mathcal{M}od_\infty$ then $\langle M, N \rangle_\infty = \langle M, N \rangle$.*

F.E.A. Johnson, *Syzygies and Homotopy Theory*, Algebra and Applications 17,
DOI 10.1007/978-1-4471-2294-4_6, © Springer-Verlag London Limited 2012

Proof It is clear that $\langle M, N \rangle_\infty \subset \langle M, N \rangle$. Suppose that M, N are countably generated Λ-modules and that $f : M \to N$ has a factorization $f = \xi \circ \eta$ through a projective $P \in \mathcal{M}od_\Lambda$ as in (6.1). Choose a projective $Q \in \mathcal{M}_\Lambda$ and a specific isomorphism $\varphi : P \oplus Q \to F_X$ where F_X is the free Λ-module on a set X. Let $i_P : P \to P \oplus Q$, $\pi_P : P \oplus Q \to P$ denote respectively the canonical inclusion and projection. Modify the factorization as $f = \hat{\xi} \circ \hat{\eta}$ where $\hat{\xi} = \xi \circ \pi_P \circ \varphi^{-1}$ and $\tilde{\eta} = \varphi \circ i_P \circ \eta : M \to F_X$. As M is countably generated there exists a countable subset $Y \subset X$ such that $\mathrm{Im}(\tilde{\eta}) \subset F_Y \subset F_X$. Take $\tilde{\tilde{\xi}} = \hat{\xi}|_{F_Y}$; then we have a factorization

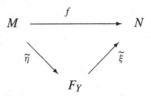

showing that $f \approx_\infty 0$.   □

Corollary 6.3 *Let M, $N \in \mathcal{M}od_\infty$; then* $\mathrm{Hom}_{\mathcal{D}er_\infty}(M, N) = \mathrm{Hom}_{\mathcal{D}er}(M, N)$.

The inclusion functor $i : \mathcal{M}od_\infty(\Lambda) \to \mathcal{M}od_\Lambda$ induces a functor $i_* : \mathcal{D}er_\infty \to \mathcal{D}er$. Then the above shows:

$$i_* : \mathcal{D}er_\infty \to \mathcal{D}er \text{ imbeds } \mathcal{D}er_\infty \text{ as a full subcategory of } \mathcal{D}er. \qquad (6.4)$$

In particular:

$$M \cong_{\mathcal{D}er_\infty} N \iff i_*(M) \cong_{\mathcal{D}er} i_*(N). \qquad (6.5)$$

For countably generated Λ-modules M, N there is no distinction between isomorphism in $\mathcal{D}er_\infty$ and isomorphism in $\mathcal{D}er$. Moreover we have a criterion for recognising isomorphic objects in $\mathcal{D}er$, namely that corresponding hyperstabilizations be isomorphic over Λ.

Proposition 6.6 *If M_1, M_2 are countably generated Λ-modules then the following statements are equivalent:*

(i) $M_1 \cong_{\mathcal{D}er} M_2$;
(ii) $M_1 \oplus P_1 \cong_\Lambda M_2 \oplus P_2$ *for some countably generated projectives P_1, P_2;*
(iii) $\widehat{M_1} \cong_\Lambda \widehat{M_2}$.

Proof (i) $\implies$ (ii): Let M_1, M_2 be countably generated Λ-modules and suppose that $M_1 \cong_{\mathcal{D}er} M_2$. In Theorem 5.9 we showed that there exist for some projective modules P_1, P_2 such that $M_1 \oplus P_1 \cong_\Lambda M_2 \oplus P_2$. However, since M_1 M_2 are countably generated it is not difficult to show that P_1, P_2 can be chosen to be countably generated. In fact, we showed in Theorem 5.9 that if there is a Λ homomorphism $f : M_1 \to M_2$ which defines an isomorphism in $\mathcal{D}er$ and if $\eta : F \to M_2$ be a surjective Λ-homomorphism where F is a free module over Λ then there is an isomorphism of Λ-modules $M_1 \oplus F \cong_\Lambda M_2 \oplus K$ where K is projective. However, since

M_2 is countably generated we may choose $F = \Lambda^\infty$. Then if M_1 is countably generated so also is $M_1 \oplus F$ and therefore so also must be K. The implication (i) $\Longrightarrow$ (ii) follows on taking $P_1 = F$ and $P_2 = K$.

(ii) $\Longrightarrow$ (iii): Suppose that $M_1 \oplus P_1 \cong_\Lambda M_2 \oplus P_2$ for some countably generated projective modules P_1, P_2. Then $M_1 \oplus P_1 \oplus \Lambda^\infty \cong_\Lambda M_2 \oplus P_2 \oplus \Lambda^\infty$. By Eilenberg's trick, $P_1 \oplus \Lambda^\infty \cong \Lambda^\infty$ so that $M_1 \oplus \Lambda^\infty \cong_\Lambda M_2 \oplus \Lambda^\infty$.

(iii) $\Longrightarrow$ (i): Now follows directly from Theorem 5.9 since $\widehat{M_i} = M_i \oplus \Lambda^\infty$ and Λ^∞ is projective. $\qquad\qquad\qquad\qquad\qquad\qquad\qquad\qquad\qquad\qquad\qquad\qquad\qquad\square$

6.2 The Category $\mathcal{D}er_{\text{fin}}$

At the outset we imposed the condition that the ring Λ should be weakly coherent, with the consequence that the category $\mathcal{M}od_\infty$ of countably generated Λ-modules and Λ-homomorphisms is abelian. Our constructions so far, though entirely universal, come with the standard drawback to universality namely that they are essentially infinite in character. In this section we begin the task of imposing finiteness conditions on our constructions.

We shall denote by $\mathcal{M}od_{\text{fin}}$ ($= \mathcal{M}od_{\text{fin}}(\Lambda)$) the category of *finitely generated* Λ-modules and Λ-homomorphisms. In general, and in many cases of interest, the condition of weak coherence is not strong enough to guarantee that the category $\mathcal{M}od_{\text{fin}}$ is abelian. There would seem to be two ways to proceed; either

(a) to strengthen the hypothesis on Λ from weak coherence and so force the category $\mathcal{M}od_{\text{fin}}$ to be abelian or

(b) to maintain the hypothesis of weak coherence but instead analyse the consequent limitations to imposing finiteness conditions upon our constructions.

We choose (b) to avoid eliminating many cases of interest. (See Appendix D.)

If $f : M \to N$ is a Λ-homomorphism between finitely generated Λ modules M, N we write '$f \approx_{\text{fin}} 0$', when f can be written as a composite $f = \xi \circ \eta$, for some *finitely generated* projective module $P \in \mathcal{M}od_{\text{fin}}$, and some Λ-homomorphisms $\eta : M_1 \to P$ and $\xi : P \to M_2$ thus:

$$M \xrightarrow{\quad f \quad} N$$

$$\eta \searrow \qquad \nearrow \xi$$

$$P$$

For $M, N \in \mathcal{M}od_{\text{fin}}$ we define $\langle M, N \rangle_{\text{fin}} = \{f \in \text{Hom}_\Lambda(M, N) : f \approx_{\text{fin}} 0\}$. The proof of the following is very similar to that of Proposition 6.2; we leave changes to the reader.

$$\text{If } M, N \in \mathcal{M}od_{\text{fin}} \text{ then } \langle M, N \rangle_{\text{fin}} = \langle M, N \rangle. \qquad (6.7)$$

Corollary 6.8 *Let* $M, N \in \mathcal{M}od_{\text{fin}}$; *then* $\text{Hom}_{\mathcal{D}er_{\text{fin}}}(M_1, M_2) = \text{Hom}_{\mathcal{D}er}(M_1, M_2)$.

The inclusion functor $i : \mathcal{Mod}_{\text{fin}} \to \mathcal{Mod}_\Lambda$ induces a functor $i_* : \mathcal{Der}_{\text{fin}} \to \mathcal{Der}$. Then the above shows:

$$i_* : \mathcal{Der}_{\text{fin}} \to \mathcal{Der} \text{ imbeds } \mathcal{Der}_{\text{fin}} \text{ as a full subcategory of } \mathcal{Der}. \qquad (6.9)$$

In particular:

$$\text{If } M, N \in \mathcal{Mod}_{\text{fin}} \text{ then } M \cong_{\mathcal{Der}_{\text{fin}}} N \iff i_*(M) \cong_{\mathcal{Der}} i_*(N). \qquad (6.10)$$

For finitely generated Λ-modules M_1, M_2 there is thus no distinction between isomorphism in $\mathcal{Der}_{\text{fin}}$, isomorphism in $\mathcal{Der}$ or, indeed, isomorphism in $\mathcal{Der}_\infty$. We have given a criterion for recognising isomorphic objects in $\mathcal{Der}_\infty$, namely that corresponding hyperstabilizations be isomorphic over Λ. Although the same isomorphism criterion then holds for objects in $\mathcal{Der}_{\text{fin}}$, it has the disadvantage of being expressed outside the category $\mathcal{Der}_{\text{fin}}$. A slight strengthing of the argument of Proposition 6.6 however gives a criterion which is intrinsic to $\mathcal{Der}_{\text{fin}}$:

Proposition 6.11 *Let M_1, $M_2 \in \mathcal{Mod}_{\text{fin}}$; then the following conditions are equivalent*:

(i) $M_1 \cong_{\mathcal{Der}} M_2$;
(ii) $M_1 \oplus P_1 \cong_\Lambda M_2 \oplus P_2$ *for some finitely generated projective modules P_1, P_2*;
(iii) *there exists an integer $a \geq 1$ and a finitely generated projective module P such that $M_1 \oplus \Lambda^a \cong_\Lambda M_2 \oplus P$.*

Proof It is evident from Theorem 5.9 and Proposition 6.6 that (iii) $\implies$ (ii) $\implies$ (i).

Suppose that M_1, M_2 are finitely generated Λ-modules, let $f : M_1 \to M_2$ be a Λ-homomorphism which defines an isomorphism in $\mathcal{Der}$ and let $\eta : \Lambda^a \to M_2$ be a surjective Λ-homomorphism. Then, as in the proof of Theorem 5.9, the exact sequence

$$0 \to K \to M_1 \oplus \Lambda^a \overset{(f,\eta)}{\to} M_2 \to 0$$

splits and so $M_1 \oplus \Lambda^a \cong M_2 \oplus K$. It follows that K is finitely generated, and, as again in the proof of Theorem 5.9, K is projective. This shows (i) $\implies$ (iii) and completes the proof. $\qquad\square$

6.3 Finiteness Conditions and Syzygies

If $M \in \mathcal{Mod}_\infty$ then M is finitely generated if and only if there exists an exact sequence of Λ-homomorphisms

$$F_0 \overset{\eta}{\to} M \to 0,$$

where F_0 is finitely generated and free over Λ. In this case we shall also say that M is of *finite type in dimension zero*, abbreviated to $FT(0)$. More generally, if $n \geq 1$

we say that M is of *finite type up to dimension* n, abbreviated to $FT(n)$, when there exists an exact sequence of Λ-homomorphisms

$$F_n \xrightarrow{\partial_n} F_{n-1} \xrightarrow{\partial_{n-1}} \cdots \xrightarrow{\partial_2} F_1 \xrightarrow{\partial_1} F_0 \xrightarrow{\eta} M \to 0,$$

where F_r is finitely generated free over Λ for $0 \leq r \leq n$. The following is easily verified:

Proposition 6.12 *Let* $M \in \mathcal{M}od_{\mathrm{fin}}$ *and let* $n \geq 1$. *Then* M *satisfies condition* $FT(n)$ *if and only if there exists an exact sequence of* Λ-*homomorphisms*

$$P_n \xrightarrow{\partial_n} P_{n-1} \xrightarrow{\partial_{n-1}} \cdots \xrightarrow{\partial_2} P_1 \xrightarrow{\partial_1} P_0 \xrightarrow{\eta} M \to 0,$$

in which each P_r *is finitely generated projective over* Λ.

Proposition 6.13 *Let* M, N *be right* Λ-*modules such that* $N \in [M]$; *then as* Λ-*modules,* N *is finitely generated if and only if* M *is finitely generated.*

If M is a finitely generated Λ-module, then, by Schanuel's Lemma, there is a well defined stable module $\Omega_1(M)$ determined by the rule that $\Omega_1(M) = [\Omega]$ if there exists an exact sequence of Λ-modules of the form

$$0 \to \Omega \to \Lambda^a \to M \to 0,$$

where a is a positive integer. It may or may not be true that Ω is finitely generated. However, this depends only on M, and not the particular representative Ω of $\Omega_1(M)$ which is chosen. We say that $\Omega_1(M)$ is *finitely generated* when Ω is finitely generated for *at least one* (and therefore for *any*) representative $\Omega \in \Omega_1(M)$. Otherwise we say that $\Omega_1(M)$ is *infinitely generated*. Regardless of whether or not $\Omega_1(\mathbf{Z})$ is finitely generated, the stable module $\Omega_1(M)$ is called the *first syzygy* of the finitely generated module M.

We say that M is *finitely presented* when both M and $\Omega_1(M)$ are finitely generated; equivalently, M is finitely presented when there exists an exact sequence of Λ-modules of the form

$$0 \to \Omega \to \Lambda^b \to \Lambda^a \to M \to 0.$$

We then write $\Omega_2(M)$ for the stability class of Ω; again by Schanuel's Lemma, the stability class $[\Omega]$ of Ω depends only upon M, and we write

$$\Omega_2(M) = [\Omega].$$

As before, we say that $\Omega_2(M)$ is finitely generated when at least one element $\Omega \in \Omega_2(M)$ is finitely generated (then every $\Omega \in \Omega_2(M)$ is finitely generated).

Otherwise we say that $\Omega_2(M)$ is infinitely generated and again, in any case, when M is finitely presented the stable module $\Omega_2(M)$ is called the *second syzygy* of M.

In general, if M is a finitely generated module and the stable modules $\Omega_1(M)$, $\ldots, \Omega_{m-1}(M)$ are defined and finitely generated, then there exists an exact sequence

$$0 \to \Omega \to \Lambda^{e_{m-1}} \to \cdots \to \Lambda^{e_1} \to \Lambda^{e_0} \to M \to 0,$$

and the stable class $[\Omega]$ then defines the nth-syzygy $\Omega_m(M)$, which again may or may not be finitely generated. We note the following:

Theorem 6.14 *The following conditions are equivalent for any ring Λ:*

  (i) *if M is a finitely presented Λ-module and $\Omega \in \Omega_1(M)$ then Ω is also finitely presented;*
 (ii) *if M is a finitely presented Λ-module then $\Omega_n(M)$ is defined and finitely generated for all $n \geq 2$;*
(iii) *if M is a finitely generated Λ-module such that $\Omega_1(M)$ is finitely generated then $\Omega_n(M)$ is defined and finitely generated for all $n \geq 2$;*
 (iv) *in any exact sequence of Λ-modules $0 \to \Omega \to \Lambda^b \to \Lambda^a \to M \to 0$ in which a, b are positive integers, the module Ω is finitely generated;*
  (v) *in any exact sequence of Λ-modules $0 \to \Omega \to \Lambda^b \to \Lambda^a$ in which a, b are positive integers, the module Ω is finitely generated.*

A ring Λ which satisfies any of these conditions (i)–(v) is said to be *coherent*. Otherwise we shall say that Λ is *incoherent*. It is straightforward to see that when G is finite its integral group ring $\mathbf{Z}[G]$ is coherent. By contrast (see Appendix D) many familiar finitely presented infinite groups have incoherent group rings.

We note that if $\Omega_n(M)$ is defined then it represents $D_n(M)$ in the sense that

$$J \in \Omega_n(M) \quad \Longrightarrow \quad J \in D_n(M).$$

It is useful to think of $\Omega_n(M)$ as a 'polarized state' of $D_n(M)$. A more exact relation between them is considered in Proposition 6.17 below.

6.4 Graph Structure on Hyperstabilizations and the Action of $\widetilde{K_0}$

Let $\widetilde{M} \in \mathcal{M}od_\infty$ be hyperstable. We say that $\widetilde{M}$ is *finitely determined* when there exists a finitely generated Λ-module M_0 such that

$$\widetilde{M} \cong M_0 \oplus \Lambda^\infty.$$

M_0 is then said to be a *finite form* of $\widetilde{M}$. If $\widetilde{M}$ is a finitely determined hyperstable module we write $\langle \widetilde{M} \rangle$ for the set of isomorphism classes of finite forms of $\widetilde{M}$; that is:

$$\langle \widetilde{M} \rangle = \{N \in \mathcal{M}od_{\mathrm{fin}} : \widehat{N} \cong_\Lambda \widehat{M}\}/\cong .$$

We then have $\langle \widehat{M} \rangle = \langle M \rangle$. It follows also that if M is of type $FT(1)$ then the generalised syzygy $D_1(M)$ is finitely determined. More generally:

Proposition 6.15 *Let* $M \in \mathcal{M}od_{\mathrm{fin}}$; *if* M *satisfies condition* $FT(n)$ *then* $D_r(M)$ *is finitely determined for* $1 \le r \le n$.

We note that the converse to Proposition 6.15 is false. If Λ is a ring with the ∞-kernel property (see Appendix D) then there exists an exact sequence

$$0 \to P \overset{i}{\to} \Lambda^a \overset{f}{\to} \Lambda^b,$$

where a, b are positive integers and P is a projective module of infinite rank. Taking $M = \mathrm{Im}(f)$, we get an exact sequence

$$0 \to P \overset{i}{\to} \Lambda^a \overset{f}{\to} M \to 0,$$

where M is finitely generated, and hence also an exact sequence

$$0 \to P \oplus \Lambda^\infty \overset{i \oplus \mathrm{Id}}{\to} \Lambda^\infty \overset{\pi \circ f}{\to} M \to 0.$$

Thus $D_1(M) \cong P \oplus \Lambda^\infty \cong \Lambda^\infty$ by Eilenberg's trick. Thus $D_1(M)$ is finitely determined (by Λ^n for any positive integer n, or even by 0). However, M does not satisfy $FT(1)$, for otherwise there would be an exact sequence $\Lambda^c \overset{\partial_1}{\to} \Lambda^d \overset{\eta}{\to} M \to 0$ and so we would have an exact sequence

$$0 \to \mathrm{Im}(\partial_1) \to \Lambda^d \overset{\eta}{\to} M \to 0,$$

in which $\mathrm{Im}(\partial_1)$ is finitely generated. However, comparison of this with the exact sequence

$$0 \to P \to \Lambda^a \overset{f}{\to} M \to 0,$$

would show, by Schanuel, that $P \oplus \Lambda^d \cong \mathrm{Im}(\partial_1) \oplus \Lambda^a$, and since $\mathrm{Im}(\partial_1)$ is finitely generated, P would also be finitely generated. Contradiction.

By Proposition 6.6 the isomorphism class of a finitely generated Λ-module M in $\mathcal{D}er$ is entirely determined by the isomorphism class in $\mathcal{M}od_\Lambda$ of its hyperstabilization $\widehat{M}$. We put

$$\langle M \rangle = \{N \in \mathcal{M}od_{\mathrm{fin}} : \widehat{N} \cong_\Lambda \widehat{M}\}/\cong .$$

In Sect. 1.2 we considered the stable module $[M]$ as a directed graph. The construction generalises straightforwardly to $\langle M \rangle$; that is, we take $\langle M \rangle$ to be the directed graph in which the vertices are isomorphism classes of finitely generated Λ-modules N for which $\widehat{N} \cong \widehat{M}$ and where we draw an arrow $N \to N \oplus \Lambda$ for every isomorphism type N.

Proposition 6.16 *Let* M *be a finitely generated* Λ-*module; then the graph* $\langle M \rangle$ *is a disjoint union of trees each of which has the form* $[M \oplus P]$ *where* P *is a finitely generated projective module.*

Proof If N is a finitely generated Λ-module such that $\widehat{N} \cong \widehat{M}$ then $[N]$ is contained as a subgraph in $\langle M \rangle$. In particular, if P is a finitely generated projective module then $\widehat{M \oplus P} \cong \widehat{M}$ so that $[M \oplus P]$ is contained in $\langle M \rangle$. Furthermore, if $\widehat{N} \cong \widehat{M}$ then, by Propositions 6.11 and 6.6, $N \sim M \oplus P$ for some finitely generated projective module P and so $N \in [M \oplus P]$. It follows that

$$\langle M \rangle = \bigcup_P [M \oplus P],$$

where P runs through the isomorphism classes of finitely generated projective modules. However, if P, Q are finitely generated projectives then either $M \oplus P \sim M \oplus N$, in which case $[M \oplus P] = [M \oplus Q]$, or $M \oplus P \not\sim M \oplus N$, in which case $[M \oplus P] \cap [M \oplus Q] = \emptyset$. Thus $\langle M \rangle$ is a disjoint union of trees

$$\langle M \rangle = \coprod_{P_i} [M \oplus P_i],$$

where P_i runs through a set of isomorphism classes of finitely generated projectives for which the graphs $[M \oplus P_i]$ are pairwise distinct. $\qquad\qquad\square$

We note the following:

Proposition 6.17 *Let M be a finitely generated Λ-module with nth generalized syzygy $D_n(M)$; if $\Omega_n(M)$ is defined and finitely generated then $\Omega_n(M)$ comprises a connected component of $\langle D_n(M) \rangle$.*

Let $\pi_0(M)$ denote the set of connected components of $\langle M \rangle$. The stable class $[N] \in \pi_0(M)$ is thus the connected component of $\langle M \rangle$ which contains N; there is an action of the reduced projective class group $\widetilde{K}_0(\Lambda)$ on the set $\pi_0(M)$ given by:

$$\bullet : \widetilde{K}_0(\Lambda) \times \pi_0(M) \longrightarrow \pi_0(M)$$
$$[P] \bullet [N] \quad = \quad [P \oplus N]$$

It follows from Proposition 6.11 that this action is transitive. Moreover, as $\widetilde{K}_0(\Lambda)$ is abelian any two points in $\pi_0(M)$ have the same stability group. In particular, putting

$$\mathrm{Stab}(M) = \{ [P] \in \widetilde{K}_0(\Lambda) : M \oplus P \sim M \}.$$

We obtain:

If M is a finitely generated Λ-module then there is a $1 - 1$ correspondence

$$\pi_0(M) \longleftrightarrow \widetilde{K}_0(\Lambda) / \mathrm{Stab}(M). \tag{6.18}$$

$\mathrm{Stab}(M)$ is thus an invariant of the hyperstability class of M. In particular, $|\widetilde{K}_0(\Lambda)|$ is an upper bound on the number of connected components of $\langle M \rangle$. Moreover, in the simplest case, namely $M = 0$, then $\mathrm{Stab}(0) = \{0\}$ and the number of connected components of $\langle 0 \rangle$ is precisely $|\widetilde{K}_0(\Lambda)|$. We give an example taken from the calculations of Swan [94]:

Example 6.19 Take $\Lambda = \mathbf{Z}[Q(36)]$ where $Q(36) = \langle x, y | x^9 = y^2, xyx = y \rangle$; then $\widetilde{K}_0(\Lambda) \cong \mathbf{Z}/2 \oplus \mathbf{Z}/2$ so that $\pi_0([0])$ has precisely four connected components; these are drawn below, the height function being the normalized $\mathbf{Z}$-rank of the Λ-modules involved:

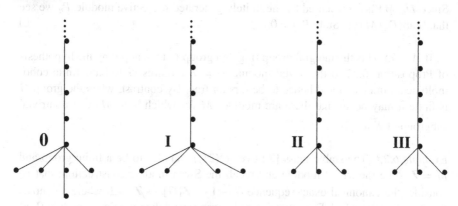

Here the tree labelled **0** is simply the stable class of the zero module; that is, the tree of stably free modules over $\mathbf{Z}[Q(36)]$. There is a useful relationship between stability groups:

Proposition 6.20 *If M is a module of type FT(1) then* $\operatorname{Stab}(M) \subset \operatorname{Stab}(D_1(M))$.

Proof Let $\mathcal{E} = (0 \to K \to \Lambda^a \to M \to 0)$ be an exact sequence and suppose that P is a finitely generated projective module such that $M \oplus P \sim M$. We show that $K \oplus P \sim K$.

Interpreting the relation $M \oplus P \sim M$ as meaning $M \oplus P \oplus \Lambda^b \cong_\Lambda M \oplus \Lambda^c$ for some $b, c \geq 0$ then we may modify $\mathcal{E}$ to give two new sequences

$$\mathcal{E}_1 = (0 \to K \to \Lambda^{a+b} \oplus P \to M \oplus P \oplus \Lambda^b \to 0);$$

$$\mathcal{E}_2 = (0 \to K \to \Lambda^{a+c} \to M \oplus \Lambda^c \to 0).$$

Since $M \oplus P \oplus \Lambda^b \cong_\Lambda M \oplus \Lambda^c$ then $K \oplus P \oplus \Lambda^{a+b} \cong_\Lambda K \oplus \Lambda^{a+c}$ by Schanuel's Lemma and so $K \oplus P \sim K$ as claimed. $\square$

More generally, if M is of type $FT(n)$ then $D_r(M)$ is finitely determined for $1 \leq r \leq n$; $\operatorname{Stab}(D_r(M))$ is then defined also for $1 \leq r \leq n$ and we have the following:

Proposition 6.21 *Let M be a module of type FT(n) then*

$$\operatorname{Stab}(M) \subset \operatorname{Stab}(D_1(M)) \subset \cdots \subset \operatorname{Stab}(D_{n-1}(M)) \subset \operatorname{Stab}(D_n(M)).$$

Proposition 6.22 *Let* $M \in \mathcal{M}\mathrm{od}_{\mathrm{fin}}$ *be a module of type FP(n); that is, suppose there exists an exact sequence* $0 \to P_n \to P_{n-1} \to \cdots \to P_1 \to P_0 \to M \to 0$ *where each P_r is finitely generated free over Λ; then* $\operatorname{Stab}(M) = 0$.

Proof By Proposition 6.21 we have inclusions

$$\text{Stab}(M) \subset \text{Stab}(D_1(M)) \subset \cdots \subset \text{Stab}(D_{n-1}(M)) \subset \text{Stab}(D_n(M)).$$

Since $D_n(M)$ is represented by the finitely generated projective module P_n we see that $\text{Stab}(D_n(M)) = \text{Stab}(P_n) = 0$. $\square$

If $\Lambda = \mathbf{Z}[G]$ is the integral group ring of a group G then applying the hypothesis of Proposition 6.22 to the trivial module $M = \mathbf{Z}$ requires G to have finite cohomological dimension and hence to be torsion free. By contrast, when the group G is finite it may occur that there are modules M for which $\text{Stab}(M)$ is a nontrivial subgroup of $\widetilde{K}_0(\mathbf{Z}[G])$.

Example 6.23 (The stable class $[\mathbf{Z}]$ over $Q(2^n)$) Take G to be a finite group and $M = \mathbf{Z}$ to be the trivial module and recall the Swan module construction; that is, consider the canonical exact sequence $0 \to I \to \mathbf{Z}[G] \xrightarrow{\epsilon} \mathbf{Z} \to 0$ where I denotes the augmentation ideal. For any nonzero integer r we define the Swan module (I, r) by

$$(\text{I}, r) = \epsilon^{-1}(r)$$

so that (I, r) occurs in an exact sequence $0 \to I \to (\text{I}, r) \xrightarrow{\epsilon/r} \mathbf{Z} \to 0$. It follows by dualising Swan's projectivity criterion Sect. 5.7 that (I, r) is projective exactly when r is coprime to $|G|$. In that case, since $\text{Ext}^r(\mathbf{Z}, \mathbf{Z}[G]) = 0$ for all $r \geq 1$, we may apply the dual Schanuel Lemma to deduce that:

Proposition 6.24 *If G is a finite group and r is coprime to $|G|$ then*

$$\mathbf{Z} \oplus (\text{I}, r) \cong \mathbf{Z} \oplus \mathbf{Z}[G].$$

In the case where G is the quaternion group $Q(2^n)$ of order 2^n ($n \geq 3$), it is known [94] that $(\text{I}, 3)$ represents the nontrivial class in $\widetilde{K}_0(\mathbf{Z}[Q(2^n)]) \cong \mathbf{Z}/2\mathbf{Z}$ so that for any finitely generated projective module P over $\mathbf{Z}[Q(2^n)]$ by Proposition 6.24 $\mathbf{Z} \oplus P \sim \mathbf{Z}$. It follows that over $G = Q(2^n)$:

Proposition 6.25 $\text{Stab}(\mathbf{Z}) = \widetilde{K}_0(\mathbf{Z}[Q(2^n)]) \cong \mathbf{Z}/2\mathbf{Z}$.

In the same way that Proposition 6.20 is deduced from Schanuel's Lemma, the following is obtained from the dual version of Schanuel's Lemma.

Proposition 6.26 *Let $M \in \mathcal{M}od_{\text{fin}}$; if $\text{Ext}^1(M, \Lambda) = 0$ then $\text{Stab}(M) = \text{Stab}(D_1(M))$.*

Proof Let $\mathcal{E} = (0 \to K \to P \to M \to 0)$ be an exact sequence with P finitely generated projective. It follows from Proposition 6.20 that $\text{Stab}(M) \subset \text{Stab}(K)$. We must show that $\mathrm{Stab}(K) \subset \mathrm{Stab}(M)$. Suppose that Q is a finitely generated projective module such that $K \oplus Q \sim K$. We show that $M \oplus Q \sim M$.

Interpreting the relation $K \oplus Q \sim K$ as meaning $K \oplus Q \oplus \Lambda^b \cong_\Lambda K \oplus \Lambda^c$ for some $b, c \geq 0$ then we may modify $\mathcal{E}$ to give two new sequences

$$\mathcal{E}_1 = (0 \to K \oplus Q \oplus \Lambda^b \to P \oplus Q \oplus \Lambda^b \to M \to 0);$$

$$\mathcal{E}_2 = (0 \to K \oplus \Lambda^c \to P \oplus \Lambda^c \to M \to 0).$$

Since $K \oplus Q \oplus \Lambda^b \cong_\Lambda K \oplus \Lambda^c$ then from the dual version of Schanuel's Lemma

$$M \oplus P \oplus \Lambda^c \cong_\Lambda M \oplus P \oplus Q \oplus \Lambda^b. \tag{$*$}$$

Choose a finitely generated projective module P_1 such that for some $d \geq 0$ $P \oplus P_1 \cong \Lambda^d$. Then applying $\oplus P_1$ to each side of $(*)$ we obtain

$$M \oplus \Lambda^{c+d} \cong_\Lambda M \oplus Q \oplus \Lambda^{b+d} \tag{$**$}$$

and so $M \oplus Q \sim M$ as claimed.   □

In the same way that Proposition 6.21 follows from Proposition 6.20 we have

Proposition 6.27 *Let $M \in \mathcal{M}od_{\mathrm{fin}}(\Lambda)$; if $\mathrm{Ext}^r(M, \Lambda) = 0$ for $1 \leq r \leq n$ then*

$$\mathrm{Stab}(M) = \mathrm{Stab}(D_1(M)) = \cdots = \mathrm{Stab}(D_{n-1}(M)) = \mathrm{Stab}(D_n(M)).$$

We say that a Λ-module M has *projective cancellation* when for projective modules P, Q

$$M \oplus P \cong_\Lambda M \oplus Q \implies P \cong_\Lambda Q.$$

If M is a right (resp. left) Λ-module we denote by M^* the dual left (resp. right) Λ-module $M^* = \mathrm{Hom}_\Lambda(M, \Lambda)$.

Proposition 6.28 *If $M^* = 0$ then M has the projective cancellation property.*

Proof Suppose that $M \oplus P \cong M \oplus Q$ where P, Q are projective. Then $M^* \oplus P^* \cong M^* \oplus Q^*$. However, since $M^* = 0$, then $P^* \cong Q^*$ and so $P^{**} \cong Q^{**}$. Thus $P \cong Q$.   □

More particularly, the set $\mathcal{SF}_+$ of nonzero stably free Λ modules is a semigroup under $\oplus$ and for any Λ-module M there is a semigroup action of $\mathcal{SF}_+$ on the stable module $[M]$

$$\mathcal{SF}_+ \times [M] \longrightarrow [M]$$

$$(S, J) \longmapsto S \oplus J$$

This action clearly has implications for the tree structure in $[M]$. For example, when M has the projective cancellation property then the tree $\mathcal{SF}_+$ imbeds in the stable class $[M]$ via $S \mapsto M \oplus S$. Thus from Proposition 6.28 we obtain:

Corollary 6.29 *If $M^* = 0$ then the tree $\mathcal{SF}_+(\Lambda)$ imbeds in the stable class $[M]$.*

Example 6.30 (The stable class $[\mathbf{Z}]$ over an infinite group G) Let Λ denote the integral group ring $\Lambda = \mathbf{Z}[G]$ of a group G. Furthermore, we denote by $\mathbf{Z}$ the (right) module over $\mathbf{Z}[G]$ whose underlying abelian group is $\mathbf{Z}$ on which G acts trivially. First note that:

Proposition 6.31 *If G is infinite then $\mathrm{Hom}_\Lambda(\mathbf{Z}, \Lambda) = 0$.*

Proof Recall that $\Lambda = \mathbf{Z}[G]$ consists of functions $\alpha : G \to \mathbf{Z}$ with finite support. The element $\alpha \in \Lambda$ then has the canonical representation $\alpha = \sum_{g \in G} \alpha_g \widehat{g}$ where $\widehat{g} : G \to \mathbf{Z}$ is the characteristic function of g

$$\widehat{g}(\gamma) = \begin{cases} 1 & \text{if } \gamma = g, \\ 0 & \text{if } \gamma \neq g. \end{cases}$$

Let $f : \mathbf{Z} \to \Lambda$ be a Λ-homomorphism; to show $f = 0$ it suffices to show that $f(1) = 0$. Write $f(1) = \alpha = \sum_{g \in G} \alpha_g \widehat{g}$; then for $h \in G$,

$$f(1 \cdot h^{-1}) = f(1) \cdot \widehat{h^{-1}}$$
$$= \sum_{g \in G} \alpha_g \widehat{gh^{-1}}$$
$$= \sum_{\gamma \in G} \alpha_{\gamma h} \widehat{\gamma}.$$

However, $\mathbf{Z}$ is a trivial module over Λ so that $f(1) = f(1 \cdot h^{-1})$ for all $h \in G$. Comparing the coefficient of $\widehat{1}$ in the two expressions we see that $\alpha_h = \alpha_1$ for all $h \in G$. Thus $\alpha_g = \alpha_h$ for all $g, h \in G$. However, α has finite support and G is infinite, so choosing $h \notin \mathrm{Supp}(\alpha)$ we see that $\alpha_g = \alpha_h = 0$ for all $g \in G$. That is, $f(1) = 0$ and so $f = 0$ as required. $\qquad\qquad\square$

It follows that:

Corollary 6.32 *If G is infinite then the tree $\mathcal{SF}_+$ of nonzero stably free modules over $\mathbf{Z}[G]$ imbeds in the stable class $[\mathbf{Z}]$ of the trivial module.*

Similarly, if $\mathbf{Z}_\sigma$ is the $\mathbf{Z}[G]$-module structure on $\mathbf{Z}$ where G acts via a group homomorphism $\sigma : G \to \{\pm 1\} = \mathrm{Aut}(\mathbf{Z})$ then again $\mathrm{Hom}_\Lambda(\mathbf{Z}_\sigma, \Lambda) = 0$ and $\mathcal{SF}_+$ likewise imbeds in $[\mathbf{Z}_\sigma]$.

Chapter 7
The Swan Mapping

In his fundamental paper on group cohomology Swan [91] defined, for any finite group G, a homomorphism $(\mathbf{Z}/|G|)^* \to \widetilde{K}_0(\mathbf{Z}[G])$ which, in this restricted context, has since been used extensively both in the classification of projective modules [94] and the algebraic homotopy theory of finite complexes [52]. We extend the definition so that, for suitable modules J over reasonably general rings Λ, it takes the form $S_J : \mathrm{Aut}_{\mathcal{D}\mathrm{er}}(J) \to \widetilde{K}_0(\Lambda)$.

7.1 The Structure of Projective 0-Complexes

We consider projective 0-complexes; that is, exact sequences

$$(0 \to J \to P \to M \to 0)$$

where P is projective. Such a complex is said to be *of finite type* when both M and J are finitely generated and to be *stably free* when P is stably free.

The set $\mathcal{SF}$ of isomorphism classes of finitely generated stably free modules is precisely the same as [0], the stable class of the zero module. It follows from Sect. 1.3 that $\mathcal{SF}$ has a well defined height function. In this case the height of $S \in \mathcal{SF}$ is called the *rank*, $\mathrm{rk}(S)$; it may be calculated as follows:

$$\mathrm{rk}(S) = r \quad \Longleftrightarrow \quad S \oplus \Lambda^a \cong \Lambda^{r+a}.$$

We first prove:

Theorem 7.1 *For $i = 1, 2$ let $\mathcal{E}_i = (0 \to J_i \to S_i \to M \to 0)$ be a stably free 0-complex of finite type in which M is coprojective; then*

$$h(J_1) = h(J_2) \quad \Longleftrightarrow \quad \mathrm{rk}(S_1) = \mathrm{rk}(S_2).$$

F.E.A. Johnson, *Syzygies and Homotopy Theory*, Algebra and Applications 17, DOI 10.1007/978-1-4471-2294-4_7, © Springer-Verlag London Limited 2012

Proof ($\Longrightarrow$) As $J_1, J_2 \in \Omega_1(M)$ and $h(J_1) = h(J_2)$ then $J_1 \oplus \Lambda^N \cong J_2 \oplus \Lambda^N$ for some positive integer N. Stabilizing by Λ^N we get

$$\Sigma_+^N(\mathcal{E}_i) = (0 \to J_i \oplus \Lambda^N \to S_i \oplus \Lambda^N \to M \to 0).$$

Comparison of stabilizations gives $M \oplus S_1 \oplus \Lambda^N \cong M \oplus S_2 \oplus \Lambda^N$ by the dual version of Schanuel's Lemma. As S_1, S_2 are finitely generated stably free then $S_i \oplus \Lambda^m \cong \Lambda^{n_i}$ for some positive integers m, n_1, n_2. Thus $M \oplus \Lambda^{N+n_1} \cong M \oplus \Lambda^{N+n_2}$ and so $n_1 = n_2$. However $\mathrm{rk}(S_i) + m = n_i$ so that $\mathrm{rk}(S_1) = \mathrm{rk}(S_2)$ as claimed.

The proof of ($\Longleftarrow$) is similar but uses Schanuel's Lemma rather than its dual.

$\square$

For a positive integer n, as abbreviation we write $\Sigma_+^n(\gamma)$ in place of $\Sigma_+^{\Lambda^n}(\gamma)$; that is:

$$\Sigma_+^n(\gamma) = (0 \to N \oplus \Lambda^n \overset{i \oplus \mathrm{Id}}{\to} P \oplus \Lambda^n \overset{p \circ \pi}{\to} M \to 0).$$

Proposition 7.2 *Let* $\alpha_r = (0 \to J_r \overset{j_r}{\to} P_r \overset{p_r}{\to} M_r \to 0)$ *be projective 0-complexes of finite type for* $r = 1, 2$ *and let* $h : M_1 \to M_2$ *be an isomorphism; then for some* $m \in \mathbf{Z}_+$ *there exists a morphism over* h

$$\begin{matrix} \Sigma_+^m(\alpha_1) \\ \downarrow \psi \\ \alpha_2 \end{matrix} = \begin{pmatrix} 0 \to J_1 \oplus \Lambda^m \overset{j_1 \oplus \mathrm{Id}}{\to} P_1 \oplus \Lambda^m \overset{p_1 \pi}{\to} M_1 \to 0 \\ \quad\quad \downarrow \psi_+ \quad\quad\quad \downarrow \psi_0 \quad\quad \downarrow h \\ 0 \to \quad J_2 \quad \overset{j_2}{\to} \quad P_2 \quad \overset{p_2}{\to} M_2 \to 0 \end{pmatrix}$$

in which ψ_+, ψ_0 *are surjective.*

Proof Suppose given a morphism of exact sequences thus;

$$\begin{matrix} 0 \to A \overset{i}{\to} B \overset{p}{\to} C \to 0 \\ \quad \downarrow \gamma_+ \quad \downarrow \gamma_0 \quad \downarrow g \\ 0 \to A' \overset{i'}{\to} B' \overset{p'}{\to} C' \to 0 \end{matrix}$$

if $g : C \to C'$ is an isomorphism then

(i) $i : \mathrm{Ker}(\gamma_+) \to \mathrm{Ker}(\gamma_0)$ is an isomorphism and
(ii) γ_0 is surjective $\Leftrightarrow$ γ_+ is surjective.

With our hypotheses, as P_1 is projective there exists a morphism $f : \alpha_1 \to \alpha_2$ over h:

$$\begin{matrix} \alpha_1 \\ \downarrow f \\ \alpha_2 \end{matrix} = \begin{pmatrix} 0 \to J_1 \overset{j_1}{\to} P_1 \overset{p_1}{\to} M_1 \to 0 \\ \quad \downarrow f_+ \quad \downarrow f_0 \quad \downarrow h \\ 0 \to J_2 \overset{j_2}{\to} P_2 \overset{p_2}{\to} M_2 \to 0 \end{pmatrix}.$$

As J_2 is finitely generated there exists a surjective Λ-homomorphism $\mu : \Lambda^m \to J_2$. Defining $\psi_+ : J_1 \oplus \Lambda^m \to J_2$, $\psi_0 : P_1 \oplus \Lambda^m \to P_2$ by

$$\psi_+ \begin{pmatrix} x \\ y \end{pmatrix} = f_+(x) + \mu(y); \qquad \psi_0 \begin{pmatrix} z \\ y \end{pmatrix} = f_0(z) + j' \circ \mu(y)$$

the following diagram commutes

$$
\begin{array}{ccccccccc}
0 \to & J_1 \oplus \Lambda^m & \overset{j_1 \oplus \mathrm{Id}}{\to} & P_1 \oplus \Lambda^m & \overset{p_1 \pi}{\to} & M_1 & \to 0 \\
& \downarrow \psi_+ & & \downarrow \psi_0 & & \downarrow h \\
0 \to & J_2 & \overset{j_2}{\to} & P_2 & \overset{p_2}{\to} & M_2 & \to 0
\end{array}
$$

and gives a morphism $\psi : \Sigma^m_+(\alpha_1) \to \alpha_2$ in which $\psi_+ : J_1 \oplus \Lambda^m \to J_2$ is surjective by construction; thus $\psi_0 : P_1 \oplus \Lambda^m \to P_2$ is surjective by our initial remark. $\qquad \square$

Fix a morphism $\psi : \alpha_1 \to \alpha_2$ between projective 0-complexes of finite type

$$
\begin{array}{cc}
\alpha_1 \\
\downarrow \psi \\
\alpha_2
\end{array}
=
\left(
\begin{array}{ccccccc}
0 \to & J_1 & \overset{j_1}{\to} & P_1 & \overset{p_1}{\to} & M_1 \to 0 \\
& \downarrow \psi_+ & & \downarrow \psi_0 & & \downarrow h \\
0 \to & J_2 & \overset{j_2}{\to} & P_2 & \overset{p_2}{\to} & M_2 \to 0
\end{array}
\right),
$$

where $h : M_1 \to M_2$ is an isomorphism and $\psi_+ : J_1 \to J_2$ is surjective. Then $\psi_0 : P_1 \to P_2$ is surjective and, as in Proposition 7.2, $j_1 : \mathrm{Ker}(\psi_+) \to \mathrm{Ker}(\psi_0)$ $(= K)$ is an isomorphism.

Proposition 7.3 *With this notation there exists an isomorphism over h thus*

$$
\begin{array}{cc}
\alpha_1 \\
\downarrow \varphi \\
\Sigma^K_+(\alpha_2)
\end{array}
=
\left(
\begin{array}{ccccccc}
0 \to & J_1 & \overset{j_1}{\to} & P_1 & \overset{p_1}{\to} & M_1 \to 0 \\
& \downarrow \varphi_+ & & \downarrow \varphi_0 & & \downarrow h \\
0 \to & J_2 \oplus K & \overset{j_2 \oplus \mathrm{Id}_K}{\to} & P_2 \oplus K & \overset{p_2}{\to} & M_2 \to 0
\end{array}
\right).
$$

Proof Since P_2 is projective there exists a homomorphism $r : P_1 \to K$ which splits the exact sequence $0 \to K \to P_1 \overset{\psi_0}{\to} P_2 \to 0$ on the left. Define $\varphi_+ : J_1 \to J_2 \oplus K$, $\varphi_+ : P_1 \to P_2 \oplus K$ by

$$\varphi_+(x) = \begin{pmatrix} \psi_+(x) \\ r \circ j_1(x) \end{pmatrix}; \qquad \varphi_0(y) = \begin{pmatrix} \psi_0(x) \\ r(y) \end{pmatrix}.$$

Then the following diagram commutes

$$
\begin{array}{ccccccc}
0 \to & J_1 & \overset{j_1}{\to} & P_1 & \overset{p_1}{\to} & M_1 \to 0 \\
& \downarrow \varphi_+ & & \downarrow \varphi_0 & & \downarrow h \\
0 \to & J_2 \oplus K & \overset{j_2 \oplus \mathrm{Id}_K}{\to} & P_2 \oplus K & \overset{p_2}{\to} & M_2 \to 0
\end{array}
$$

Since φ_0 is an isomorphism it follows that φ_+ is also an isomorphism. $\qquad \square$

Theorem 7.4 *Let* $\alpha_r = (0 \to J_r \xrightarrow{j_r} P_r \xrightarrow{p_r} M_r \to 0)$ *be projective* 0-*complexes of finite type* $(r = 1, 2)$ *and suppose that* $h : M_1 \to M_2$ *is an isomorphism. If* $[P_1] = [P_2] \in \widetilde{K}_0(\Lambda)$ *then there exist* $n_1, n_2 \in \mathbf{Z}_+$ *such that* $\Sigma_+^{n_1}(\alpha_1) \cong_h \Sigma_+^{n_2}(\alpha_2)$.

Proof By Proposition 7.2 we first construct a morphism $\psi : \Sigma_+^m(\alpha_1) \to \alpha_2$ in which both ψ_+ and ψ_0 are surjective. Then by Proposition 7.3 there exists an isomorphism $\varphi : \Sigma_+^m(\alpha_1) \to \Sigma_+^K(\alpha_2)$ where $K = \mathrm{Ker}(\psi_0)$ $(\cong \mathrm{Ker}(\psi_+))$. As $[P_1] = [P_2] \in \widetilde{K}_0(\Lambda)$ it follows that K is stably free. In particular, $K \oplus \Lambda^a \cong \Lambda^b$ for some a, b. Applying Σ_+^a to φ gives an isomorphism over h:

$$\Sigma_+^a(\varphi) : \Sigma_+^a(\Sigma_+^m(\alpha_1)) \to \Sigma_+^a(\Sigma_+^K)(\alpha_2).$$

However, $\Sigma_+^a(\Sigma_+^m(\alpha_1)) \cong_{\mathrm{Id}_M} \Sigma_+^{a+m}(\alpha_1)$ and since $K \oplus \Lambda^a \cong \Lambda^b$ we may identify $\Sigma_+^a(\Sigma_+^K)(\alpha_2) \cong_{\mathrm{Id}_M} \Sigma_+^b(\alpha_2)$. $\Sigma_+^a(\varphi)$ now gives the required isomorphism over h on taking $n_1 = a + m$ and $n_2 = b$. □

7.2 Endomorphism Rings

If $\alpha = (0 \to J_\alpha \to P_\alpha \to M_\alpha \to 0)$ and $\beta = (0 \to J_\beta \to P_\beta \to M_\beta \to 0)$ are projective 0-complexes and $f : M_\alpha \to M_\beta$ is a Λ-homomorphism it is a consequence of the universal property of projective modules that then there exists a morphism $\widetilde{f_{\beta\alpha}}$ over f

$$
\begin{array}{c} \alpha \\ \downarrow \widetilde{f_{\beta\alpha}} = \\ \beta \end{array}
\begin{pmatrix}
0 \to J_\alpha \xrightarrow{j_\alpha} P_\alpha \xrightarrow{p_\alpha} M_\alpha \to 0 \\
\downarrow f_{\beta\alpha} \quad \downarrow \widetilde{f_{\beta\alpha}} \quad \downarrow f \\
0 \to J_\beta \xrightarrow{j_\beta} P_\beta \xrightarrow{p_\beta} M_\beta \to 0
\end{pmatrix}.
$$

Although the homomorphism $f_{\beta\alpha} : J_\alpha \to J_\beta$ need not be unique, it becomes unique if we work instead in the category $\mathcal{D}er$; if $\widetilde{g_{\beta\alpha}} : P_\alpha \to P_\beta$ is a lifting of $g : M_\alpha \to M_\beta$ then:

$$f \approx g \quad \Longrightarrow \quad f_{\beta\alpha} \approx g_{\beta\alpha}. \tag{7.5}$$

For any projective 0-complex α, the correspondence $\rho_\alpha(f) = f_{\alpha\alpha}$ determines a ring homomorphism $\rho_\alpha : \mathrm{End}_{\mathcal{D}er}(M_\alpha) \to \mathrm{End}_{\mathcal{D}er}(J_\alpha)$. When $M_\alpha = M_\beta$ $(= M$ say) we denote any morphism $J_\alpha \to J_\beta$ obtained from a lifting of Id_M by $k_{\beta\alpha}$ thus

$$
\begin{array}{c} \alpha \\ \downarrow \widetilde{k_{\beta\alpha}} = \\ \beta \end{array}
\begin{pmatrix}
0 \to J_\alpha \xrightarrow{j_\alpha} P_\alpha \xrightarrow{p_\alpha} M \to 0 \\
\downarrow k_{\beta\alpha} \quad \downarrow \widetilde{k_{\beta\alpha}} \quad \downarrow \mathrm{Id} \\
0 \to J_\beta \xrightarrow{j_\beta} P_\beta \xrightarrow{p_\beta} M \to 0
\end{pmatrix}.
$$

It is straightforward to see that

$$\rho_\beta(f) = k_{\beta\alpha}\rho_\alpha(f)k_{\alpha\beta}. \tag{7.6}$$

Note that the natural map $\nu : \mathrm{End}_\Lambda(M) \to \mathrm{End}_{\mathcal{D}\mathrm{er}}(M)$ is also a ring homomorphism with $\mathrm{Ker}(\nu) = \mathcal{P}(M, M) = \{f \in \mathrm{End}_\Lambda(M) : f \approx 0\}$. It follows from (7.5) that $\mathrm{Ker}(\nu) \subset \mathrm{Ker}(\rho_\alpha)$ so that $\rho_\alpha : \mathrm{End}_\Lambda(M) \to \mathrm{End}_{\mathcal{D}\mathrm{er}}(J)$ descends to a ring homomorphism denoted by the same symbol

$$\rho_\alpha : \mathrm{End}_{\mathcal{D}\mathrm{er}}(M) \to \mathrm{End}_{\mathcal{D}\mathrm{er}}(J).$$

In general ρ_α need not be injective. It is not immediately clear whether ρ_α is necessarily surjective; however it becomes clear as soon as M is coprojective.

Theorem 7.7 *Let* $\alpha = (0 \to J \xrightarrow{j} P \xrightarrow{p} M \to 0)$ *be a projective 0-complex; if M is coprojective then* $\rho_\alpha : \mathrm{End}_{\mathcal{D}\mathrm{er}}(M) \to \mathrm{End}_{\mathcal{D}\mathrm{er}}(J)$ *is an isomorphism of rings.*

Proof It suffices to show that

 (I) $\rho_\alpha : \mathrm{End}_\Lambda(M) \to \mathrm{End}_{\mathcal{D}\mathrm{er}}(J)$ is surjective; and
(II) $\mathrm{Ker}(\rho_\alpha) = \mathcal{P}(M, M) = \{\alpha \in \mathrm{Hom}_\Lambda(M, M) : \alpha \approx 0\}$.

Clearly (I) implies that ρ_α is surjective whilst (II) implies ρ_α is injective. To show (I), observe that $\mathrm{Hom}_{\mathcal{D}\mathrm{er}}(P, J) = \mathrm{Ext}^1(P, J) = 0$ so that the modified exact sequence of Sect. 5.4,

$$\mathrm{Hom}_{\mathcal{D}\mathrm{er}}(P, J) \xrightarrow{j^*} \mathrm{Hom}_{\mathcal{D}\mathrm{er}}(J, J) \xrightarrow{\delta_*} \mathrm{Ext}^1(M, J) \xrightarrow{p^*} \mathrm{Ext}^1(P, J)$$

reduces to an isomorphism $\delta_* : \mathrm{End}_{\mathcal{D}\mathrm{er}}(J) \xrightarrow{\simeq} \mathrm{Ext}^1(M, J)$. Since M is coprojective then $\mathrm{Ext}^1(M, P) = 0$. From the exact sequence

(III) $\mathrm{Hom}_\Lambda(M, P) \xrightarrow{p_*} \mathrm{Hom}_\Lambda(M, M) \xrightarrow{\partial} \mathrm{Ext}^1(M, J) \xrightarrow{j_*} 0 \ (= \mathrm{Ext}^1(M, P))$

it follows that $\mathrm{End}_\Lambda(M) \xrightarrow{\partial} \mathrm{Ext}^1(M, J)$ is surjective. Given $[g] \in \mathrm{End}_{\mathcal{D}\mathrm{er}}(J)$ there exists $f \in \mathrm{End}_\Lambda(M)$ such that $\partial(f) = \delta_*([g])$; that is, there is a congruence $c : g_*(\alpha) \equiv f^*(\alpha)$

$$
\begin{array}{ccc}
g_*(\alpha) \\
\downarrow c \\
f^*(\alpha)
\end{array}
=
\left(
\begin{array}{ccc}
0 \longrightarrow J \longrightarrow \varinjlim(g, j) \longrightarrow M \longrightarrow 0 \\
\quad\quad \downarrow \mathrm{Id}_J \quad\quad\quad \downarrow c \quad\quad\quad \downarrow \mathrm{Id}_M \\
0 \longrightarrow J \longrightarrow \varprojlim(p, f) \longrightarrow M \longrightarrow 0
\end{array}
\right).
$$

In addition, there are natural maps $\nu_1 : \alpha \to g_*(\alpha)$ and $\nu_2 : f^*(\alpha) \to \alpha$ thus

$$
\begin{array}{ccc}
\alpha \\
\downarrow \nu_1 \\
g_*(\alpha)
\end{array}
=
\left(
\begin{array}{ccc}
0 \longrightarrow J \xrightarrow{j} P \xrightarrow{p} M \longrightarrow 0 \\
\quad\quad \downarrow g \quad\quad \downarrow \nu_1 \quad\quad \downarrow \mathrm{Id}_M \\
0 \longrightarrow J \longrightarrow \varinjlim(g, j) \longrightarrow M \longrightarrow 0
\end{array}
\right);
$$

$$
\begin{array}{ccc}
f^*(\alpha) \\
\downarrow \nu_2 \\
\alpha
\end{array}
=
\left(
\begin{array}{ccc}
0 \longrightarrow J \longrightarrow \varprojlim(p, f) \longrightarrow M \longrightarrow 0 \\
\quad\quad \downarrow \mathrm{Id}_J \quad\quad \downarrow \nu_2 \quad\quad \downarrow f \\
0 \longrightarrow J \xrightarrow{j} P \xrightarrow{p} M \longrightarrow 0
\end{array}
\right).
$$

In the composition $\mu = \nu_2 \circ c \circ \nu_1 : \alpha \to \alpha$, the mapping $g : J \to J$ is induced over $f : M \to M$ showing that $\rho(f) = [g]$ as required. This proves (I).

For (II), observe that $\rho = \delta_*^{-1} \circ \partial$ where δ_* is the isomorphism already noted and $\partial : \mathrm{Hom}_\Lambda(M, M) \to \mathrm{Ext}^1(M, J)$ is the boundary map of the exact sequence

$$\cdots \xrightarrow{i_*} \mathrm{Hom}_\Lambda(M, P) \xrightarrow{p_*} \mathrm{Hom}_\Lambda(M, M) \xrightarrow{\partial} \mathrm{Ext}^1(M, D) \xrightarrow{i_*} \mathrm{Ext}^1(M, P) \xrightarrow{p_*} \cdots .$$

Thus $\mathrm{Ker}(\rho) = \mathrm{Ker}(\partial)$, so it suffices to show that $\mathrm{Ker}(\partial) = \mathcal{P}(M, M)$. From (III), in addition to the surjectivity of ∂ we note that $\mathrm{Ker}(\partial) = \mathrm{Im}(p_*) \subset \mathcal{P}(M, M)$. Suppose, conversely, that $\alpha \in \mathcal{P}(M, M)$; that is, α factors through a projective Q;

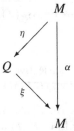

Since p is surjective then by the universal property of the projective Q there exists a homomorphism $\widehat{\xi} : Q \to P$ such that $\xi = p \circ \widehat{\xi}$:

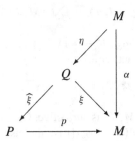

Hence $\alpha = \xi \circ \eta = p \circ \widehat{\xi} \circ \eta$ and $\alpha = p_*(\widehat{\xi} \circ \eta) \in \mathrm{Im}(p_*)$. Thus $\mathcal{P}(M, M) \subset \mathrm{Im}(p_*)$ and $\mathrm{Ker}(\rho) = \mathrm{Im}(p_*) = \mathcal{P}(M, M)$ as required. This shows (II) and completes the proof.   □

7.3 The Dual Swan Mapping

Fix a projective 0-complex $\alpha = (0 \to J \xrightarrow{j} P \xrightarrow{p} M \to 0)$ and let $f : M' \to M$ be a Λ-homomorphism.

Proposition 7.8 *If f defines an isomorphism in $\mathcal{D}\mathrm{er}$ then $f^*(\alpha)$ is a projective 0-complex.*

Proof Consider the natural morphism

$$
\begin{array}{ccc}
f^*(\alpha) \\
\downarrow \nu \\
\alpha
\end{array}
=
\left(
\begin{array}{ccccccccc}
0 & \longrightarrow & J & \overset{i}{\longrightarrow} & \varprojlim(p,f) & \overset{\pi}{\longrightarrow} & M' & \longrightarrow & 0 \\
& & \downarrow \mathrm{Id} & & \downarrow \nu & & \downarrow f & & \\
0 & \longrightarrow & J & \overset{j}{\longrightarrow} & P & \overset{p}{\longrightarrow} & M & \longrightarrow & 0
\end{array}
\right).
$$

For brevity we write $\varprojlim = \varprojlim(p,f)$. Given a Λ-module N on applying the co-homology exact sequence extending $\mathrm{Hom}_\Lambda(-,N)$ we obtain a commutative ladder with exact rows

$$
\begin{array}{ccccccc}
\mathrm{Ext}^1(M,N) & \overset{j_*}{\to} & \mathrm{Ext}^1(P,N) & \overset{p_*}{\to} & \mathrm{Ext}^1(J,N) & \overset{\partial}{\to} & \mathrm{Ext}^2(M,N) \\
\downarrow f^* & & \downarrow \nu^* & & \downarrow \mathrm{Id} & & \downarrow f^* \\
\mathrm{Ext}^1(M',N) & \overset{i_*}{\to} & \mathrm{Ext}^1(\varprojlim,N) & \overset{\pi_*}{\to} & \mathrm{Ext}^1(J,N) & \overset{\partial}{\to} & \mathrm{Ext}^2(M',N)
\end{array}
$$

Since f^* and Id are isomorphisms it follows that $\nu^* : \mathrm{Ext}^1(P,N) \to \mathrm{Ext}^1(\varprojlim,N)$ is surjective. However, $\mathrm{Ext}^1(P,N) = 0$ since P is projective. Thus $\mathrm{Ext}^1(\varprojlim,N) = 0$ for arbitrary N, and so, by Proposition 5.5, $\varprojlim(p,f)$ is projective as claimed. $\square$

Let $\alpha = (0 \to J_\alpha \overset{j_\alpha}{\to} P_\alpha \overset{p_\alpha}{\to} M \to 0)$ be a projective 0-complex of finite type and $f : M \to M$ a Λ-homomorphism which defines an automorphism of M in the category $\mathcal{D}\mathrm{er}$; then by Proposition 7.8 above,

$$
f^*(\alpha) = (0 \to J_\alpha \overset{i}{\to} \varprojlim(p_\alpha,f) \overset{\pi}{\to} M \to 0)
$$

is a projective 0-complex, also of finite type, and hence defines a class $[\varprojlim(p_\alpha,f)] \in \widetilde{K}_0(\Lambda)$. The correspondence

$$
\widehat{S}^\alpha(f) = [\varprojlim(p_\alpha,f)]
$$

defines a mapping $\widehat{S}^\alpha : \mathrm{Aut}_{\mathcal{D}\mathrm{er}}(M) \to \widetilde{K}_0(\Lambda)$, the *dual Swan mapping associated* with α. Evidently congruent projective 0-complexes give the same dual Swan mapping; that is:

Proposition 7.9 *Let α, β be projective 0-complexes of finite type; if $\alpha \equiv \beta$ then $\widehat{S}^\alpha = \widehat{S}^\beta$.*

If α is a projective 0-complex of finite type and K is a finitely generated projective module then $\Sigma^K_+(\alpha)$ is also a projective 0-complex of finite type. Moreover, the stabilization operator Σ^K_+ commutes with pullback; that is:

Proposition 7.10 $\Sigma^K_+(f^*(\alpha)) \equiv f^*(\Sigma^K_+(\alpha))$.

Furthermore, if $f : M \to M$ is a Λ-homomorphism and $\widetilde{f} : \alpha \to \alpha$ is a lifting of f to α then there is a lifting $\Sigma_+^K(\widetilde{f}) : \Sigma_+^K(\alpha) \to \Sigma_+^K(\alpha)$ constructed as follows:

$$\begin{matrix} \Sigma_+^K(\alpha) \\ \downarrow \Sigma_+^K(\widetilde{f}) = \\ \Sigma_+^K(\alpha) \end{matrix} \begin{pmatrix} 0 \longrightarrow J_\alpha \oplus K \xrightarrow{i \oplus \mathrm{Id}} P_\alpha \oplus K \xrightarrow{p_\alpha \pi} M \longrightarrow 0 \\ \downarrow \widetilde{f} \oplus \mathrm{Id} \qquad \downarrow \widetilde{f} \oplus \mathrm{Id} \qquad \downarrow f \\ 0 \longrightarrow J_\alpha \oplus K \xrightarrow{i \oplus \mathrm{Id}} P_\alpha \oplus K \xrightarrow{p_\alpha \pi} M \longrightarrow 0 \end{pmatrix}.$$

It now follows easily that:

Proposition 7.11 $\widehat{S}^{\Sigma_+^K(\alpha)}(f) = \widehat{S}^\alpha(f) + [K]$.

In the special case where $K = \Lambda^n$ we have $[\Lambda^n] = 0 \in \widetilde{K_0(\Lambda)}$ so that by Proposition 7.11:

Proposition 7.12 $\widehat{S}^{\Sigma_+^n(\alpha)} = \widehat{S}^\alpha$.

Let $\alpha = (0 \to J_\alpha \xrightarrow{j_\alpha} P_\alpha \xrightarrow{p_\alpha} M \to 0)$, $\beta = (0 \to J_\beta \xrightarrow{j_\beta} P_\beta \xrightarrow{p_\beta} M \to 0)$ be projective 0-complexes of finite type; we write $\alpha \cong_M \beta$ when there exists a commutative diagram

$$\begin{matrix} \alpha \\ \downarrow \varphi = \\ \beta \end{matrix} \begin{pmatrix} 0 \to J_\alpha \xrightarrow{j_\alpha} P_\alpha \xrightarrow{p_\alpha} M \to 0 \\ \downarrow \varphi_+ \quad \downarrow \varphi \quad \downarrow \mathrm{Id} \\ 0 \to J_\beta \xrightarrow{j_\beta} P_\beta \xrightarrow{p_\beta} M \to 0 \end{pmatrix}.$$

φ is then said to be an *isomorphism over the identity*; with this notation:

$$\alpha \cong_M \beta \quad \Longrightarrow \quad \widehat{S}^\alpha = \widehat{S}^\beta. \qquad (7.13)$$

Proof Let $\varphi : \alpha \to \beta$ be an isomorphism over Id_M:

$$\begin{matrix} \alpha \\ \downarrow \varphi = \\ \beta \end{matrix} \begin{pmatrix} 0 \to J_\alpha \xrightarrow{j_\alpha} P_\alpha \xrightarrow{p_\alpha} M \to 0 \\ \downarrow \varphi_+ \quad \downarrow \varphi \quad \downarrow \mathrm{Id} \\ 0 \to J_\beta \xrightarrow{j_\beta} P_\beta \xrightarrow{p_\beta} M \to 0 \end{pmatrix}.$$

Then $\varphi \times \mathrm{Id} : P_\alpha \times M \to P_\beta \times M$ is an isomorphism of Λ-modules whose restriction is a Λ-isomorphism $\varphi \times \mathrm{Id} : \varprojlim(p_\alpha, f) \xrightarrow{\cong} \varprojlim(p_\beta, f)$ for each $f \in \mathrm{End}_{\mathcal{D}er}(M)$. Hence $\widehat{S}^\alpha(f) = \widehat{S}^\beta(f)$. $\qquad \square$

Corollary 7.14 *Let*

$$\alpha = (0 \to J_\alpha \xrightarrow{j_\alpha} P_\alpha \xrightarrow{p_\alpha} M \to 0), \qquad \beta = (0 \to J_\beta \xrightarrow{j_\beta} P_\beta \xrightarrow{p_\beta} M \to 0)$$

be projective 0-complexes of finite type. Then

$$\widehat{S}^\alpha = \widehat{S}^\beta \quad \Longleftrightarrow \quad [P_\alpha] = [P_\beta] \in \widetilde{K_0}(\Lambda).$$

Proof If $\widehat{S}^\alpha = \widehat{S}^\beta$ then in particular $\widehat{S}^\alpha(\mathrm{Id}) = \widehat{S}^\beta(\mathrm{Id})$. The implication ($\Longrightarrow$) now follows since $\widehat{S}^\alpha(\mathrm{Id}) = [P_\alpha]$ and $\widehat{S}^\beta(\mathrm{Id}) = [P_\beta]$. Suppose conversely that $[P_\alpha] = [P_\beta]$. By Theorem 7.4, $\Sigma^n(\alpha) \cong_M \Sigma^m(\beta)$ for some n, m. Thus $\widehat{S}^{\Sigma^n(\alpha)} = \widehat{S}^{\Sigma^m(\beta)}$ by (7.13). The implication ($\Longleftarrow$) now follows from Proposition 7.12. □

Corollary 7.15 *For* $r = 1, 2$ *let* $\mathbf{F}_r = (0 \to J_r \xrightarrow{j_r} \Lambda^{n_r} \xrightarrow{p_r} M \to 0)$ *be free* 0-*complexes over the finitely presented module* M; *then* $\widehat{S}^{\mathbf{F}_1} = \widehat{S}^{\mathbf{F}_2}$.

Proof $[\Lambda^{n_1}] = [\Lambda^{n_2}] = 0 \in \widetilde{K}_0(\Lambda)$ so the conclusion follows by Corollary 7.14. □

If $\mathbf{F}$ is a free 0-complex of finite type over the finitely presented module M then writing

$$\widehat{S^M} = \widehat{S}^{\mathbf{F}}$$

allows us to define the *dual Swan mapping* $\widehat{S^M} : \mathrm{Aut}_{\mathcal{D}er}(M) \to \widetilde{K}_0(\Lambda)$; that is, if

$$\mathbf{F} = (0 \to J \xrightarrow{j_{\mathbf{F}}} \Lambda^n \xrightarrow{p_{\mathbf{F}}} M \to 0)$$

then

$$\widehat{S^M}(f) = [\varprojlim(p_{\mathbf{F}}, f)].$$

By Corollary 7.15 the definition depends only upon M and not upon the particular choice of $\mathbf{F}$.

We now proceed to show that $\widehat{S^M}$ is a homomorphism; for the remainder of this section we fix a finitely presented module M and write $\widehat{S} = \widehat{S^M}$. We first establish the following composition property:

Proposition 7.16 *Let* α *be a projective* 0-*complex of finite type; then for all* $f, g \in \mathrm{Aut}_{\mathcal{D}er}(M)$

$$\widehat{S}^\alpha(fg) = \widehat{S}^{f^*(\alpha)}(g).$$

Proof We compute that $\widehat{S}^\alpha(fg) = [\varprojlim(p, fg)]$ using the commutative diagram

$$
\begin{array}{cc}
\begin{array}{c} (fg)^*(\alpha) \\ \downarrow v_{fg} \\ \alpha \end{array} & =
\left(
\begin{array}{ccccccc}
0 & \longrightarrow & J_\alpha & \xrightarrow{\ i\ } & \varprojlim(p, fg) & \xrightarrow{\ \pi\ } & M & \longrightarrow 0 \\
 & & \downarrow \mathrm{Id} & & \downarrow v_{fg} & & \downarrow fg & \\
0 & \longrightarrow & J_\alpha & \xrightarrow{\ j\ } & P_\alpha & \xrightarrow{\ p\ } & M & \longrightarrow 0
\end{array}
\right) .
\end{array}
$$

Likewise $\widehat{S}^{f^*(\alpha)}(g)$ is computed using the commutative diagram

$$
\begin{array}{ccccccccc}
g^*f^*(\alpha) & & 0 \longrightarrow J_\alpha \xrightarrow{\ i\ } \varprojlim(\pi_1,g) \xrightarrow{\ \pi_2\ } M \longrightarrow 0 \\
\downarrow v_g & & \quad\ \ \downarrow \mathrm{Id} \qquad\quad \downarrow v_g \qquad\quad \downarrow g \\
f^*(\alpha) & = & 0 \longrightarrow J_\alpha \xrightarrow{\ i\ } \varprojlim(p,f) \xrightarrow{\ \pi_1\ } M \longrightarrow 0 \\
\downarrow v_f & & \quad\ \ \downarrow \mathrm{Id} \qquad\quad \downarrow v_f \qquad\quad \downarrow f \\
\alpha & & 0 \longrightarrow J_\alpha \xrightarrow{\ j\ } \quad P_\alpha \xrightarrow{\ p\ } M \longrightarrow 0
\end{array}
$$

which contains the two pullback squares

$$
\begin{array}{ccc}
\varprojlim(p,f) \xrightarrow{\ \pi_1\ } M \\
\ \downarrow v_f \qquad\quad \downarrow f\,; \\
\ P_\alpha \xrightarrow{\ p\ } M
\end{array}
\qquad\qquad
\begin{array}{ccc}
\varprojlim(\pi_1,g) \xrightarrow{\ \pi_2\ } M \\
\ \downarrow v_g \qquad\quad \downarrow g \\
\ \varprojlim(p,f) \xrightarrow{\ \pi_1\ } M
\end{array}
$$

and gives $\widehat{S}^{f^*(\alpha)}(g) = [\varprojlim(\pi_1,g)]$. As the following square is also a pullback

$$
\begin{array}{ccc}
\varprojlim(\pi_1,g) \xrightarrow{\ \pi_2\ } M \\
\ \downarrow v_f v_g \qquad\quad \downarrow fg \\
\ P_\alpha \xrightarrow{\ p\ } M
\end{array}
$$

the conclusion follows from the resulting isomorphism $\varprojlim(p,fg) \cong \varprojlim(\pi_1,g)$. □

Proposition 7.17 *Let* $\alpha = (0 \to J_\alpha \xrightarrow{j_\alpha} P_\alpha \xrightarrow{p_\alpha} M \to 0)$ *be a projective 0-complex of finite type; then for each* $f \in \mathrm{Aut}_{\mathcal{D}\mathrm{er}}(M)$

$$\widehat{S}^\alpha(f) = \widehat{S}(f) + [P_\alpha].$$

Proof Write $\alpha = (0 \to J_\alpha \xrightarrow{j} P_\alpha \xrightarrow{p} M \to 0)$ and let $\mathbf{F} = (0 \to J \to \Lambda^n \to M \to 0)$ be a free 0-complex. Then $\Sigma^{P_\alpha}\mathbf{F} = (0 \to J \oplus P_\alpha \to \Lambda^n \oplus P_\alpha \to M \to 0)$ has the property that $[\Lambda^n \oplus P_\alpha] = [P_\alpha]$. and so, by Corollary 7.14, $\widehat{S}^\alpha = \widehat{S}^{\Sigma^{P_\alpha}(\mathbf{F})}$. Now by Proposition 7.11 we see that $\widehat{S}^{\Sigma^{P_\alpha}(\mathbf{F})}(f) = \widehat{S}^{\mathbf{F}}(f) + [P_\alpha]$. The conclusion follows as $\widehat{S}^{\mathbf{F}} = \widehat{S}$. □

Corollary 7.18 $\widehat{S} : \mathrm{Aut}_{\mathcal{D}\mathrm{er}}(M) \to \widetilde{K}_0(\Lambda)$ *is a homomorphism.*

Proof Let $f, g \in \mathrm{Aut}_{\mathcal{D}\mathrm{er}}(M)$ and let 0-complex $\mathbf{F} = (0 \to J \to \Lambda^n \xrightarrow{p} M \to 0)$ be a free 0-complex. Then, by Proposition 7.16,

$$\widehat{S}(fg) = \widehat{S}^{f^*(\mathbf{F})}(g).$$

However, $f^*(\mathbf{F}) = (0 \to J \to \varprojlim(p,f) \to M \to 0)$ so that, by Proposition 7.17,

$$\widehat{S}^{f^*(\mathbf{F})}(g) = \widehat{S}(g) + [\varprojlim(p,f)] = [\varprojlim(p,f)] + \widehat{S}(g).$$

However, $\widehat{S}(f) = [\varprojlim(p, f)]$ so that $\widehat{S}^{f^*(\mathbf{F})}(g) = \widehat{S}(f) + \widehat{S}(g)$, and hence

$$\widehat{S}(fg) = \widehat{S}(f) + \widehat{S}(g). \qquad \square$$

7.4 Representing Projective 0-Complexes

Let M be a finitely generated coprojective module of type $\mathcal{F}(1)$ and let $J \in \Omega_1(M)$; then we know that there exists an exact sequence $\alpha = (0 \to J \xrightarrow{j} P \xrightarrow{p} M \to 0)$ where P is finitely generated projective; in fact, P may be taken to be stably free. Given any Λ-homomorphism $g : J \to J$ we recall that $g_*(\alpha)$ is the exact sequence

$$g_*(\alpha) = \left(0 \to J \xrightarrow{j} \varinjlim(g, j) \xrightarrow{\pi} M \to 0\right),$$

where $\varinjlim(g, j)$ is the pushout $\varinjlim(g, j) = (J \oplus P)/\mathrm{Im}(g \times -j)$. To simplify the discussion we adopt some notation; α will denote a fixed projective 0-complex as above; $\beta = (0 \to J \xrightarrow{i} E \xrightarrow{\pi} M \to 0)$ will denote an otherwise arbitrary element of $\mathbf{Ext}^1(M, J)$; g, h will denote Λ-homomorphisms $g, h : J \to J$. With notation as above:

$$\beta \equiv g_*(\alpha) \quad \text{for some } g \in \mathrm{End}_\Lambda(J); \tag{7.19}$$

$$g_*(\alpha) \equiv h_*(\alpha) \quad \Longleftrightarrow \quad [g] = [h] \in \mathrm{End}_{\mathcal{D}\mathrm{er}}(J); \tag{7.20}$$

$$g_*(\alpha) \text{ is a projective 0-complex} \quad \Longleftrightarrow \quad [g] \in \mathrm{Aut}_{\mathcal{D}\mathrm{er}}(J); \tag{7.21}$$

$$\beta \cong_M \alpha \quad \Longleftrightarrow \quad \beta \equiv g_*(\alpha) \quad \text{for some } g \in \mathrm{Aut}_\Lambda(J). \tag{7.22}$$

Proofs of (7.19) and (7.20) Consider the exact sequence

$$\mathrm{Hom}_{\mathcal{D}\mathrm{er}}(P, J) \xrightarrow{j^*} \mathrm{Hom}_{\mathcal{D}\mathrm{er}}(J, J) \xrightarrow{\delta_*} \mathrm{Ext}^1(M, J) \xrightarrow{p^*} \mathrm{Ext}^1(P, J).$$

Since P is projective then $\mathrm{Hom}_{\mathcal{D}\mathrm{er}}(P, J) = \mathrm{Ext}^1(P, J) = 0$ so that we have a isomorphism $\delta_* : \mathrm{Hom}_{\mathcal{D}\mathrm{er}}(J, J) \to \mathrm{Ext}^1(M, J)$ given by $\delta_*([g]) = g_*(\alpha)$. Equations (7.19) and (7.20) follow immediately. $\square$

Proof of (7.21) Consider the natural homomorphism

$$\begin{array}{c} \alpha \\ \downarrow \nu_g \\ g_*(\alpha) \end{array} = \left(\begin{array}{ccccccccc} 0 & \longrightarrow & J & \xrightarrow{j} & P & \xrightarrow{p} & M & \longrightarrow & 0 \\ & & \downarrow g & & \downarrow \nu & & \downarrow \mathrm{Id} & & \\ 0 & \longrightarrow & J & \xrightarrow{i} & \varinjlim(g, j) & \xrightarrow{\pi_1} & M & \longrightarrow & 0 \end{array} \right).$$

The exact sequence from $\mathrm{Hom}_{\mathcal{D}\mathrm{er}}(-, N)$ gives a ladder with exact rows

$$\begin{array}{ccccccc} \mathrm{Hom}_{\mathcal{D}\mathrm{er}}(J, N) & \xrightarrow{\partial} & \mathrm{Ext}^1(M, N) & \xrightarrow{\pi^*} & \mathrm{Ext}^1(\varinjlim, N) & \xrightarrow{i^*} & \mathrm{Ext}^1(J, N) \\ \downarrow g^* & & \downarrow \mathrm{Id} & & \downarrow \nu^* & & \downarrow g^* \\ \mathrm{Hom}_{\mathcal{D}\mathrm{er}}(J, N) & \xrightarrow{\partial} & \mathrm{Ext}^1(M, N) & \xrightarrow{p^*} & \mathrm{Ext}^1(P, N) & \xrightarrow{j^*} & \mathrm{Ext}^1(J, N) \end{array}$$

By hypothesis, g^* is an isomorphism; so also is Id. It follows that v^* is injective. However $\text{Ext}^1(P, N) = 0$ so that also $\text{Ext}^1(\varinjlim, N) = 0$. Since this is true for arbitrary N it follows that $\varinjlim(g, j)$ is projective. This proves ($\Longleftarrow$).

To show ($\Longrightarrow$) note that $\text{Hom}_{\mathcal{D}\text{er}}(P, N) = 0$ since P is projective. If $\varinjlim(g, j)$ is also projective then $\text{Ext}^1(P, N) = \text{Ext}^1(\varinjlim, N) = 0$ for any Λ-module N. Since M is coprojective, we may apply the (modified) reverse exact sequence from $\text{Hom}_{\mathcal{D}\text{er}}(-, J)$ to obtain a commutative diagram with exact rows:

$$0 \to \text{Hom}_{\mathcal{D}\text{er}}(J, J) \overset{\partial}{\to} \text{Ext}^1(M, J) \to 0$$
$$\downarrow g^* \qquad\qquad\qquad \downarrow \text{Id}$$
$$0 \to \text{Hom}_{\mathcal{D}\text{er}}(J, J) \overset{\partial}{\to} \text{Ext}^1(M, J) \to 0$$

In particular, $g^* : \text{Hom}_{\mathcal{D}\text{er}}(J, J) \to \text{Hom}_{\mathcal{D}\text{er}}(J, J)$ is bijective. We may choose a Λ-homomorphism $h : J \to J$ such that $g^*(h) = h \circ g \approx \text{Id}_J$. Then

$$g^*(g \circ h) = (g \circ h) \circ g$$
$$= g \circ (h \circ g)$$
$$\approx g \circ \text{Id}$$
$$\approx g$$

whilst evidently $g^*(\text{Id}) \approx g$. Since $g^* : \text{Hom}_{\mathcal{D}\text{er}}(J, J) \to \text{Hom}_{\mathcal{D}\text{er}}(J, J)$ is injective then $g \circ h \approx \text{Id}$, and so $g : J \to J$ is an isomorphism in $\mathcal{D}\text{er}$. This completes the proof. $\qquad\qquad\qquad\qquad\qquad\qquad\qquad\qquad\qquad\qquad\qquad\qquad\qquad\qquad\square$

Proof of (7.22) Suppose that $\beta \cong_M \alpha$; then there is a commutative diagram

$$\begin{matrix} \alpha \\ \downarrow \hat{g} = \\ \beta \end{matrix} \begin{pmatrix} 0 \to J \overset{j}{\to} P \overset{p}{\to} M \to 0 \\ \downarrow g \quad \downarrow g_0 \quad \downarrow \text{Id} \\ 0 \to J \overset{\iota}{\to} E \overset{p}{\to} M \to 0 \end{pmatrix},$$

where g, g_0 are Λ-isomorphisms. Putting $g_\#(\alpha) = (0 \to J \overset{jg^{-1}}{\to} S \overset{p}{\to} M \to 0)$ then the above commutative diagram may be modified to give a congruence

$$\begin{matrix} g_\#(\alpha) \\ \downarrow c = \\ \beta \end{matrix} \begin{pmatrix} 0 \to J \overset{jg^{-1}}{\to} P \overset{p}{\to} M \to 0 \\ \downarrow \text{Id} \quad \downarrow g_0 \quad \downarrow \text{Id} \\ 0 \to J \overset{\iota}{\to} E \overset{p}{\to} M \to 0 \end{pmatrix}.$$

There is also a conguence $c' : g_*(\alpha) \to g_\#(\alpha)$ given by $c'[x, y] = jh^{-1}(x) + y$ as follows;

$$\begin{matrix} g_*(\alpha) \\ \downarrow c' = \\ g_\#(\alpha) \end{matrix} \begin{pmatrix} 0 \longrightarrow J \overset{i}{\longrightarrow} \varinjlim(h, j) \overset{\pi}{\longrightarrow} M \longrightarrow 0 \\ \downarrow \text{Id} \qquad\quad \downarrow c' \qquad\quad \downarrow \text{Id} \\ 0 \longrightarrow J \overset{jg^{-1}}{\longrightarrow} P \overset{p}{\longrightarrow} M \longrightarrow 0 \end{pmatrix}.$$

The composite $c \circ c' : g_*(\alpha) \to \beta$ gives the required congruence $g_*(\alpha) \equiv \beta$.

Conversely, if $\beta \equiv g_*(\alpha)$ for some $g \in \mathrm{Aut}_\Lambda(J)$ then since $g_\#(\alpha) \equiv g_*(\alpha)$ there is also a congruence $c : g_\#(\alpha) \to \beta$ which we write thus:

$$
\begin{array}{c} g_\#(\alpha) \\ \downarrow c \\ \beta \end{array} = \left(\begin{array}{ccccccc} 0 \to & J & \xrightarrow{jg^{-1}} & P & \xrightarrow{p} & M \to 0 \\ & \downarrow \mathrm{Id} & & \downarrow g_0 & & \downarrow \mathrm{Id} \\ 0 \to & J & \xrightarrow{\iota} & E & \xrightarrow{p} & M \to 0 \end{array} \right).
$$

Equivalently we have an isomorphism $\hat{g} : \alpha \to \beta$ over Id_M:

$$
\begin{array}{c} \alpha \\ \downarrow \hat{g} \\ \beta \end{array} = \left(\begin{array}{ccccccc} 0 \to & J & \xrightarrow{j} & P & \xrightarrow{p} & M \to 0 \\ & \downarrow g & & \downarrow g_0 & & \downarrow \mathrm{Id} \\ 0 \to & J & \xrightarrow{\iota} & E & \xrightarrow{p} & M \to 0 \end{array} \right). \qquad \square
$$

7.5   The Swan Mapping Proper

A finitely generated module J is said to be *tame* when $J \in \Omega_1(M)$ for some finitely generated coprojective module M. Observe also that this forces M to be finitely presented and so the dual Swan mapping $\widehat{S^M}$ is well defined, as in Sect. 7.3. We first note:

Proposition 7.23 *If M is a finitely presented coprojective module and $J \in \Omega_1(M)$ then J then occurs in a 0-complex $\alpha = (0 \to J \xrightarrow{j} E \xrightarrow{p} M \to 0)$ in which E is stably free.*

Proof The proof is a refinement of Proposition 5.31. Let

$$
S = (0 \to K \xrightarrow{i} \Lambda^n \xrightarrow{p} M \to 0)
$$

be an exact sequence. Since $J \in \Omega_1(M)$ then $J \oplus \Lambda^a \cong K \oplus \Lambda^b$. for some integers $a, b \geq 0$. We may now modify S to get an exact sequence

$$
0 \to K \oplus \Lambda^b \xrightarrow{i'} \Lambda^{n+b} \xrightarrow{p'} M \to 0.
$$

Choosing an isomorphism $\varphi : J \oplus \Lambda^a \xrightarrow{\cong} K \oplus \Lambda^b$, put $j = i' \circ \varphi$; we get an exact sequence

$$
0 \to J \oplus \Lambda^a \xrightarrow{j} \Lambda^{n+b} \xrightarrow{p'} M \to 0.
$$

Define $E = \Lambda^{n+b}/j(\Lambda^a)$; then we get an exact sequence

$$
\alpha = (0 \to J \to E \to M \to 0)
$$

and E is projective by Proposition 5.17. Hence the sequence

$$0 \to \Lambda^a \to \Lambda^{n+b} \to E \to 0$$

defining E splits so that $E \oplus \Lambda^a \cong \Lambda^{n+b}$; that is, E is stably free.   □

A 0-complex α as in Proposition 7.23 is also said to be *tame*; note that then both M and J are finitely geneated. We shall associate to each tame module J a mapping $S_J : \mathrm{Aut}_{\mathcal{D}\mathrm{er}}(J) \to \widetilde{K}_0(\Lambda)$, the *Swan mapping* of J. S_J may be regarded as the dual to $\widehat{S^M}$, the coprojectivity hypothesis on M being necessary for the duality.

Thus let $\alpha = (0 \to J \xrightarrow{j} E \xrightarrow{p} M \to 0)$ be a tame 0-complex. In Sect. 7.2 we constructed a ring isomorphism $\rho_\alpha : \mathrm{End}_{\mathcal{D}\mathrm{er}}(M) \to \mathrm{End}_{\mathcal{D}\mathrm{er}}(J)$; we denote by $\mu_\alpha : \mathrm{End}_{\mathcal{D}\mathrm{er}}(J) \to \mathrm{End}_{\mathcal{D}\mathrm{er}}(M)$ the inverse isomorphism; with this notation:

Proposition 7.24 *For any $g \in \mathrm{Aut}_{\mathcal{D}\mathrm{er}}(J)$ $g_*(\alpha) \equiv \mu_\alpha(g)^*(\alpha)$.*

Proof An easy chase of definitions shows that the following diagram commutes

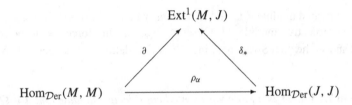

that is, $\rho_\alpha(f)_*(\alpha) \equiv f^*(\alpha)$ for $f \in \mathrm{Hom}_{\mathcal{D}\mathrm{er}}(M, M)$. Equivalently, $g_*(\alpha) \equiv \mu_\alpha(g)^*(\alpha)$ for $g \in \mathrm{Hom}_{\mathcal{D}\mathrm{er}}(J, J)$.   □

Suppose that $\alpha = (0 \to J \xrightarrow{j} E \xrightarrow{p} M \to 0)$ is a tame 0-complex and let $g_1, g_2 : J \to J$ be Λ-homomorphisms; if $g_1 \approx g_2$ then, as M is coprojective, the exact sequences $(g_1)_*(\alpha)$ and $(g_2)_*(\alpha)$ are congruent so that $\varinjlim(g_1, j) \cong \varinjlim(g_2, j)$. By (7.21) above, if g defines an isomorphism in the derived module category then $\varinjlim(g, j)$ is projective. As both M and J are finitely generated then $\varinjlim(g, j)$ is also finitely generated and we obtain a function

$$S_\alpha : \mathrm{Aut}_{\mathcal{D}\mathrm{er}}(J) \to \tilde{K}_0(\Lambda); \qquad S_\alpha(g) = [\varinjlim(g, j)].$$

Clearly $S_\alpha = S_\beta$ when α is congruent to β. Hence directly from Proposition 7.11 we obtain:

Proposition 7.25 *For any $f \in \mathrm{Aut}_{\mathcal{D}\mathrm{er}}(J)$, $S_\alpha(f) = \widehat{S^\alpha}(\mu_\alpha(f))$.*

We proceed to show that, in a tame 0-complex α of this type, S_α depends only on J_α. The proof is a not entirely obvious dual version of Corollary 7.14. First we consider the dual notion of stabilization; if a is a nonnegative integer the co-stabilization $\Sigma^a_-(\alpha)$ of α is the extension

$$\Sigma^a_-(\alpha) = (0 \to J \xrightarrow{\hat{j}} E \oplus \Lambda^a \xrightarrow{p \oplus \mathrm{Id}} M \oplus \Lambda^a \to 0),$$

where $\hat{j} : J \to E \oplus \Lambda^a$ is the composition of j with the canonical inclusion $E \to E \oplus \Lambda^a$. One may check directly that the dual result to Proposition 7.12 holds, namely:

Proposition 7.26 $S_{\Sigma^a_-(\alpha)} = S_\alpha$.

Let $\alpha = (0 \to J \xrightarrow{j_\alpha} E_\alpha \xrightarrow{p_\alpha} M \to 0)$, $\beta = (0 \to J \xrightarrow{j_\beta} E_\beta \xrightarrow{p_\beta} M \to 0)$ be tame 0-complexes in which $J_\alpha = J_\beta = J$. We write $\alpha \underset{J}{\cong} \beta$ when there exists a commutative diagram of the form

$$\begin{array}{cc} \alpha \\ \downarrow \varphi = \\ \beta \end{array} \begin{pmatrix} 0 \to J \xrightarrow{j_\alpha} E_\alpha \xrightarrow{p_\alpha} M_\alpha \to 0 \\ \quad \downarrow \mathrm{Id} \quad \downarrow \varphi \quad\;\; \downarrow \varphi_- \\ 0 \to J \xrightarrow{j_\beta} E_\beta \xrightarrow{p_\beta} M_\beta \to 0 \end{pmatrix}$$

φ is then said to be an *isomorphism over the kernel*.

Proposition 7.27 *If α, β are tame 0-complexes such that $\alpha \underset{J}{\cong} \beta$ then $S_\alpha = S_\beta$.*

Proof Let $\varphi : \alpha \to \beta$ be an isomorphism over J:

$$\begin{array}{cc} \alpha \\ \downarrow \varphi = \\ \beta \end{array} \begin{pmatrix} 0 \to J \xrightarrow{j_\alpha} E_\alpha \xrightarrow{p_\alpha} M_\alpha \to 0 \\ \quad \downarrow \mathrm{Id} \quad \downarrow \varphi \quad\;\; \downarrow \varphi_- \\ 0 \to J \xrightarrow{j_\beta} E_\beta \xrightarrow{p_\beta} M_\beta \to 0 \end{pmatrix}.$$

Then $\mathrm{Id} \times \varphi : J \times P_\alpha \to J \times P_\beta$ is an isomorphism of Λ modules and for each $g \in \mathrm{End}_{\mathcal{D}\mathrm{er}}\Lambda(J)$ restricts to a Λ-module isomorphism $\mathrm{Id} \times \varphi : \mathrm{Im}(g \times -j_\alpha) \to \mathrm{Im}(g \times -j_\beta)$ inducing an isomorphism $\varprojlim(g, j_\alpha) \xrightarrow{\approx} \varprojlim(g, j_\beta)$. Thus $S_\alpha(g) = S_\beta(g)$ as claimed. $\qquad\qquad\square$

Theorem 7.28 *Let α, β be tame 0-complexes with $J_\alpha = J_\beta$; then $S_\alpha = S_\beta$.*

Proof Write

$$\alpha = (0 \to J \xrightarrow{j_\alpha} E_\alpha \xrightarrow{p_\alpha} M_\alpha \to 0) \quad \text{and} \quad \beta = (0 \to J \xrightarrow{j_\beta} E_\beta \xrightarrow{p_\beta} M_\beta \to 0),$$

where E_α, E_β are stably free. We consider a sequence of decreasingly special cases.

Case I Firstly suppose that $M_\alpha = M_\beta = M$ (say) and let $\widehat{S} = \widehat{S^M}$ be the dual Swan homomorphism of Sect. 7.3. As E_α, E_β are stably free it follows from Corollary 7.14 that $\widehat{S^\alpha} = \widehat{S^\beta} = \widehat{S}$. In particular, $\widehat{S^\alpha}$ is a homomorphism by Corollary 7.18. By (7.6), $\rho_\beta(f) = k_{\beta\alpha}\rho_\alpha(f)k_{\alpha\beta}$ for $f \in \text{Aut}_{\text{Der}}(M)$. Since $\mu_\alpha = \rho_\alpha^{-1}$ it follows for $g \in \text{Aut}_{\text{Der}}(J)$ that

$$\mu_\alpha(g) = \lambda_{\beta\alpha}\mu_\beta(g)\lambda_{\alpha\beta},$$

where $\lambda_{\alpha\beta} = \mu_\beta(k_{\alpha\beta}) \in \text{Aut}_{\text{Der}}(M)$. Then $\widehat{S^\alpha}(\lambda_{\beta\alpha}) = -\widehat{S^\alpha}(\lambda_{\alpha\beta})$ since $\lambda_{\beta\alpha}\lambda_{\alpha\beta} = \text{Id}$ and $\widehat{S^\alpha}$ is a homomorphism. Thus

$$\begin{aligned}
S_\alpha(g)) &= \widehat{S^\alpha}(\mu_\alpha(g))\\
&= \widehat{S^\alpha}(\lambda_{\beta\alpha}\mu_\beta(g)\lambda_{\alpha\beta})\\
&= \widehat{S^\alpha}(\lambda_{\beta\alpha}) + \widehat{S^\alpha}(\mu_\beta(g)) + \widehat{S^\alpha}(\lambda_{\alpha\beta})\\
&= \widehat{S^\alpha}(\lambda_{\beta\alpha}) + \widehat{S^\alpha}(\mu_\beta(g)) - \widehat{S^\alpha}(\lambda_{\beta\alpha})\\
&= \widehat{S^\alpha}(\mu_\beta(g)).
\end{aligned}$$

However, $\widehat{S^\alpha} = \widehat{S^\beta}$, so that $S_\alpha(g)) = \widehat{S^\beta}(\mu_\beta(g)) = S_\beta(g))$ and $S_\alpha = S_\beta$ as claimed.

Case II Next suppose that there is an isomorphism of Λ-modules $f : M_\alpha \xrightarrow{\approx} M_\beta$ then f induces a canonical isomorphism over the kernel $f^*(\beta) \cong_J \beta$, so that $S_{f^*(\beta)} = S_\beta$ by Proposition 7.27. However, there is a congruence $\alpha \equiv f^*(\beta)$ so that $S_\alpha = S_{f^*(\beta)}$ and the result follows.

General Case Then

$$\alpha = (0 \to J \to E_\alpha \to M_\alpha \to 0), \qquad \beta = (0 \to J \to E_\beta \to M_\beta \to 0).$$

Put $\alpha_1 = \Sigma_-^{E_\beta}(\alpha)$ thus $\alpha_1 = (0 \to J \to E_\alpha \oplus E_\beta \to M_\alpha \oplus E_\beta \to 0)$. Since E_α, E_β are finitely generated stably free, so also is $E_\alpha \oplus E_\beta$. Moreover, we may choose $N \geq 1$ such that $E_\alpha \oplus \Lambda^N$, $E_\beta \oplus \Lambda^N$ are both free.

$$\alpha_2 = \Sigma_-^N(\alpha_1) = (0 \to J \to E_\alpha \oplus E_\beta \oplus \Lambda^N \to M_\alpha \oplus E_\beta \oplus \Lambda^N \to 0).$$

Then $S_{\alpha_2} = S_{\alpha_1}$ by Proposition 7.26. However $\alpha_2 = \Sigma_-^a(\alpha)$ where $E_\beta \oplus \Lambda^N \cong \Lambda^a$ so that $S_\alpha = S_{\alpha_2}$ by Proposition 7.26. Thus $S_\alpha = S_{\alpha_1}$. Similarly $S_\beta = S_{\beta_1}$ where $\beta_1 = \Sigma_-^{E_\alpha}(\beta)$. However, since M_α, M_β are coprojective then $M_\alpha \oplus E_\beta \cong M_\beta \oplus E_\alpha$ by the Dual Schanuel Lemma, Thus $S_{\alpha_1} = S_{\beta_1}$ by Case II above, and so $S_\alpha = S_\beta$ as required. $\square$

Now let J be a tame module occurring in a tame 0-complex

$$\alpha = (0 \to J \xrightarrow{j} E \xrightarrow{p} M \to 0);$$

we associate to J a mapping

$$S_J : \mathrm{Aut}_{\mathcal{D}\mathrm{er}}(J) \to \widetilde{K}_0(\Lambda)$$

by the rule $S_J = S_\alpha$. S_J is the *Swan mapping* of J. By Theorem 7.28, S_J is independent of the tame 0-complex α in which it occurs. Finally we establish:

Theorem 7.29 $S_J : \mathrm{Aut}_{\mathcal{D}\mathrm{er}}(J) \to \widetilde{K}_0(\Lambda)$ *is a homomorphism.*

Proof Start with a tame 0-complex $\alpha = (0 \to J \to E \to M \to 0)$. As E is stably free then $E \oplus \Lambda^a \cong \Lambda^n$ for some integers $a, n \geq 0$. Thus we may co-stabilize to get

$$\Sigma_-^a(\alpha) = \mathbf{F} = (0 \to J \to \Lambda^n \to N \to 0),$$

where $F = E \oplus \Lambda^a$ and $N = M \oplus \Lambda^a$. As $S_J = S_\alpha = S_{\mathbf{F}}$ it suffices to show that $S_{\mathbf{F}}$ is a homomorphism. However, from Proposition 7.25 we get that $S_{\mathbf{F}}(f) = \widehat{S^{\mathbf{F}}}(\mu_{\mathbf{F}}(f))$; whilst from the definition immediately following Corollary 7.15 we see that $\widehat{S^{\mathbf{F}}} = \widehat{S^N}$. Thus

$$S_{\mathbf{F}}(f) = \widehat{S^N}(\mu_{\mathbf{F}}(f)).$$

Thus for all $f, g \in \mathrm{Aut}_{\mathcal{D}\mathrm{er}}(J)$ we have $S_{\mathbf{F}}(fg) = \widehat{S^N}(\mu_{\mathbf{F}}(fg)) = \widehat{S^N}(\mu_{\mathbf{F}}(f)\mu_{\mathbf{F}}(g))$. However, $\widehat{S^N}$ is a homomorphism by Corollary 7.18, so that

$$\widehat{S^N}(\mu_{\mathbf{F}}(f)\mu_{\mathbf{F}}(g)) = \widehat{S^N}(\mu_{\mathbf{F}}(f)) + \widehat{S^N}(\mu_{\mathbf{F}}(g)) = S_{\mathbf{F}}(f) + S_{\mathbf{F}}(g).$$

Hence $S_{\mathbf{F}}(fg) = S_{\mathbf{F}}(f) + S_{\mathbf{F}}(g)$ and this completes the proof. $\square$

7.6 Stabilising Endomorphisms

Let J, P be modules over Λ; we represent elements of the direct sum $J \oplus P$ as column vectors $\binom{\mathbf{x}}{\mathbf{y}}$ where $\mathbf{x} \in J$ and $\mathbf{y} \in P$. There are obvious inclusions and projections

$$i_J : J \to J \oplus P; \qquad i_P : J \to J \oplus P;$$
$$i_J(\mathbf{x}) = \begin{pmatrix} \mathbf{x} \\ 0 \end{pmatrix}; \qquad i_P(\mathbf{y}) = \begin{pmatrix} 0 \\ \mathbf{y} \end{pmatrix};$$
$$\pi_J : J \oplus P \to J; \qquad \pi_P : J \oplus P \to P;$$
$$\pi_J \begin{pmatrix} \mathbf{x} \\ \mathbf{y} \end{pmatrix} = \mathbf{x}; \qquad \pi_P \begin{pmatrix} \mathbf{x} \\ \mathbf{y} \end{pmatrix} = \mathbf{y}.$$

Any endomorphism $\alpha : J \oplus P \to J \oplus P$ may be uniquely represented by a 2×2 matrix

$$\alpha = \begin{pmatrix} \alpha_{JJ} & \alpha_{JP} \\ \alpha_{PJ} & \alpha_{PP} \end{pmatrix},$$

where $\alpha_{RS} = \pi_R \alpha i_S$. There is a ring homomorphism $\mathbf{e} : \mathrm{End}_\Lambda(J) \to \mathrm{End}_\Lambda(J \oplus P)$

$$\mathbf{e}(\gamma) = \begin{pmatrix} \gamma & 0 \\ 0 & \mathrm{Id}_P \end{pmatrix}.$$

When P is projective then $\mathrm{Id}_P \approx 0$; it then follows that $\mathbf{e}$ descends to a ring homomorphism $\mathbf{e}_* : \mathrm{End}_{\mathcal{D}\mathrm{er}}(J) \to \mathrm{End}_{\mathcal{D}\mathrm{er}}(J \oplus P)$. Moreover, the diagram below commutes:

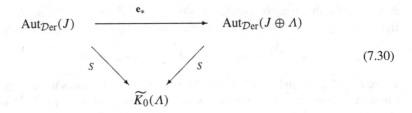

$$(7.30)$$

Proposition 7.31 *If P is projective then* $\mathbf{e}_* : \mathrm{End}_{\mathcal{D}\mathrm{er}}(J) \to \mathrm{End}_{\mathcal{D}\mathrm{er}}(J \oplus P)$ *is an isomorphism of rings.*

Proof It suffices to show $\mathbf{e}_*$ is bijective. The mapping $\rho : \mathrm{End}_\Lambda(J \oplus P) \to \mathrm{End}_\Lambda(J)$ defined by $\rho(\alpha) = \alpha_{JJ}$ is a surjective homomorphism of abelian groups and clearly $\rho\mathbf{e} = \mathrm{Id}$. Observe that $\mathrm{Id}_{J \oplus P} = i_J \pi_J + i_P \pi_P \,(= \mathrm{Id})$. Expanding $\mathrm{Id} \circ \alpha \circ \mathrm{Id}$ for $\alpha \in \mathrm{End}_\Lambda(J \oplus P)$ gives the familiar identity

$$\alpha = i_J \alpha_{JJ} \pi_J + i_J \alpha_{JP} \pi_P + i_P \alpha_{PJ} \pi_J + i_P \alpha_{PP} \pi_P. \qquad (*)$$

We claim that:

$$\alpha \approx 0 \iff \alpha_{JJ} \approx 0. \qquad (**)$$

Observe that if $\alpha \approx 0$ then $\pi_J \alpha i_J \approx 0$ which, since $\alpha_{JJ} = \pi_J \alpha i_J$, proves ($\Longrightarrow$).

Conversely, since P is projective then $\alpha_{JP} \approx 0$, $\alpha_{PJ} \approx 0$ and $\alpha_{PP} \approx 0$ so that by ($*$)

$$\alpha \approx i_J \alpha_{JJ} \pi_J. \qquad (***)$$

If $\alpha_{JJ} \approx 0$ then $\alpha \approx 0$. This proves ($**$) from which it follows that ρ descends to a homomorphism of abelian groups $\rho_* : \mathrm{End}_{\mathcal{D}\mathrm{er}}(J \oplus P) \to \mathrm{End}_{\mathcal{D}\mathrm{er}}(J);\ [\alpha] \mapsto [\alpha_{JJ}]$. Since $\rho\mathbf{e} = \mathrm{Id}$ then $\rho_* \mathbf{e}_* = \mathrm{Id}$ so that ρ_* is surjective. However, ρ_* is also injective by ($**$). Thus ρ_* is bijective. Hence $\mathbf{e}_* = \rho_*^{-1}$ is also bijective as required. $\square$

7.7 Full Modules

Let J be a tame module and let $\alpha = (0 \to J \xrightarrow{j} E \to M \to 0)$ be a tame 0-complex with E stably free. The Swan mapping $S_J : \mathrm{Aut}_{\mathcal{D}\mathrm{er}}(J) \to \widetilde{K}_0(\Lambda)$ may be described by

$$S_J(f) = [\varinjlim(f, j)].$$

Note that there is a canonical morphism

$$
\begin{matrix}
\alpha \\
\downarrow \nu_f \\
f_*(\alpha)
\end{matrix}
=
\begin{pmatrix}
0 \to J \overset{j}{\to} & E & \to M \to 0 \\
\downarrow f & \downarrow \nu & \downarrow \mathrm{Id} \\
0 \to J \to \varinjlim(f, j) \to M \to 0
\end{pmatrix}.
$$

If $f : J \to J$ is an isomorphism of Λ-modules then, by the Five Lemma, $\nu : E \to \varinjlim(f, j)$ is also an isomorphism of Λ-modules. Hence $[\varinjlim(f, j)] = [E] \in \widetilde{K}_0(\Lambda)$. However, E represents 0 in $\widetilde{K}_0(\Lambda)$ so that:

Proposition 7.32 *If $f \in \mathrm{Aut}_\Lambda(J)$ then $S_J(f) = 0 \in \widetilde{K}_0(\Lambda)$.*

There is natural homomorphism $\nu^J : \mathrm{Aut}_\Lambda(J) \to \mathrm{Aut}_{\mathcal{D}\mathrm{er}}(J)$ which assigns to $f \in \mathrm{Aut}_\Lambda(J)$ its class modulo '$\approx$'. Thus we see from Proposition 7.32 that:

Proposition 7.33 *If J is a tame module then $\mathrm{Im}(\nu^J) \subset \mathrm{Ker}(S_J)$.*

We shall say that J is *full* when $\mathrm{Im}(\nu^J) = \mathrm{Ker}(S_J)$. In this section we will show that when J is tame then some stabilization $J \oplus \Lambda^N$ is full. Firstly we write

$$r(J) = \min\{s \geq 1 : \text{there exists a surjective}\,\Lambda\text{-homomorphism}\, \Lambda^s \to J\}.$$

A ring Λ is said to have a *stable range* when there exists an integer $n \geq 0$ such that there is no nontrivial finitely generated stably free module S with $n < \mathrm{rk}(S)$. We write $\sigma(\Lambda) = n$ when there is a nontrivial stably free module of rank $= n$ but none of rank $> n$. This definition of $\sigma(\Lambda)$ is normalized so that $\sigma(\mathbf{F}) = 0$ when $\mathbf{F}$ is a field. Moreover, if Λ has no stable range we write $\sigma(\Lambda) = +\infty$.

Theorem 7.34 *Let J be tame; if $\max\{r(J), \sigma(\Lambda) + 1\} \leq n$ then $J \oplus \Lambda^n$ is full.*

Proof In essence the proof repeats Theorem 7.4 with more precise book-keeping. Since J is tame we may first choose a free 0-complex of finite type

$$\mathcal{E} = (0 \to J \overset{j}{\to} \Lambda^m \overset{p}{\to} M \to 0).$$

Let n be a positive integer such that $\max\{r(J), \sigma(\Lambda) + 1\} \leq n$ and choose a Λ-homomorphism $g : J \oplus \Lambda^n \to J \oplus \Lambda^n$ such that $[g] \in \mathrm{Ker}(S_{J \oplus \Lambda^n})$. Write g as a matrix with respect to the direct sum $J \oplus \Lambda^n$

$$g = \begin{pmatrix} g_{00} & g_{01} \\ g_{10} & g_{11} \end{pmatrix}$$

and put $f = g_{00} : J \to J$. Then $[g] = \mathbf{e}_*([f])$ so that $[f] \in \mathrm{Ker}(S)$ by commutativity of (7.30). It will suffice to show that:

There is an element $F \in \mathrm{Aut}_\Lambda(J \oplus \Lambda^n)$ of the form $F = \begin{pmatrix} f & f_{01} \\ f_{10} & f_{11} \end{pmatrix}$. (7.35)

For then $[F] = \mathbf{e}_*([f]) = [g]$ and so $[g] \in \mathrm{Im}(\nu^J)$ as required since F is a Λ-automorphism of $J \oplus \Lambda^n$. Given f obtained from g as above, construct the pushout sequence $f_*(\mathcal{E})$ together with the canonical morphism $\nu_f : \mathcal{E} \to f_*(\mathcal{E})$;

$$
\begin{array}{c} \mathcal{E} \\ \downarrow \nu_f \\ f_*(\mathcal{E}) \end{array} = \left(\begin{array}{ccccccc} 0 \longrightarrow & J & \xrightarrow{\ j\ } & \Lambda^m & \xrightarrow{\ p\ } & M & \longrightarrow 0 \\ & \downarrow f & & \downarrow f_0 & & \downarrow \mathrm{Id}_M & \\ 0 \longrightarrow & J & \xrightarrow{\ i\ } & \varinjlim(f,j) & \xrightarrow{\ \pi\ } & M & \longrightarrow 0 \end{array} \right).
$$

Since $r(J) \leq n$ choose a surjective Λ-homomorphism $f_{01} : \Lambda^n \to J$ and construct a morphism of exact sequences thus

$$
\begin{array}{c} \Sigma_+^n(\mathcal{E}) \\ \downarrow \hat{\varphi} \\ f_*(\mathcal{E}) \end{array} = \left(\begin{array}{ccccccc} 0 \longrightarrow & J \oplus \Lambda^n & \xrightarrow{j \oplus \mathrm{Id}} & \Lambda^m \oplus \Lambda^n & \xrightarrow{p\pi} & M & \longrightarrow 0 \\ & \downarrow \varphi & & \downarrow \varphi_0 & & \downarrow \mathrm{Id}_M & \\ 0 \longrightarrow & J & \xrightarrow{\ i\ } & \varinjlim(f,j) & \xrightarrow{\ \pi\ } & M & \longrightarrow 0 \end{array} \right),
$$

where $\varphi = (f, f_{01})$ and $\varphi_0 = (f_0, i \circ f_{01})$. As f_{01} is surjective so also is φ_0 and we have an exact sequence $\mathcal{K} = (0 \longrightarrow \mathrm{Ker}(\varphi_0) \xrightarrow{\kappa} \Lambda^m \oplus \Lambda^n \xrightarrow{\varphi_0} \varinjlim(f,j) \longrightarrow 0)$. As $f \in \mathrm{Aut}_{\mathcal{D}\mathrm{er}}(J)$ then $\varinjlim(f,j)$ is projective, so that on splitting $\mathcal{K}$

$$
\mathrm{Ker}(\varphi_0) \oplus \varinjlim(f,j) \cong \Lambda^m \oplus \Lambda^n
$$

and $\mathrm{Ker}(\varphi_0)$ is also projective. Since $f \in \mathrm{Ker}(S)$ then $\varinjlim(f,j)$ is stably free, so that $\mathrm{Ker}(\varphi_0)$ is also stably free. Furthermore, $\mathrm{rk}(\mathrm{Ker}(\varphi_0)) + \mathrm{rk}(\varinjlim(f,j)) = m + n$. On comparing $\mathcal{E}$ with $f_*(\mathcal{E})$ we see from Theorem 7.1 that $\mathrm{rk}(\varinjlim(f,j) = \mathrm{rk}(\Lambda^m) = m$. It follows that $\mathrm{rk}(\mathrm{Ker}(\varphi_0)) = n$. However, since $\sigma(\Lambda) + 1 \leq n$ then $\mathrm{Ker}(\varphi_0) \cong \Lambda^n$. Choosing a specific isomorphism $\psi : \Lambda^n \to \mathrm{Ker}(\varphi_0)$ we may re-write $\mathcal{K}$ to give an exact sequence of the form

$$
\mathcal{K}' = (0 \longrightarrow \Lambda^n \xrightarrow{\kappa\psi} \Lambda^m \oplus \Lambda^n \xrightarrow{\varphi_0} \varinjlim(f,j) \longrightarrow 0).
$$

Let $r = (r_0, r_1) : \Lambda^m \oplus \Lambda^n \to \Lambda^n$ be a left splitting of $\mathcal{K}'$. Taking

$$
F = \begin{pmatrix} f & f_{01} \\ r_0 j & r_1 \end{pmatrix}, \qquad F_0 = \begin{pmatrix} f_0 & i f_{01} \\ r_0 & r_1 \end{pmatrix}
$$

we may construct a morphism $\hat{F} : \Sigma_+^n(\mathcal{E}) \to \Sigma_+^n(f_*(\mathcal{E}))$ as follows:

$$
\begin{array}{c}
\Sigma_+^n(\mathcal{E}) \\
\downarrow \hat{F} \\
\Sigma_+^n(f_*(\mathcal{E}))
\end{array}
=
\begin{pmatrix}
0 \longrightarrow J \oplus \Lambda^n \xrightarrow{j \oplus \mathrm{Id}} \Lambda^m \oplus \Lambda^n \xrightarrow{p\pi} M \longrightarrow 0 \\
\qquad\qquad \downarrow F \qquad\qquad\quad \downarrow F_0 \qquad\qquad \downarrow \mathrm{Id}_M \\
0 \longrightarrow J \oplus \Lambda^n \xrightarrow{i \oplus \mathrm{Id}} \varinjlim(f, j) \oplus \Lambda^n \xrightarrow{\pi} M \longrightarrow 0
\end{pmatrix}.
$$

Now F_0 is a Λ-isomorphism since it is constructed from a splitting of $\mathcal{K}'$. Extending by zeroes on the left, it follows from the Five Lemma that F is also a Λ-isomorphism; taking F to be the fulfilment of the promise at (7.35), the proof is now complete. $\qquad\square$

we may construct a morphism

$$\begin{pmatrix} & \end{pmatrix}$$

Chapter 8
Classification of Algebraic Complexes

By an *algebraic n-complex* over $\mathbf{Z}[G]$ we mean an exact sequence of $\mathbf{Z}[G]$-modules

$$E_* = (0 \to J \to E_n \xrightarrow{\partial_n} E_{n-1} \xrightarrow{\partial_{n-1}} \cdots \xrightarrow{\partial_2} E_1 \xrightarrow{\partial_1} E_0 \to \mathbf{Z} \to 0)$$

in which each E_r is finitely generated and stably free over $\mathbf{Z}[G]$. The notion is an abstraction from a cell complex X with $\pi_1(X) = G$ for which $\pi_r(\tilde{X}) = 0$ for $0 < r < n$. In this chapter we use the Swan homomorphism of Chap. 7 to classify algebraic n-complexes up to homotopy equivalence.

8.1 Algebraic n-Complexes

An algebraic n-complex may be described in two slightly different ways; the naive definition, which for operational purposes we regard as secondary, views an algebraic n-complex $\mathcal{E}$ as a chain complex of stably free Λ-modules

$$0 \to E_n \xrightarrow{\partial_n} E_{n-1} \xrightarrow{\partial_{n-1}} \cdots \xrightarrow{\partial_2} E_2 \xrightarrow{\partial_2} E_1 \xrightarrow{\partial_1} E_0 \to 0 \qquad (*)$$

which is exact at E_r for $0 < r < n$. The condition 'stably free' rather than the more obvious 'free' is made for technical convenience. For $n \geq 2$ there is, up to congruence [68, p. 84], no difference between the two. By analogy with the Hurewicz Theorem, it is useful to think of $H_n(\mathbf{E})$ as an 'algebraic π_n'; putting

$$\pi_n(\mathbf{E}) = \mathrm{Ker}(\partial_n : E_n \to E_{n-1})$$

enables us to write $\mathbf{E}$ in augmented, co-augmented form as an exact sequence

$$\mathbf{E} = (0 \to \pi_n(\mathbf{E}) \to E_n \xrightarrow{\partial_n} \cdots \xrightarrow{\partial_2} E_1 \xrightarrow{\partial_1} E_0 \to H_0(\mathbf{E}) \to 0).$$

Thus in this description an algebraic n-complex is an exact sequence

$$0 \to J \xrightarrow{j} E_n \xrightarrow{\partial_n} \cdots \xrightarrow{\partial_2} E_1 \xrightarrow{\partial_1} E_0 \xrightarrow{\epsilon} M \to 0 \qquad (**)$$

F.E.A. Johnson, *Syzygies and Homotopy Theory*, Algebra and Applications 17, DOI 10.1007/978-1-4471-2294-4_8, © Springer-Verlag London Limited 2012

of Λ-homomorphisms in which all modules are finitely generated and each E_r is stably free.

More general is the notion of a *projective n-complex* by which we mean an exact sequence

$$\mathbf{P} = (0 \to \pi_n(\mathbf{P}) \to P_n \xrightarrow{\partial_n} \cdots \xrightarrow{\partial_2} P_1 \xrightarrow{\partial_1} P_0 \to H_0(\mathbf{P}) \to 0)$$

of Λ-homomorphisms in which each P_r is projective.

Corresponding to the two descriptions are two different notions of homotopy. In the geometric version, if $f, g : C_* \to D_*$ are chain maps then f is homotopic to g (written $f \simeq g$) when there are Λ-homomorphisms $\eta_n : C_n \to D_{n+1}$ such that, for each n:

$$f_n - g_n = \partial_{n+1}\eta_n + \eta_{n-1}\partial_n.$$

As is well known if $f \simeq g : C_* \to D_*$ are homotopic chain maps then $H_r(f) = H_r(g)$ for all r. In particular $\pi_n(\mathbf{P})$ is an invariant of chain homotopy type.

The appropriate notion of homotopy for the augmented/co-augmented description is *weak homotopy equivalence*. A chain mapping $h : \mathbf{P} \to \mathbf{Q}$ between projective n-complexes $\mathbf{P}, \mathbf{Q}$ is a weak homotopy equivalence when the induced maps $h_* : H_0(\mathbf{P}) \to H_0(\mathbf{Q})$ and $h_* : \pi_n(\mathbf{P}) \to \pi_n(\mathbf{Q})$ are isomorphisms. By hypothesis, $H_r(\mathbf{P}) = H_r(\mathbf{Q}) = 0$ for $1 \leq r \leq n-1$ so the mappings $h_* : H_r(\mathbf{P}) \to H_r(\mathbf{Q})$ are automatically isomorphisms. The following is essentially due to J.H.C. Whitehead (cf. [52, 100, p. 177]):

Proposition 8.1 *If $f : C_* \to D_*$ is a chain map between (nonnegative) projective chain complexes then the following statements are equivalent;*

(i) *$H_*(f)$ is an isomorphism for each n;*
(ii) *there exists a chain mapping $g : D_* \to C_*$ such that $f \circ g \simeq \mathrm{Id}$ and $g \circ f \simeq \mathrm{Id}$.*

For projective n-complexes both weak homotopy equivalence and elementary congruence are equivalence relations; this is false for more general extensions. The question now being essentially a matter of taste, we choose, for practical purposes, the description which includes augmentation and co-augmentation as primary. Given a module M we denote by $\mathbf{Alg}_n(M)$ the set of homotopy equivalence classes of algebraic n-complexes over Id_M. We have seen, in Sect. 1.2, how the stable module $[M]$ has the structure of a tree. $\mathbf{Alg}_n(M)$ also acquires the structure of an infinite directed graph on drawing arrows of the form

$$\mathbf{E} \longrightarrow \Sigma_+(\mathbf{E}) = (0 \to J \oplus \Lambda \to E_n \oplus \Lambda \to E_{n-1} \to \cdots \to E_1 \to E_0 \to M \to 0).$$

The correspondence $\mathbf{E} \to \pi_n(\mathbf{E})$ gives a mapping of graphs $\pi_n : \mathbf{Alg}_n(M) \to \Omega_{n+1}(M)$ from which it is easy to see that $\mathbf{Alg}_n(M)$ contains no loops. In Sect. 8.3 we shall prove that $\mathbf{Alg}_n(M)$ is connected and hence is a tree. We note that the classification theorem which we establish assumes a much simpler form for hyperstable complexes [55].

8.2 A Cancellation Theorem for Chain Homotopy Equivalences

The following cancellation theorem is perhaps well known but as there is no convenient reference in the literature we give a proof:

Theorem 8.2 *Let* $\mathbf{Q}$ *be a projective m-complex and for* $r = 1, 2$ *let* $\mathbf{P}(r)$ *be a projective* $(n + m)$-*complex decomposed as a Yoneda product* $\mathbf{P}(r) = \mathbf{P}_+(r) \circ \mathbf{Q}$ *for some projective* $(n - 1)$-*complexes* $\mathbf{P}_+(r)$; *then* $\mathbf{P}(1) \simeq \mathbf{P}(2) \Longrightarrow \mathbf{P}_+(1) \simeq \mathbf{P}_+(2)$.

Yoneda product,[1] or cutting and splicing, is treated at greater length in, for example the book by Maclane [68, p. 229]. However, its meaning here should be entirely clear from context. Thus given a projective n-complex

$$\mathbf{P} = (0 \to J \to P_n \to \cdots \to P_0 \to M \to 0)$$

we decompose it thus $\mathbf{P} = \mathbf{P}_+ \circ \mathbf{P}_-$ where

$$\mathbf{P}_+ = (0 \to J \to P_n \to \cdots \to P_1 \to \Omega \to 0);$$
$$\mathbf{P}_- = (0 \to \Omega \to P_0 \to M \to 0).$$

Likewise, given a chain mapping $f : \mathbf{P} \to \mathbf{Q}$ between projective n-complexes

$$
\begin{matrix}
\mathbf{P} \\
\downarrow f = \\
\mathbf{Q}
\end{matrix}
\begin{pmatrix}
0 \to J \to & P_n \to & \cdots \to & P_0 \to & M \to 0 \\
\downarrow f_J & \downarrow f_n & & \downarrow f_0 & \downarrow f_M \\
0 \to J' \to & Q_n \to & \cdots \to & Q_0 \to & M' \to 0
\end{pmatrix}
$$

we decompose it as $f = f_+ \circ f_- : \mathbf{P}_+ \circ \mathbf{P}_- \to \mathbf{Q}_+ \circ \mathbf{Q}_-$

$$
\begin{matrix}
\mathbf{P}_- \\
\downarrow f_- = \\
\mathbf{Q}_-
\end{matrix}
\begin{pmatrix}
0 \to \Omega \to & P_0 \to & M \to 0 \\
\downarrow f_\Omega & \downarrow f_0 & \downarrow f_M \\
0 \to \Omega' \to & Q_0 \to & M' \to 0
\end{pmatrix};
$$

$$
\begin{matrix}
\mathbf{P}_+ \\
\downarrow f_+ = \\
\mathbf{Q}_+
\end{matrix}
\begin{pmatrix}
0 \to J \to & P_n \to & \cdots \to & P_1 \to & \Omega \to 0 \\
\downarrow f_J & \downarrow f_n & & \downarrow f_1 & \downarrow f_\Omega \\
0 \to J' \to & Q_n \to & \cdots \to & Q_1 \to & \Omega' \to 0
\end{pmatrix}.
$$

Starting from (7.5) as base, an obvious inductive argument shows that when $\mathbf{P}, \mathbf{Q}$ are projective:

If $g, h : \mathbf{P} \to \mathbf{Q}$ are chain mappings then $h_M \approx g_M \Longrightarrow h_J \approx g_J$.   (8.3)

On taking $\mathbf{F} = \mathbf{P}$ and $g = \mathrm{Id}$ we obtain the useful special case of (8.3):

If $h : \mathbf{P} \to \mathbf{P}$ is a chain mapping then $h_M \approx \mathrm{Id}_M \Longrightarrow h_J \approx \mathrm{Id}_J$.   (8.4)

---

[1]An interpretation of Yoneda product as composition in the derived module category is given in the thesis of Gollek [33].

Now let $h : \mathbf{P} \to \mathbf{Q}$ be a chain mapping between projective n-complexes over $h_M : M \to M$ and suppose that $h_M \approx \mathrm{Id}_M$. We shall produce a chain mapping $\hat{g} : \mathbf{P} \to \mathbf{Q}$ over Id_M in which $g_J = h_J$. In the case $n = 0$, choose a factorization $\mathrm{Id}_M - h_M = \xi \circ \eta$ through a projective module X. Then since $q : Q_0 \to M$ is surjective and X is projective there exists a homomorphism $\hat{\xi} : X \to Q_0$ such that $\xi = q \circ \hat{\xi}$ so that the diagram below commutes:

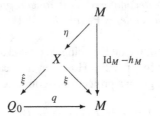

Put $g_0 = h_0 + \hat{\xi} \circ \eta \circ p$. It is straightforward to check that the diagram below commutes as required:

$$\begin{matrix} \mathbf{P} \\ \downarrow g = \\ \mathbf{Q} \end{matrix} \begin{pmatrix} 0 \to J \xrightarrow{j} P_0 \xrightarrow{p} M \to 0 \\ \downarrow h_J \quad \downarrow g_0 \quad \downarrow \mathrm{Id}_M \\ 0 \to J' \xrightarrow{i} Q_0 \xrightarrow{q} M \to 0 \end{pmatrix}.$$

When $n > 0$, decompose both $\mathbf{P}$ and $\mathbf{Q}$ as Yoneda products $\mathbf{P} = \mathbf{P}_+ \circ \mathbf{P}_-$ and $\mathbf{Q} = \mathbf{Q}_+ \circ \mathbf{Q}_-$, and also decompose h as $h = h_+ \circ h_-$; by the case $n = 0$ already proved there exists a chain mapping $f : \mathbf{P}_- \to \mathbf{Q}_-$ of the form

$$\begin{matrix} \mathbf{P}_- \\ \downarrow f = \\ \mathbf{Q}_- \end{matrix} \begin{pmatrix} 0 \to \Omega \to P_0 \to M \to 0 \\ \downarrow h_\Omega \quad \downarrow f_0 \quad \downarrow \mathrm{Id}_M \\ 0 \to \Omega' \to Q_0 \to M \to 0 \end{pmatrix}.$$

Since $f_\Omega = h_\Omega$, the spliced mapping $g = h_+ \circ f$ is well defined and takes the form:

$$\begin{matrix} \mathbf{P} \\ \downarrow g = \\ \mathbf{Q} \end{matrix} \begin{pmatrix} 0 \to J \to P_n \to \cdots \to P_1 \to P_0 \to M \to 0 \\ \downarrow h_J \quad \downarrow h_n \qquad\quad \downarrow h_1 \quad \downarrow f_0 \quad \downarrow \mathrm{Id}_M \\ 0 \to J' \to Q_n \to \cdots \to Q_1 \to Q_0 \to M \to 0 \end{pmatrix}.$$

For a chain mapping $h : \mathbf{P} \to \mathbf{Q}$ between projective n-complexes over M we have: If $h_M \approx \mathrm{Id}_M$ then there exists a chain mapping $g : \mathbf{P} \to \mathbf{Q}$ over Id_M

$$\begin{matrix} \mathbf{P} \\ \downarrow g = \\ \mathbf{Q} \end{matrix} \begin{pmatrix} 0 \to J \to P_n \to \cdots \to P_0 \to M \to 0 \\ \downarrow g_J \quad \downarrow g_n \qquad\quad \downarrow g_0 \quad \downarrow \mathrm{Id}_M \\ 0 \to J' \to Q_n \to \cdots \to Q_0 \to M \to 0 \end{pmatrix} \qquad (8.5)$$

in which $g_J = h_J$.

Proof of Theorem 8.2 Write $\mathbf{Q} = (0 \to K \to Q_m \to \cdots \to Q_0 \to M \to 0)$ and $\mathbf{P}_+(r) = (0 \to J(r) \to P_{m+n}(r) \to \cdots \to P_{m+1}(r) \to K \to 0)$ and suppose given a homotopy equivalence $h : \mathbf{P}_+(1) \circ \mathbf{Q} \overset{\sim}{\to} \mathbf{P}_+(2) \circ \mathbf{Q}$. Decompose h as $h = h_+ \circ h_-$:

$$
\begin{array}{c}
\mathbf{Q} \\
\downarrow h_- = \\
\mathbf{Q}
\end{array}
\left(
\begin{array}{ccccccc}
0 \to & K & \to Q_m \to & \cdots & \to Q_0 & \to M & \to 0 \\
 & \downarrow h_K & \downarrow h_m & & \downarrow h_0 & \downarrow \mathrm{Id}_M & \\
0 \to & K & \to Q_m \to & \cdots & \to Q_0 & \to M & \to 0
\end{array}
\right);
$$

$$
\begin{array}{c}
\mathbf{P}_+(1) \\
\downarrow h_+ = \\
\mathbf{P}_+(2)
\end{array}
\left(
\begin{array}{ccccccc}
0 \to & J(1) & \to P_{m+n}(1) \to & \cdots & \to P_{m+1}(1) & \to K & \to 0 \\
 & \downarrow h_J & \downarrow h_{m+n} & & \downarrow h_{m+1} & \downarrow h_K & \\
0 \to & J(2) & \to P_{m+n}(2) \to & \cdots & \to P_{m+1}(2) & \to K & \to 0
\end{array}
\right).
$$

Evidently $J(r) = \pi_{m+n}(\mathbf{P}_+(r) \circ \mathbf{Q})$. Since h is a homotopy equivalence and $h_J : J(1) \to J(2)$ is just the induced map

$$
\pi_{m+n}(h) : \pi_{m+n}(\mathbf{P}_+(1) \circ \mathbf{Q}) \to \pi_{m+n}(\mathbf{P}_+(2) \circ \mathbf{Q})
$$

then h_J is an isomorphism of Λ-modules; (8.4) applied to h_- shows that $h_K \approx \mathrm{Id}_K$; (8.5) applied to h_+ gives a chain mapping $g : \mathbf{P}_+(1) \to \mathbf{P}_+(2)$ in which $g_J = h_J$ thus:

$$
\begin{array}{c}
\mathbf{P}_+(1) \\
\downarrow g = \\
\mathbf{P}_+(2)
\end{array}
\left(
\begin{array}{ccccccc}
0 \to & J(1) & \to P_{m+n}(1) \to & \cdots & \to P_{m+1}(1) & \to K & \to 0 \\
 & \downarrow h_J & \downarrow g_{m+n} & & \downarrow g_{m+1} & \downarrow \mathrm{Id}_K & \\
0 \to & J(2) & \to P_{m+n}(2) \to & \cdots & \to P_{m+1}(2) & \to K & \to 0
\end{array}
\right).
$$

By construction $g_J = h_J$ so g_J is a Λ-isomorphism. Since $\mathbf{P}_+(1)$, $\mathbf{P}_+(2)$ are projective $(n-1)$-complexes then $g : \mathbf{P}_+(1) \to \mathbf{P}_+(2)$ is a homotopy equivalence. $\square$

8.3 Connectivity of $\mathbf{Alg}_n(M)$

If $\mathbf{E} = (0 \to J \to E_n \to E_{n-1} \to \cdots \to E_0 \to M \to 0)$ is an algebraic n-complex with $n \geq 0$ we denote by Σ_+, Σ_- the 'external' stabilization operators

$$
\Sigma_+(\mathbf{E}) = (0 \to J \oplus \Lambda \to E_n \oplus \Lambda \to E_{n-1} \to \cdots \to E_0 \to M \to 0);
$$

$$
\Sigma_-(\mathbf{E}) = (0 \to J \to E_n \to \cdots \to E_1 \to E_0 \oplus \Lambda \to M \oplus \Lambda \to 0).
$$

For $n \geq 1$ there is also an 'internal' stabilization operator between the $(n-1)$th and nth stages, denoted by L thus:

$$
L(\mathbf{E}) = (0 \to J \to E_n \oplus \Lambda \to E_{n-1} \oplus \Lambda \to E_{n-2} \to \cdots \to M \to 0).
$$

Evidently $L(\mathbf{E})$ differs from $\mathbf{E}$ by an elementary congruence; that is:

Proposition 8.6 $L(\mathbf{E}) \equiv \mathbf{E}$.

For $n \geq 1$ we decompose algebraic n-complexes systematically as Yoneda products $\mathbf{E} = \mathbf{E}_+ \circ \mathbf{E}_-$ where

$$\mathbf{E}_- = (0 \to K \to E_{n-1} \to \cdots \to E_0 \to M \to 0) \quad \text{and}$$

$$\mathbf{E}_+ = (0 \to J \to E_n \to K \to 0).$$

Writing $L(\mathbf{E}) = L(\mathbf{E})_+ \circ L(\mathbf{E})_-$ forces identities amongst the various constructions:

$$L(\mathbf{E})_+ = \Sigma_-(\mathbf{E}_+), \tag{8.7}$$

$$L(\mathbf{E})_- = \Sigma_+(\mathbf{E}_-). \tag{8.8}$$

We denote by Σ_+^k, Σ_-^k and L^k the k-fold iterates of the respective operators. We proceed to prove:

Theorem 8.9 *Let* $\mathbf{E}, \mathbf{F}$ *be algebraic n-complexes over modules* M, M' *respectively and suppose that* $h : M \to M'$ *is an isomorphism. Then for some positive integers a, b there exists a homotopy equivalence* $\eta : \Sigma_+^a(\mathbf{E}) \xrightarrow{\sim}_h \Sigma_+^b(\mathbf{F})$ *over h.*

Proof We prove it by induction, noting that the induction base, the case $n = 0$, is already established as Theorem 7.4. Write

$$\mathbf{E}_+ = (0 \to J \to E_n \to K \to 0);$$

$$\mathbf{E}_- = (0 \to K \to E_{n-1} \to \cdots \to E_0 \to M \to 0);$$

$$\mathbf{F}_+ = (0 \to J' \to F_n \to K' \to 0);$$

$$\mathbf{F}_- = (0 \to K' \to F_{n-1} \to \cdots \to F_0 \to M' \to 0);$$

so that $\mathbf{E} = \mathbf{E}_+ \circ \mathbf{E}_-$ and $\mathbf{E} = \mathbf{F}_+ \circ \mathbf{F}_-$. By induction choose a homotopy equivalence $\xi_- : \Sigma_+^c(\mathbf{E}_-) \xrightarrow{\sim}_h \Sigma_+^d(\mathbf{F}_-)$ over h. This gives an isomorphism $\xi_K : K \oplus \Lambda^c \to K' \oplus \Lambda^d$. Apply the case $n = 0$ to $\Sigma_-^c(\mathbf{E}_+)$, $\Sigma_-^d(\mathbf{F}_+)$ to get an isomorphism of 0-complexes over ξ_K

$$\xi_+ : \Sigma_+^a \Sigma_-^c(\mathbf{E}_+) \xrightarrow{\simeq}_{\xi_K} \Sigma_+^b \Sigma_-^d(\mathbf{F}_+).$$

We may now splice ξ_+ together with ξ_- along ξ_K to obtain a homotopy equivalence over h

$$\xi : \Sigma_+^a L^c(\mathbf{E}) \xrightarrow{\sim}_h \Sigma_+^b L^d(\mathbf{F}).$$

However $\mathbf{E} \equiv L^c(\mathbf{E})$ so that $\Sigma_+^a(\mathbf{E}) \equiv \Sigma_+^a(L^c(\mathbf{E}))$ and likewise $\Sigma_+^b(\mathbf{F}) \equiv \Sigma_+^b(L^d(\mathbf{F}))$. Let $\iota : \Sigma_+^a(\mathbf{E}) \to \Sigma_+^a(L^c(\mathbf{E}))$ and $\pi : \Sigma_+^b(L^d(\mathbf{F})) \to \Sigma_+^b(\mathbf{F})$ be congruences; the required homotopy equivalence over h is given by $\eta = \pi \circ \xi \circ \iota : \Sigma_+^a(\mathbf{E}) \xrightarrow{\sim}_h \Sigma_+^b(\mathbf{F})$. $\qquad\square$

It follows from Theorem 8.9 that $\mathbf{Alg}_n(M)$ is connected, so that:

Corollary 8.10 $\mathbf{Alg}_n(M)$ *is a tree for each $n \geq 0$.*

8.4 Comparison of Trees

For a finitely presented Λ-module M denote by $\mathbf{Stab}_0(M, -)$ the collection of exact sequences of the form $\mathbf{S} = (0 \to J \to S \to M \to 0)$ where S is finitely generated stably free. By Schanuel's Lemma J is finitely generated and $J \in \Omega_1(M)$.

If $\mathbf{S}' = (0 \to J' \to S' \to M \to 0)$ also belongs to $\mathbf{Stab}_0(M, -)$ write $\mathbf{S} \cong_{\mathrm{Id}_M} \mathbf{S}'$ when there exists a morphism $\varphi : \mathbf{S} \to \mathbf{S}'$ of the following form:

$$\begin{pmatrix} 0 \to J \to S \to & M \to 0 \\ \downarrow \varphi_+ \ \downarrow \varphi_0 & \downarrow \mathrm{Id}_M \\ 0 \to J' \to S' \to & M \to 0 \end{pmatrix}.$$

Then $\cong_{\mathrm{Id}_M}$ is an equivalence relation on $\mathbf{Stab}_0(M, -)$; we denote by $\mathbf{Alg}_0(M)$ the set of equivalence classes of $\mathbf{Stab}_0(M, -)$ under $\cong_{\mathrm{Id}_M}$. $\mathbf{Alg}_0(M)$ has the structure of a directed tree on drawing $\mathbf{S} \to \Sigma_+(\mathbf{S})$. For $\mathbf{S} = (0 \to J \to S \to M \to 0) \in \mathbf{Stab}_0(M, -)$ we have seen that $J \in \Omega_1(M)$; the correspondence

$$\mathbf{S} = (0 \to J \to S \to M \to 0) \mapsto J(= \natural(\mathbf{S}))$$

induces a mapping of directed trees $\natural : \mathbf{Alg}_0(M) \to \Omega_1(M)$. From Proposition 7.23 we have:

Proposition 8.11 *If M is coprojective then $\natural : \mathbf{Alg}_0(M) \to \Omega_1(M)$ is surjective.*

To elucidate the structure of $\mathbf{Alg}_0(M)$ in terms of $\Omega_1(M)$ it is necessary to compute the size of the fibres $\natural^{-1}(J)$ for $J \in \Omega_1(M)$. We first prove:

Theorem 8.12 *If $\mathbf{E} = (0 \to J \xrightarrow{j} E \xrightarrow{p} M \to 0), \mathbf{E}' = (0 \to J' \xrightarrow{j'} E' \xrightarrow{p'} M \to 0)$ are extensions then $\mathbf{E} \cong_{\mathrm{Id}_M} \mathbf{E}'$ if and only if $h_*(\mathbf{E}) \equiv \mathbf{E}'$ for some Λ-isomorphism $h : J \to J'$.*

Proof To prove ($\Longrightarrow$) suppose given an isomorphism over the identity

$$\begin{matrix} \mathbf{E} \\ \downarrow \hat{h} = \\ \mathbf{E}' \end{matrix} \begin{pmatrix} 0 \to J \xrightarrow{j} E \xrightarrow{p} M \to 0 \\ \downarrow h \quad \downarrow h_0 \quad \downarrow \mathrm{Id} \\ 0 \to J' \xrightarrow{j'} E' \xrightarrow{p'} M \to 0 \end{pmatrix}$$

so that, in particular, $h : J \to J'$ is a Λ-isomorphism. By definition, $h_*(\mathbf{E})$ is the extension obtained by the pushout construction

$$\begin{matrix} \mathbf{E} \\ \downarrow v_h \ = \\ h_*(\mathbf{E}) \end{matrix} \begin{pmatrix} 0 \to J \xrightarrow{j} & E \xrightarrow{p} & M \to 0 \\ \downarrow h & \downarrow v_0 & \downarrow \mathrm{Id} \\ 0 \to J' \xrightarrow{\iota} \varinjlim(h, j) \xrightarrow{\pi} & M \to 0 \end{pmatrix}.$$

Note that by the Five Lemma, since h is a Λ-isomorphism then the pushout mapping $\nu_0 : E \to \varinjlim(h, j)$ is also a Λ-isomorphism. Then the following diagram commutes:

$$\begin{matrix} h_*(\mathbf{E}) \\ \downarrow \hat{h}\nu_h^{-1} = \\ \mathbf{E}' \end{matrix} \begin{pmatrix} 0 \to J' \overset{\iota}{\to} \varinjlim(h, j) \overset{\pi}{\to} M \to 0 \\ \quad\; \downarrow \mathrm{Id} \qquad \downarrow h_0 \qquad \downarrow \mathrm{Id} \\ 0 \to J' \overset{j'}{\to} \quad E' \; \overset{p'}{\to} \quad M \to 0 \end{pmatrix} ;$$

that is, $\hat{h}\nu_h^{-1}$ is the required congruence $h_*(\mathbf{E}) \equiv \mathbf{E}'$. Conversely, let $h : J \to J'$ be a Λ-isomorphism and suppose that c is a congruence $h_*(\mathbf{E}) \equiv \mathbf{E}'$:

$$\begin{matrix} h_*(\mathbf{E}) \\ \downarrow c = \\ \mathbf{E}' \end{matrix} \begin{pmatrix} 0 \to J' \overset{\iota}{\to} \varinjlim(h, j) \overset{\pi}{\to} M \to 0 \\ \quad\; \downarrow \mathrm{Id} \qquad \downarrow c_0 \qquad \downarrow \mathrm{Id} \\ 0 \to J' \overset{j'}{\to} \quad E' \; \overset{p'}{\to} \quad M \to 0 \end{pmatrix} .$$

Recall the natural pushout morphism $\nu_h : \mathbf{E} \to h_*(\mathbf{E})$;

$$\begin{matrix} \mathbf{E} \\ \downarrow \nu_h = \\ h_*(\mathbf{E}) \end{matrix} \begin{pmatrix} 0 \to J \overset{j}{\to} \quad E \; \overset{p}{\to} \quad M \to 0 \\ \quad \downarrow h \qquad \downarrow \nu_0 \qquad \downarrow \mathrm{Id} \\ 0 \to J' \overset{\iota}{\to} \varinjlim(h, j) \overset{\pi}{\to} M \to 0 \end{pmatrix} .$$

Then $c \circ \nu_h : \mathbf{E} \to \mathbf{E}'$ is an isomorphism over Id_M as required.   □

Now assume that M is coprojective and let $J \in \Omega_1(M)$. The relation $\cong_M$ (isomorphism over Id) is an equivalence relation on $\mathbf{Stab}_0(M, J)$. Let $\langle \mathbf{S} \rangle$ denote the equivalence class of $\mathbf{S}$ under $\cong_M$. It is tautological that the elements of $\pi_n^{-1}(J)$ are precisely the distinct equivalence classes $\langle \mathbf{S} \rangle$ as $\mathbf{S}$ runs through $\mathbf{Stab}_0(M, J)$.

Choose a specific stably free 0-complex of finite type

$$\mathbf{E} = (0 \to J \overset{i}{\to} E \overset{q}{\to} M \to 0)$$

and define a mapping $\kappa : \mathrm{Ker}(S_J) \to \pi_0^{-1}(J)$ by

$$\kappa([f]) = \langle f_*(\mathbf{E}) \rangle.$$

It is straightforward to verify that κ is well defined; moreover, from (7.19) we have:

$$\kappa \text{ is surjective.} \tag{8.13}$$

If $\mathbf{F} = (0 \to J \overset{j}{\to} F \overset{\pi}{\to} M \to 0)$ and $h \in \mathrm{Aut}_\Lambda(J)$ then $\langle h_*(\mathbf{F}) \rangle = \langle \mathbf{F} \rangle$ by (7.22). Now suppose that $f \in \mathrm{Ker}(S_J)$. Then $f_*(\mathbf{E})$ is a stably free 0-complex of finite type. Taking $\mathbf{F} = f_*(\mathbf{E})$ we see that $\langle h_* f_*(\mathbf{E}) \rangle = \langle f_*(\mathbf{E}) \rangle$; that is:

$$\kappa([h][f]) = \kappa([f]) \quad \text{for all } h \in \text{Aut}_\Lambda(J) \text{ and all } [f] \in \text{Ker}(S_J). \tag{8.14}$$

Thus there is a surjective map $\kappa_* : \text{Im}(\nu_J) \backslash \text{Ker}(S_J) \to \pi_0^{-1}(J)$ given by

$$\kappa_*(\text{Im}(\nu_J)[f]) = \langle f_*(\mathbf{E}) \rangle.$$

Suppose that $\kappa_*(\text{Im}(\nu_J)[f]) = \kappa_*(\text{Im}(\nu_J)[g])$. Then $f_*(\mathbf{E}) \cong_{\text{Id}_M} g_*(\mathbf{E})$. By (7.22) there exists $h \in \text{Aut}_\Lambda(J)$ such that $h_*(f_*(\mathbf{E})) \equiv g_*(\mathbf{E})$. Thus $[hf] = [g] \in \text{Aut}_{\mathcal{D}\text{er}}(J)$ by (7.20); that is $[h][f] = [g]$. However $[h] \in \text{Im}(\nu_J)$ so that $\text{Im}(\nu_J)[f] = \text{Im}(\nu_J)[g]$. Thus $\kappa_* : \text{Im}(\nu_J) \backslash \text{Ker}(S_J) \to \pi_0^{-1}(J)$ is injective and so also bijective. We have shown:

Theorem 8.15 *Let M be a finitely presented coprojective module; then for each $J \in \Omega_1(M)$ there is a bijection $\kappa_* : \text{Im}(\nu_J) \backslash \text{Ker}(S_J) \longleftrightarrow \pi_0^{-1}(J)$.*

It is now immediate from the definition that:

Proposition 8.16 *Let M be a finitely presented coprojective Λ-module and $J \in \Omega_1(M)$; then $|\pi_0^{-1}(J)| = 1 \iff J$ is full.*

If M is a finitely presented module we denote by $r_1(M)$ the minimal value of $r(J)$ as J runs through $\Omega_1(M)$. It follows from Theorem 7.34 that:

Proposition 8.17 *Let M be a finitely presented coprojective Λ-module; if $\Omega_1(M)$ does not branch at levels $\geq n$ then $\mathbf{Alg}_0(M)$ does not branch at levels $\geq \max\{n, r_1(M), \sigma(\Lambda) + 1\}$.*

Proposition 8.18 *Let M be a module of type $FT(n+1)$ such that $\text{Ext}^{n+1}(M, \Lambda) = 0$; then $\pi_n : \mathbf{Alg}_n(M) \to \Omega_{n+1}(M)$ is surjective.*

Proof Let $\mathbf{E} = (0 \to J' \xrightarrow{j} E_n \xrightarrow{\partial_n} \cdots \xrightarrow{\partial_2} E_1 \xrightarrow{\partial_1} E_0 \xrightarrow{\epsilon} M \to 0)$ be an algebraic n-complex and let $J \in \Omega_{n+1}(M)$. Decompose $\mathbf{E}$ as a Yoneda product $\mathbf{E} = \mathbf{G} \circ \mathbf{F}$ where

$$\mathbf{G} = (0 \to J' \xrightarrow{j} E_n \xrightarrow{\partial_n} K \to 0);$$

$$\mathbf{F} = (0 \to K \xrightarrow{\iota} E_{n-1} \xrightarrow{\partial_{n-1}} \cdots \xrightarrow{\partial_1} E_0 \xrightarrow{\epsilon} M \to 0)$$

with $K = \text{Ker}(\partial_{n-1}) = \text{Im}(\partial_n)$ and where ι is the inclusion of K in E_{n-1}. Since $J, J' \in \Omega_{n+1}(M)$ there exist positive integers a, b and a Λ-isomorphism $\psi : J \oplus \Lambda^a \xrightarrow{\cong} J' \oplus \Lambda^b$. Consider the stabilization

$$\Sigma_+^b(\mathbf{G}) = (0 \to J' \oplus \Lambda^b \xrightarrow{j \oplus \text{Id}} E_n \oplus \Lambda^b \xrightarrow{\partial_n \pi} K \to 0)$$

and modify it to $\mathbf{G}_1 = (0 \to J \oplus \Lambda^a \xrightarrow{\iota} E_n \oplus \Lambda^b \xrightarrow{\partial_n \pi} K \to 0)$ where $\iota = (j \oplus \mathrm{Id}) \circ \psi$. By dimension shifting $\mathrm{Ext}^1(K, \Lambda) \cong \mathrm{Ext}^{n+1}(M, \Lambda) = 0$ so that K is coprojective. By the de-stabilization lemma of Proposition 5.17 there exists a stably free 0-complex

$$\mathbf{G}_2 = (0 \to J \xrightarrow{i} E'_n \xrightarrow{\partial_n \pi} K \to 0)$$

such that $\Sigma^a_+(\mathbf{G}_2) \equiv \mathbf{G}_1$. Putting $\mathbf{E}' = \mathbf{G}_2 \circ \mathbf{F}$ we see it takes the form

$$\mathbf{E}' = (0 \to J \to E'_n \to \cdots \to E_1 \to E_0 \to M \to 0).$$

Now $\mathbf{E}'$ is an algebraic n-complex with $\pi_n(\mathbf{E}') = J$ and π_n is surjective as claimed. $\square$

8.5 Counting the Fibres of π_n

Let $\mathbf{F}_- = (0 \to K \xrightarrow{\iota} E_m \xrightarrow{\partial_{m-1}} \cdots \xrightarrow{\partial_1} E_0 \xrightarrow{\epsilon} M \to 0)$ be a stably free $(m - 1)$-complex and let $\mathbf{F}_+ = (0 \to J \xrightarrow{j} E_{m+k} \xrightarrow{\partial_{m+k}} \cdots \xrightarrow{\partial_{m+1}} E_m \xrightarrow{\epsilon} K \to 0$ be a stably free k-complex so that the Yoneda product

$$\mathbf{F}_+ \circ \mathbf{F}_- = (0 \to J \xrightarrow{j} E_{m+k} \xrightarrow{\partial_{m+k}} \cdots \xrightarrow{\partial_2} E_1 \xrightarrow{\partial_1} E_0 \xrightarrow{\epsilon} M \to 0)$$

is a stably free $m + k$-complex. Here $\partial_m = \iota \circ \epsilon$. There is an additive functor again given by Yoneda product, $\# : \mathbf{Ext}^{k+1}(K, J) \to \mathbf{Ext}^{m+k+1}(M, J); \mathbf{G} \mapsto \mathbf{G} \circ \mathbf{F}_-$. By standard dimension shifting the mapping $\# : \mathrm{Ext}^{k+1}(K, J) \to \mathrm{Ext}^{m+k+1}(M, J)$ defined by $\#[\mathbf{G}] = [\mathbf{G} \circ \mathbf{F}_-]$ is an isomorphism of abelian groups. Let $\mathrm{Stab}_c(A, B)$ denote the subset of $\mathrm{Ext}^{c+1}(A, B)$ represented by exact sequences of the form

$$(0 \to B \xrightarrow{\iota} E_c \xrightarrow{\partial_c} \cdots \xrightarrow{\partial_1} E_0 \xrightarrow{\epsilon} A \to 0)$$

where E_r is stably free:

Proposition 8.19 *The mapping* $\# : \mathrm{Ext}^{k+1}(K, J) \to \mathrm{Ext}^{m+k+1}(M, J)$ *given by* $\#[\mathbf{G}] = [\mathbf{G} \circ \mathbf{F}_-]$ *induces a bijection* $\# : \mathrm{Stab}_k(K, J) \to \mathrm{Stab}_{m+k}(M, J)$.

Let $\mathbf{G}_1, \mathbf{G}_2 \in \mathrm{Stab}_k(K, J)$; if $\mathbf{G}_1 \simeq \mathbf{G}_2$ then clearly $\mathbf{G}_1 \circ \mathbf{F}_- \simeq \mathbf{G}_2 \circ \mathbf{F}_-$. It follows that there is a well defined mapping $\natural : \mathbf{Alg}_k(K, J) \to \mathbf{Alg}_{m+k}(M, J)$ given by $\natural([\mathbf{G}]) = [\mathbf{G} \circ \mathbf{F}_-]$. Moreover, it is then clear that the diagram below commutes.

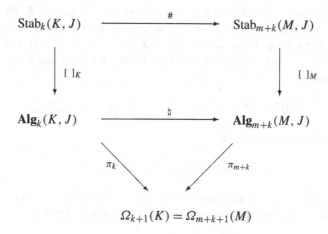

We give preference to the description $\Omega_{m+k+1}(M)$ over $\Omega_{k+1}(K)$. As both [$\,$]$_M$ and # are surjective then from the commutativity of the above diagram:

Proposition 8.20 *The mapping* $\natural : \mathbf{Alg}_k(K, J) \to \mathbf{Alg}_{m+k}(M, J)$ *is surjective for each* $J \in \Omega_{m+k+1}(M)$.

The commutativity of the bottom triangle in the above diagram shows that $\natural$ restricts to a surjective mapping $\natural : \pi_k^{-1}(J) \longrightarrow \pi_{m+k}^{-1}(J)$; in fact a stronger statement is true:

Theorem 8.21 *The mapping* $\natural : \pi_k^{-1}(J) \longrightarrow \pi_{m+k}^{-1}(J)$ *is bijective for each* $J \in \Omega_{m+k+1}(M)$.

Proof We have see that $\natural$ is surjective. However if $\mathbf{G}_1 \circ \mathbf{F}_- \simeq \mathbf{G}_2 \circ \mathbf{F}_-$ then, by Proposition 8.18, $\mathbf{G}_1 \simeq \mathbf{G}_2$ so that $\natural$ is also injective. $\square$

In the special case $k = 0$, by dimension shifting $\mathrm{Ext}^1(K, \Lambda) = \mathrm{Ext}^{m+1}(M, \Lambda) = 0$ so that K is coprojective. In this case we compute the fibre $\pi_0^{-1}(J)$ exactly by Theorem 8.15 to get a bijection $\mathrm{Im}(\nu_J)\backslash \mathrm{Ker}(S_J) \leftrightarrow \pi_0^{-1}(J)$. Composing with the bijection $\natural$ we obtain:

Theorem 8.22 *If* $\mathrm{Ext}^{m+1}(M, \Lambda) = 0$ *then for each* $J \in \Omega_{m+1}(M)$ *there is a bijective mapping* $\eta : \mathrm{Im}(\nu_J)\backslash \mathrm{Ker}(S_J) \longrightarrow \pi_m^{-1}(J)$.

8.6 Realizing Algebraic n-Complexes for $n \geq 3$

In this section we take $\Lambda = \mathbf{Z}[G]$ where G is a finitely presented group. We consider algebraic n-complexes A_* over the trivial module $\mathbf{Z}$; that is, we assume that the algebraic complexes considered have the property that $H_0(A_*) \cong \mathbf{Z}$. We will show that, when $n \geq 3$, all such algebraic n-complexes are geometrically realizable.

Suppose given a finite presentation $\mathcal{G} = \langle X_1, \ldots, X_g \mid W_1, \ldots, W_r \rangle$ of G and for each $s \geq r + 1$ make a choice of a trivial relator W_s'; for $k \geq 1$ denote by $\Sigma^k(\mathcal{G})$ the presentation

$$\Sigma^k(\mathcal{G}) = \langle X_1, \ldots, X_g \mid W_1, \ldots, W_r, W_{r+1}', \ldots, W_{r+k}' \rangle.$$

It is straightforward to see that:

$$\Sigma_+^k(C_*(\mathcal{G})) \simeq C_*(\Sigma^k(\mathcal{G})). \tag{8.23}$$

An algebraic 2-complex A_* is realizable precisely when it is homotopy equivalent to $C_*(\mathcal{G})$ for some finite presentation $\mathcal{G}$; it follows from (8.23) that:

If the algebraic 2-complex A_* is realizable then each $\Sigma^k(A_*)$ is also realizable. $\tag{8.24}$

Proposition 8.25 *Let A_* be an algebraic n-complex where $n \geq 3$; if $A_*^{(n-1)}$ is realizable then so also is A_*.*

Proof Without loss of generality we may suppose that A_n is free over Λ. Suppose that $A_*^{(n-1)}$ is realizable by the finite connected $(n-1)$-complex Y with $\pi_1(Y) = G$; then $A_*^{n-1} \simeq C_*(Y)$ where

$$C_*(Y) = (0 \to \pi_{n-1}(\widetilde{Y}) \to C_{n-1}(\widetilde{Y}) \to \cdots \to C_1(\widetilde{Y}) \to C_0(\widetilde{Y}) \to \mathbf{Z} \to 0)$$

is the cellular chain complex of $\widetilde{Y}$. Consider the factorization of the boundary map δ_n through its image K thus

By exactness of A_* and the Hurewicz Theorem $K \cong \pi_{n-1}(\widetilde{Y}) \cong \pi_{n-1}(Y)$. Let $\epsilon_1, \ldots, \epsilon_d$ be a $\mathbf{Z}[G]$-basis for A_n. Identifying K with $\pi_{n-1}(Y)$ then each $\eta(\epsilon_j)$ is represented by a mapping $\alpha_r : S^{n-1} \to Y$. Let X be the finite n-complex obtained by attaching n-cells $e_1^n, \ldots, e_d^n$ to Y via $\alpha_1, \ldots, \alpha_d$:

$$X = Y \bigcup_{\{\alpha_j\}_j} \coprod_{j=1}^{d} e_j^n.$$

Then $A_* \simeq C_*(X)$ so that A_* is realizable as claimed. $\qquad \square$

Theorem 8.26 *Every algebraic 3-complex is realizable.*

Proof Let $A_* = (0 \to J \to A_3 \to A_2 \to A_1 \to A_0 \to \mathbf{Z} \to 0)$ be an algebraic 3-complex with 2-skeleton

$$A_*^{(2)} = (0 \to K' \to A_2 \to A_1 \to A_0 \to \mathbf{Z} \to 0).$$

Choose a finite presentation $\mathcal{G}$ for G. By Theorem 8.9 $\Sigma^a(A_*^{(2)}) \simeq \Sigma^b C_*(\mathcal{G})$ for some $a, b \geq 1$. Hence by (53,1), $\Sigma^a(A_*^{(2)}) \simeq C_*(\Sigma^b \mathcal{G})$. As $C_*(\Sigma^b \mathcal{G})$ is tautologically realizable then $\Sigma^a(A_*^{(2)})$ is also realizable. Now consider $A_*' = (0 \to J \to A_3 \oplus \Lambda^a \to A_2 \oplus \Lambda^a \to A_1 \to A_0 \to \mathbf{Z} \to 0)$. Then $A_*'^{(2)} = \Sigma^a(A_*^{(2)})$. As $\Sigma^a(A_*^{(2)})$ is realizable then A_*' is realizable by Proposition 8.25. However, $A_* \simeq A_*'$ so that A_* is realizable. $\square$

From Proposition 8.25 and Theorem 8.26 we obtain immediately:

Corollary 8.27 *Every algebraic n-complex is realizable for $n \geq 3$.*

Part II
Practice

Chapter 9
Rings with Stably Free Cancellation

The ring Λ has the *stably free cancellation property* (abbreviated to SFC) when for any Λ-module S and any positive integers m, k

$$S \oplus \Lambda^m \cong \Lambda^{m+k} \implies S \cong \Lambda^k. \tag{SFC}$$

As noted in Chap. 1, the theorem of Gabel ensures that every stably free Λ-module is free precisely when Λ has the SFC property; thus we may concentrate our discussion on finitely generated modules.

There is a stronger property than stably free cancellation; the ring Λ is *projective free* when every finitely generated projective Λ-module is free. As a fundamental notion projective freeness is inconveniently restrictive; even so, it is a property too useful to be ignored and we make use of it at a number of places. For rings of Laurent polynomials the two notions are connected via the theorem of Grothendieck [4] that $\widetilde{K}_0(A[t, t^{-1}]) \cong \widetilde{K}_0(A)$ if A is a coherent ring of finite global dimension. Under this hypothesis, if $A[t, t^{-1}]$ has property SFC then

$$A[t, t^{-1}] \text{ is projective free} \iff \widetilde{K}_0(A) = 0.$$

9.1 Group Algebras and the Retraction Principle

Given a ring R and a group G we are interested in the relationship between R and $R[G]$ in the context of the SFC property. In one direction the relationship is straightforward; we say that ring A is a *retract of* B when there are ring homomorphisms $i : A \to B$ and $r : B \to A$ such that $r \circ i = \mathrm{Id}_A$. If M is a module over A it then follows that $r_* i_*(M) \cong M$. Suppose that A is a retract of B; if S is an A-module such that $S \oplus A^m \cong A^{k+m}$ then $i_*(S) \oplus B^m \cong B^{k+m}$. If B has SFC then $i_*(S) \cong B^k$ and so $S \cong r_* i_*(S) \cong A^k$; that is:

Let A be a retract of B; then B has property SFC $\implies A$ has property SFC. (9.1)

As an example, R is a retract of the group ring $R[G]$ via $i : R \to R[G]; i(a) = a.\hat{1}$ and $\epsilon : R[G] \to R; \epsilon(\sum_g a_g \hat{g}) = \sum_g a_g$ so that:

If $R[G]$ has property SFC then so also does R. (9.2)

F.E.A. Johnson, *Syzygies and Homotopy Theory*, Algebra and Applications 17, DOI 10.1007/978-1-4471-2294-4_9, © Springer-Verlag London Limited 2012

Thus we may focus our interest on:

Question Let R be a ring satisfying SFC; for which groups G is it true that $R[G]$ also satisfies SFC?

In this chapter we give some, more or less standard, examples of rings with the SFC property. In Chaps. 10, 11 and 12 we consider the above Question in detail with reference both to examples where the SFC property persists for $R[G]$ and also to cases where it fails. As a convenient class of rings with the SFC property we begin with:

9.2 Dedekind Domains

A Dedekind domain R is a commutative Noetherian integral domain which is integrally closed in its field of fractions $k(R) = k$. When R is Dedekind a theorem of Steinitz [11, 87, 88] allows us to classify finitely generated torsion free R modules.

The set $\mathcal{J}(R)$ of nonzero ideals in R is a commutative monoid under multiplication. We write $\mathcal{P}\mathrm{ic}(R)$ for the set of isomorphism classes in $\mathcal{J}(R)$. It is known that $\mathcal{P}\mathrm{ic}(R)$ is finite and acquires a group structure from the monoid structure on $\mathcal{J}(R)$. An R-module M is said to be *torsion free* when the mapping $M \to M \otimes_R k; m \mapsto m \otimes 1$ is injective; if M is a nonzero finitely generated torsion free R-module then:

(i) $M \cong J_1 \oplus \cdots \oplus J_m$ where $J_r \in \mathcal{J}(R)$.

Moreover if $M' \cong J_1' \oplus \cdots \oplus J_\mu'$ is another such direct sum then

(ii) $M' \cong M \iff \mu = m$ and $[J_1' \cdots J_\mu'] = [J_1 \cdots J_m] \in \mathcal{P}\mathrm{ic}(R)$.

It follows that:

$$\widetilde{K}_0(R) \cong \mathcal{P}\mathrm{ic}(R). \tag{9.3}$$

In general, of course, $\widetilde{K}_0(R) \neq 0$; that is, Dedekind domains are not usually projective free. However, it follows easily from the Steinitz classification that:

$$\text{Any Dedekind domain } R \text{ has the SFC property.} \tag{9.4}$$

There are two notable cases where $R[G]$ continues to have stably free cancellation:

Free Group Algebras   When A is a ring we denote by $A\langle X \rangle$ the free algebra over A on the set X. If F_X is the free group on X we may describe the group algebra $A[F_X]$ as the localization $A\langle X, X^{-1} \rangle$ of $A\langle X \rangle$ obtained by formally inverting each $x \in X$. By generalising an earlier argument of Sheshadri [85], Bass showed ([2], p. 213) that:

$$\text{If } R \text{ is a Dedekind domain then } R\langle X, X^{-1} \rangle \text{ has the SFC property.} \tag{9.5}$$

In the original context of Sheshadri's argument, R was a commutative principal ideal domain and $|X| = 1$. Bass' generalization shows rather more than we have indicated, namely that if P is a finitely generated projective module over $\Lambda = R\langle X, X^{-1}\rangle$ then $P \cong i_*(P_0) \oplus \Lambda^m$ where P_0 is an ideal (necessarily finitely generated projective) of R and $i : R \to R\langle X, X^{-1}\rangle$ is the inclusion. Taken to its logical conclusion, Sheshadri's original proof actually shows something stronger (cf. [19]):

$R\langle X, X^{-1}\rangle$ is projective free R is a commutative principal ideal domain.   (9.6)

Although it does not concern us directly we note that the same arguments establish the corresponding results for free algebras; namely:

If R is a Dedekind domain then $R\langle X\rangle$ has the SFC property.   (9.7)

If R is a commutative principal ideal domain then $R\langle X\rangle$ is projective free. (9.8)

Free Abelian Group Algebras   When A is a ring we denote by $A[T]$ the polynomial algebra over A on the finite set T. If F_T^{ab} is the free abelian group on T we may describe the group algebra $A[F_T^{ab}]$ as the localization $A[T, T^{-1}]$ of $A[T]$ obtained by formally inverting each $t \in T$. In the aftermath of the Quillen-Suslin proof of the Serre conjecture [81, 90], it was shown ([67], p.189) that:

If R is a Dedekind domain then $R[T, T^{-1}]$ has the SFC property.   (9.9)

The same considerations apply to polynomial algebras to give:

If R is a Dedekind domain then $R[T]$ has the SFC property.   (9.10)

9.3 Free Group Algebras over Division Rings

Let Γ be a commutative integral domain with field of fractions k; if Γ is a Dedekind domain then Γ is maximal in k in the sense that if Γ' is a subring of k such that $\Gamma \subset \Gamma' \subset k$ then either $\Gamma' = \Gamma$ or $\Gamma' = k$. Here of course we allow, as a special case, the possibility that $\Gamma = k$. It is natural to ask to what extent the results of Sect. 9.2 generalize if Γ is replaced by a proper subring of a noncommutative division ring. As we shall see in Chap. 12, however, the attempted generalization fails badly, even in the comparatively simple case where Γ is a maximal proper subring of the rational quaternion algebra.

In Sect. 9.2 we observed that when R is a Dedekind domain both the free group algebra $R\langle X, X^{-1}\rangle$ and the free abelian group algebra $R[T, T^{-1}]$ have the SFC property. These two cases intersect only when $|X| = |T| = 1$; that is, in the case of the group ring $R[C_\infty]$ of the infinite cyclic group C_∞.

For a ring A we may represent the group algebra $A[C_\infty]$ as the ring $A[x, x^{-1}]$ of Laurent polynomials in the variable x with coefficients in A; that is:

$$A[x, x^{-1}] = \left\{ \alpha(x) = \sum_{r=n}^{N} a_r x^r : n, N \in \mathbf{Z}, \ a_r \in A \right\}.$$

If $\alpha(x) = \sum_{r=n}^{N} a_r x^r \in A[x, x^{-1}]$ we define the *length* $\lambda(\alpha)$ to be $N - n$ provided $a_n \neq 0$ and $a_N \neq 0$. If $A = D$ is a (possibly noncommutative) division ring then λ defines a Euclidean valuation on $D[x, x^{-1}]$ allowing us to apply the division algorithm. We may thus conclude that $D[x, x^{-1}]$ is a (left and right) principal ideal domain. It follows that:

$$D[x, x^{-1}] \text{ is projective free for any division ring } D. \tag{9.11}$$

The discussion of Sect. 9.2 now bifurcates sharply. On the one hand, if $X = \{x_1, \ldots, x_n\}$ the free group algebra $D\langle X, X^{-1} \rangle$ is isomorphic to the free product of D-algebras

$$D\langle X, X^{-1} \rangle \cong D[x_1, x_1^{-1}] * \cdots * D[x_n, x_n^{-1}],$$

where the coefficients D are identified in the various copies. Then a result of Dicks and Sontag [20], generalizing earlier results of Cohn, shows that:

$$D\langle X, X^{-1} \rangle \text{ is projective free for any division ring } D. \tag{9.12}$$

Hence

$$D\langle X, X^{-1} \rangle \text{ satisfies SFC for any division ring } D. \tag{9.13}$$

The situation for *free abelian* group algebras is entirely different; one may generalize arguments of Dicks-Sontag [20], Ojurangen-Sridharan [76] and Parimala-Sridharan [79] to show that for any noncommutative division ring D

$$D[T, T^{-1}] \text{ possesses nontrivial stably free modules whenever } |T| \geq 2. \tag{9.14}$$

9.4 Local Rings and the Nakayama-Bourbaki Lemma

Denote by $\text{rad}(\Lambda)$ the *Jacobson radical* of a ring Λ (cf. [66], Chap. 2). Recall that an ideal $\mathbf{m}$ of Λ is *radical* when $\mathbf{m} \subset \text{rad}(\Lambda)$ and that the ring Λ is said to be *local* when $\Lambda/\text{rad}(\Lambda)$ is a division algebra. For the avoidance of doubt, in what follows, there is no assumption that a local ring need be commutative unless specified.

Proposition 9.15 *Let* $\mathbf{m}$ *be a two-sided radical ideal in a ring* Λ *and let* M *be a finitely presented flat* Λ*-module; then*

$$M \otimes_\Lambda (\Lambda/\mathbf{m}) \text{ is free over } \Lambda/\mathbf{m} \quad \Longrightarrow \quad M \text{ is free over } \Lambda.$$

This criterion clearly derives from Nakayama's Lemma. However, the form in which we use it is a special case of a statement from Bourbaki ([11] p. 83, Proposition 5). Any finitely generated projective Λ-module M satisfies the hypotheses of Proposition 9.15. Moreover, if M is stably free then $M \otimes_\Lambda (\Lambda/\mathbf{m})$ is stably free over $\Lambda/\mathbf{m}$; we see that:

Corollary 9.16 *Let* $\mathbf{m}$ *be a two-sided radical ideal in a ring* Λ; *then*

$$\Lambda/\mathbf{m} \text{ has SFC} \implies \Lambda \text{ has SFC}.$$

To apply this in the first instance, take Λ to be local and $\mathbf{m} = \mathrm{rad}(\Lambda)$. Then the division ring $\Lambda/\mathbf{m}$ has the SFC property so that we obtain:

Corollary 9.17 *If* Λ *is a local ring then* Λ *has* SFC.

Next suppose that A is a local ring so that $D = A/\mathrm{rad}(A)$ is a division ring. The canonical mapping $\varphi : A \to D$ induces a surjective ring homomorphism $\varphi_* : A\langle X, X^{-1}\rangle \to D\langle X, X^{-1}\rangle$ in which $\mathrm{Ker}(\varphi_*) = \mathrm{rad}(A)\langle X, X^{-1}\rangle$. In general $\mathrm{rad}(A)\langle X, X^{-1}\rangle$ is not a radical ideal in $A\langle X, X^{-1}\rangle$. In two special cases, however, it is. Firstly when $\mathrm{rad}(A)$ is nilpotent; we may then apply (9.13) and Corollary 9.16 to show that;

If A is a local ring and $\mathrm{rad}(A)$ is nilpotent then $A\langle X, X^{-1}\rangle$ has SFC. (9.18)

Secondly $\mathrm{rad}(A)\langle X, X^{-1}\rangle$ is again a radical ideal when A is complete. The same formal argument applies to show:

If A is a complete local ring then $A\langle X, X^{-1}\rangle$ has SFC. (9.19)

9.5 Matrix Rings

We briefly consider stably free cancellation over rings of matrices; fix $n \geq 1$ and let $C(n)$ and $R(n)$ denote respectively the set of $n \times 1$ and $1 \times n$ matrices over Λ;

$$C(n) = \left\{ \begin{pmatrix} \lambda_1 \\ \lambda_2 \\ \vdots \\ \lambda_n \end{pmatrix} ; \lambda_i \in \Lambda \right\}; \qquad R(n) = \{(\lambda_1, \lambda_2, \ldots, \lambda_n); \lambda_i \in \Lambda\}.$$

Then $C(n)$ is a $M_n(\Lambda)$-Λ bimodule whilst $R(n)$ is a Λ-$M_n(\Lambda)$ bimodule and we have additive functors

$$\Phi : \mathcal{Mod}_{M_n(\Lambda)} \to \mathcal{Mod}_\Lambda; \qquad \Phi(M) = M \otimes_{M_n(\Lambda)} C(n),$$

$$\Psi : \mathcal{Mod}_{M_n(\Lambda)} \to \mathcal{Mod}_\Lambda; \qquad \Psi(N) = N \otimes_\Lambda R(n).$$

Morita's Theorem is the easily verified statement that, up to equivalence, Φ and Ψ are mutually inverse. If S is a stably free module of rank k over $M_n(\Lambda)$ then $\Phi(S)$ is stably free of rank kn over Λ. Now if Λ has property SFC then $\Phi(S) \cong \Lambda^{kn}$. Likewise $\Phi(M_n(\Lambda)^k) \cong \Lambda^{kn}$. However Φ is injective on isomorphism classes so that $S \cong M_n(\Lambda)^k$ and we have shown:

$$\Lambda \text{ has property SFC} \implies M_n(\Lambda) \text{ has property SFC}. \tag{9.20}$$

9.6   Iterated Fibre Products

When A is a direct product of rings $A = A_1 \times A_2$ the projection maps $\pi_r : A \to A_r$ induce additive functors $(\pi_r)_* : \mathcal{M}_A \longrightarrow \mathcal{M}_{A_r}$ and yield an additive functor

$$(\pi_1, \pi_2) : \mathcal{M}_A \longrightarrow \mathcal{M}_{A_1} \times \mathcal{M}_{A_2}; \quad (\pi_1, \pi_2)(M) \mapsto ((\pi_1)_*(M), (\pi_2)_*(M)).$$

For $r = 1, 2$ let $i_r : A_1 \to A$ denote the canonical injection; that is:

$$i_1(a_1) = (a_1, 0); \qquad i_2(a_2) = (0, a_2).$$

Then i_1, i_2 are homomorphisms of 'rings without identity'; that is, they are additive and multiplicative but do not preserve the multiplicative identity. Even so they induce additive functors $(i_r)_* : \mathcal{M}_{A_r} \to \mathcal{M}_A$ so that we also get an additive functor

$$(i_1, i_2) : \mathcal{M}_{A_1} \times \mathcal{M}_{A_2} \longrightarrow \mathcal{M}_A; \quad (i_1, i_2)(M_1, M_2) = (i_1)_*(M_1) \oplus (i_2)_*(M_2)$$

which is an equivalence of categories with inverse (π_1, π_2). Moreover, the A-module S is stably free over A if and only if $S_i = (\pi_i)_*(S)$ is stably free over A_i. It follows easily that

$$A_1 \times A_2 \text{ has property SFC} \iff A_i \text{ has property SFC for } i = 1, 2. \quad (9.21)$$

Now suppose that $\mathbf{A}, \mathbf{B}$ are classes of rings, closed under isomorphism; for each integer $n \geq 1$ we define a class of rings $\mathcal{J}_n(\mathbf{B}, \mathbf{A})$ iteratively thus;

(i) $R \in \mathcal{J}_1(\mathbf{B}, \mathbf{A})$ if and only if $R \in \mathbf{B}$;
(ii) $R \in \mathcal{J}_n(\mathbf{B}, \mathbf{A})$ if and only if there exist $R_1 \in \mathcal{J}_k(\mathbf{B}, \mathbf{A})$, $R_2 \in \mathcal{J}_l(\mathbf{B}, \mathbf{A})$ with $k + l = n$ such that *either*
   (a) R is isomorphic to a fibre product

$$
\begin{array}{ccc}
R & \overset{\eta_2}{\to} & R_2 \\
\downarrow{\eta_1} & & \downarrow{\varphi_2} \\
R_1 & \overset{\varphi_1}{\to} & A
\end{array}
$$

   which satisfies the Milnor condition and where $A \in \mathbf{A}$ *or*
   (b) $R \cong R_1 \times R_2$.
   (Although we choose not to express it so, the condition (b) could be regarded as a degenerate case of (a) by allowing fibre products over the 'zero ring'.)

We define the class $\mathcal{J}(\mathbf{B}, \mathbf{A})$ of *iterated fibre products* with *building blocks* $\mathbf{B}$ and *amalgamations* $\mathbf{A}$ by

$$\mathcal{J}(\mathbf{B}, \mathbf{A}) = \bigcup_{n \geq 1} \mathcal{J}_n(\mathbf{B}, \mathbf{A}).$$

We note for the record that:

If $\mathbf{B}$ is a class of commutative rings then each $R \in \mathcal{J}(\mathbf{B}, \mathbf{A})$ is commutative. (9.22)

Morever, this construction is compatible with the formation of group rings. Thus if $\mathbf{A}$ is a class of rings and G is a group we denote by $\mathbf{A}[G]$ the class of rings of the form $A[G]$ where $A \in \mathbf{A}$. Then it is straightforward to verify that for any classes of rings $\mathbf{A}$, $\mathbf{B}$ and any group G

$$\mathcal{J}(\mathbf{B}, \mathbf{A})[G] = \mathcal{J}(\mathbf{B}[G], \mathbf{A}[G]). \tag{9.23}$$

By (9.21), the SFC property is closed under direct product; thus it is straightforward to iterate the conclusion of Corollary 3.54:

Theorem 9.24 *Let* $\mathbf{A}$ *be a class of weakly Euclidean commutative rings and let* $\mathbf{B}$ *be a class of commutative rings possessing the* SFC *property; then each* $R \in \mathcal{J}(\mathbf{B}, \mathbf{A})$ *has the* SFC *property.*

Let $\mathbf{L}$ denote the class of rings L of the form $L \cong L_1 \times \cdots \times L_m$ where each L_i is a commutative local ring with nilpotent radical. Then by Corollary 2.46, Proposition 2.47, each $L \in \mathbf{L}$ is weakly Euclidean. Next take $\mathbf{D}$ to be the class of Dedekind domains. Then each $D \in \mathbf{D}$ has SFC by (9.4). It now follows from Theorem 9.24 that:

$$\text{If } R \in \mathcal{J}(\mathbf{D}, \mathbf{L}) \text{ then } R \text{ has the SFC property.} \tag{9.25}$$

We can improve on this example. With the same notation, and taking $G = C_\infty$ for any $L \in \mathbf{L}$, the group ring $L[C_\infty]$ is weakly Euclidean by Proposition 2.47 and Corollary 2.52. Moreover, for $D \in \mathbf{D}$, $D[C_\infty]$ has property SFC by (9.5). Thus each $R' \in \mathcal{J}(\mathbf{D}[C_\infty], \mathbf{L}[C_\infty])$ has the SFC property. However, if $R \in \mathcal{J}(\mathbf{D}, \mathbf{L})$ then, as noted in (9.23), $R[C_\infty] \in \mathcal{J}(\mathbf{D}[C_\infty], \mathbf{L}[C_\infty])$. It follows that:

$$\text{For each } R \in \mathcal{J}(\mathbf{D}, \mathbf{L}), \ R[C_\infty] \text{ has the SFC property.} \tag{9.26}$$

Chapter 10
Group Rings of Cyclic Groups

In this chapter we begin the detailed study of the SFC property for group rings of the form $\mathbf{Z}[F_n \times \Phi]$ where F_n is the free group of rank $n \geq 1$ and Φ is finite. In the first instance we consider the rings $\mathbf{Z}[F_n \times C_m]$ where C_m is the cyclic group of order m.

As $\mathbf{Z}[\Phi]$ is a retract of $\mathbf{Z}[F_n \times \Phi]$ it follows that a prior condition for $\mathbf{Z}[F_n \times \Phi]$ to have stably free cancellation is that $\mathbf{Z}[\Phi]$ should also have this property. The question of stably free cancellation for $\mathbf{Z}[\Phi]$ and related rings has been studied extensively by Swan [92–94] and Jacobinski [46], building upon earlier work of Eichler [27].

10.1 Stably Free Modules over $\mathbf{Z}[\Phi]$

Take Φ to be a finite group; one may begin with a general finiteness statement which is a consequence of the Jordan-Zassenhaus Theorem [18, 102]:

$$\mathcal{SF}_n(\mathbf{Z}[\Phi]) \text{ is finite for each } n \geq 1. \tag{10.1}$$

A preliminary consequence of the Swan-Jacobinski generalization of Eichler's work is that a stably free module of rank ≥ 2 is necessarily free; that is:

$$|\mathcal{SF}_n(\mathbf{Z}[\Phi])| = 1 \quad \text{when } n \geq 2. \tag{10.2}$$

In consequence of which:

$$\mathbf{Z}[\Phi] \text{ has the stably free cancellation property} \quad \Longleftrightarrow \quad |\mathcal{SF}_1(\mathbf{Z}[\Phi])| = 1. \tag{10.3}$$

Swan also gives a sufficient condition for $\mathbf{Z}[\Phi]$ to have stably free cancellation. To explain this, we consider the real group ring $\mathbf{R}[\Phi]$; by Wedderburn's Theorem we have

$$\mathbf{R}[\Phi] \cong \prod_{i=1}^{m} M_{d_i}(\mathcal{D}_i),$$

F.E.A. Johnson, *Syzygies and Homotopy Theory*, Algebra and Applications 17, DOI 10.1007/978-1-4471-2294-4_10, © Springer-Verlag London Limited 2012

where $\mathcal{D}_i$ is $\mathbf{R}$, $\mathbf{C}$ or $\mathbf{H}$, the last being the division ring of Hamiltonian quaternions

$$\mathbf{H} = \left(\frac{-1, -1}{\mathbf{R}} \right).$$

We say that Φ satisfies the *Eichler condition* when $\mathbf{H}$ is *not* a factor of $\mathbf{R}[\Phi]$; that is, when the case $\mathcal{D}_i = \mathbf{H}$ and $d_i = 1$ does not occur. Swan's first theorem gives a sufficient condition for $\mathbf{Z}[\Phi]$ to have stably free cancellation [92, 94].

If Φ satisfies the Eichler condition then $|\mathcal{SF}_1(\mathbf{Z}[\Phi])| = 1$. $\qquad$ (10.4)

The Eichler condition rules out the existence of nontrivial stably free modules in very many cases and has the advantage that it is relatively easy to check. It is, however, not a necessary condition and further analysis of the problem is rather more intricate. Evidently it is the presence of quaternionic factors that causes problems and as test cases we may consider the generalized quaternion groups $Q(4n)$ $(n \geq 2)$:

$$Q(4n) = \langle x, y | x^n = y^2, xyx = y \rangle.$$

It is straightforward to check that $|Q(4n)| = 4n$; moreover

$$\mathbf{R}[Q(4n)] \cong \begin{cases} \mathbf{R}^{(4)} \times M_2(\mathbf{R})^{(n-2)/2} \times \mathbf{H}^{(n/2)} & n \text{ even,} \\ \mathbf{R}^{(2)} \times M_2(\mathbf{R})^{(n-1)/2} \times \mathbf{C} \times \mathbf{H}^{(n-1)/2} & n \text{ odd.} \end{cases}$$

Evidently each $Q(4n)$ fails the Eichler condition. In his remarkable paper [94] Swan shows:

$\mathbf{Z}[Q(4n)]$ has the SFC property if and only if $n \leq 5$. $\qquad$ (10.5)

In particular:

$\mathbf{Z}[Q(4n)]$ has at least one nontrivial stably free module whenever $n \geq 6$. $\quad$ (10.6)

In [94] Swan also gives explicit calculations of $|\mathcal{SF}_1(Q(4n))|$ for $n \leq 10$ and in addition for the cases $n = 12, 15, 21$. Moreover, employing results of Vigneras [97], he also estimates the asymptotic growth of $|\mathcal{SF}_1(Q(2^k))|$.

In studying stably free cancellation over $\mathbf{Z}[F_n \times \Phi]$ we are, in effect, attempting to generalize the results of Swan from $\mathbf{Z}[\Phi]$ to $A[\Phi]$ where $A = \mathbf{Z}[F_n]$. The extent to which this is possible is still problematic. In this chapter and the next we show that $A[\Phi]$ has SFC in the cases $\Phi = C_p, D_{2p}$ where p is an odd prime. However, as we shall also see, as Φ becomes only slightly more complicated the SFC property fails. When k is a field similar results hold for the group rings $k[C_\infty^n \times \Phi]$ although [59] with different justifications. However, the group rings $k[F_n \times \Phi]$ always have SFC.

10.2 Stably Free Cancellation for $\mathbf{Z}[F_n \times C_p]$

In this section we show that, when p is prime, the group ring $\mathbf{Z}[F_n \times C_p]$ has the SFC property. As $\mathbf{Z}[C_p]$ is a retract of $\mathbf{Z}[F_n \times C_p]$ this implies the well known statement that $\mathbf{Z}[C_p]$ has the SFC property and it is instructive first to demonstrate this point. Throughout this chapter, $c_d(x)$ will denote the cyclotomic polynomial corresponding to a primitive dth root of unity. When p is prime clearly $c_p(x) = x^{p-1} + x^{p-2} + \cdots + x + 1$. We denote by $R(p) = \mathbf{Z}[x]/c_p(x)$ the ring of integers in the cyclotomic field

$$\mathbf{Q}(\zeta_p) = \mathbf{Q}[x]/(x^{p-1} + x^{p-2} + \cdots + x + 1).$$

When p is understood we simply write $R(p) = R$. To begin we note ([38], pp. 525–526) that

$$R \text{ is a } \mathbf{Z}\text{-lattice of rank } p - 1 \text{ in } \mathbf{Q}(\zeta_p). \tag{10.7}$$

$$(\zeta - 1)R \text{ has index } p \text{ in } R. \tag{10.8}$$

Moreover, if $1 \le k \le p - 1$ then the correspondence $\zeta \mapsto \zeta^k$ induces an automorphism $R \mapsto R$. In particular $(\zeta^k - 1)R = (\zeta - 1)R$ and hence $(\zeta^k - 1)/(\zeta - 1) \in R^*$. Thus

$$1 + \zeta + \cdots + \zeta^{k-1} \in R^* \quad \text{whenever } 2 \le k \le p - 1. \tag{10.9}$$

As ζ has order p and $|\mathbf{F}_p^*| = p - 1$ then under the canonical homomorphism $R \to \mathbf{F}_p$ we see that $\zeta \mapsto 1$. In particular, given $k \in \mathbf{F}_p^*$ the unit $1 + \zeta + \cdots + \zeta^{k-1}$ maps to k under $R \to \mathbf{F}_p$; that is:

The canonical homomorphism $R \to \mathbf{F}_p$ has the lifting property for units. (10.10)

The factorization $x^p - 1 = (x - 1)c_p(x)$ gives rise to a fibre square

$$\left\{ \begin{array}{ccc} \mathbf{Z}[x]/(x^p - 1) & \xrightarrow{\pi_-} & \mathbf{Z}[x]/c_p(x) \\ \downarrow \pi_+ & & \downarrow \nu \\ \mathbf{Z}[x]/(x - 1) & \xrightarrow{\nu} & \mathbf{F}_p \end{array} \right.$$

where $\mathbf{F}_p$ is the field with p elements and ν is reduction mod p. After making the identifications $\mathbf{Z}[C_p] = \mathbf{Z}[x]/(x^p - 1)$ and $\mathbf{Z} = \mathbf{Z}[x]/(x - 1)$ this becomes

$$\mathcal{Z}(p) = \left\{ \begin{array}{ccc} \mathbf{Z}[C_p] & \xrightarrow{\pi_-} & R \\ \downarrow \pi_+ & & \downarrow \nu \\ \mathbf{Z} & \xrightarrow{\nu} & \mathbf{F}_p \end{array} \right.$$

As the ring homomorphism $R \to \mathbf{F}_p$ is surjective then $\mathcal{Z}(p)$ has the Milnor patching property. Moreover, both $\mathbf{Z}$ and R are Dedekind domains and so have the SFC property by (9.4). Thus $\mathcal{Z}(p)$ is of locally free type; furthermore $\mathbf{F}_p$ is weakly Euclidean. Finally, it follows from (10.10) that $\mathcal{Z}(p)$ is pointlike in dimension one. As the hypotheses of Corollary 3.48 are satisfied we conclude:

$$\text{If } p \text{ is prime then } \mathbf{Z}[C_p] \text{ has property SFC.} \tag{10.11}$$

Of course, (10.11) is special case of Swan's Theorem (10.4) as $\mathbf{R}[C_p] \cong \mathbf{R} \times \mathbf{C}^{(p-1)/2}$ has no quaternionic factors. Nevertheless, whereas the Swan-Jacobinski Theorem requires that the coefficients of the group ring should at least lie in a Dedekind domain, we may use the above proof as a model in more general contexts. Let F_n denote the free group of rank n. By applying the functor $- \otimes_Z \mathbf{Z}[F_n]$ to $\mathcal{Z}(p)$ we obtain another Milnor square

$$\mathcal{Z}(p)[F_n] = \begin{cases} \mathbf{Z}[F_n \times C_p] & \xrightarrow{\pi_-} & R[F_n] \\ \downarrow \pi_+ & & \downarrow \nu \\ \mathbf{Z}[F_n] & \xrightarrow{\nu} & \mathbf{F}_p[F_n] \end{cases}$$

As F_n is a $\mathcal{TUP}$ group then for any integral domain A, the group ring $A[F_n]$ has only trivial units. It follows easily from (10.10) that:

$$\text{The induced map on units } \nu : R[F_n]^* \to \mathbf{F}_p[F_n]^* \text{ is surjective.} \qquad (10.12)$$

Hence:

$$\mathcal{Z}(p)[F_n] \text{ is pointlike in dimension one.} \qquad (10.13)$$

If $X = \{x_1, \ldots, x_n\}$ is a generating set for F_n then the group ring $A[F_n]$ may be described alternatively as $A\langle X, X^{-1} \rangle = A\langle x_1, x_1^{-1}, \ldots, x_n, x_n^{-1} \rangle$, the ring of Laurent polynomials in noncommuting variables $x_1, \ldots, x_n$. Moreover, as $\mathbf{Z}$ and R are Dedekind domains then both $\mathbf{Z}\langle X, X^{-1} \rangle$ and $R\langle X, X^{-1} \rangle$ have the SFC property by (9.5). Finally $\mathbf{F}_p\langle X, X^{-1} \rangle$ is weakly Euclidean by Theorem 2.49. Thus $\mathcal{Z}(p)[F_n]$ satisfies all the hypotheses of Corollary 3.48 and so:

$$\mathbf{Z}[F_n \times C_p] \text{ has the SFC property for each prime } p. \qquad (10.14)$$

10.3 Stably Free Cancellation for $\mathbf{Z}[C_\infty \times C_m]$

Let d be a positive integer and let $\zeta_d \in \overline{\mathbf{Q}}$ be a primitive dth root of unity. The dth cyclotomic polynomial c_d is then

$$c_d(x) = \prod_{(r,d)=1} (x - \zeta_d^r).$$

Although $c_d(x)$ is ostensibly a polynomial over $\overline{\mathbf{Q}}$ it is actually defined over $\mathbf{Z}$ and is irreducible over $\mathbf{Q}$. When A is a nonempty finite set of positive integers we define

$$c_A(x) = \prod_{a \in A} c_a(x).$$

The next proposition is presumably well known but difficult to locate within the literature:

Proposition 10.15 *Let A be a finite nonempty set of positive integers and let d be a positive integer such that $d \notin A$; then for some nonzero integer N*

$$\mathbf{Z}[x]/((c_d(x)) + (c_A(x))) \cong (\mathbf{Z}/N)((c_d(x)) + (c_A(x))).$$

Proof First consider $c_d(x)$, $c_A(x)$ as polynomials over $\overline{\mathbf{Q}}$. It is clear from definitions that $c_d(x)$, $c_A(x)$ have no common factor over $\overline{\mathbf{Q}}$. A fortiori they have no common factor over $\mathbf{Q}$. Hence there are rational polynomials $\alpha(x)$, $\beta(x)$ such that

$$\alpha(x)c_d(x) + \beta(x)c_A(x) = 1.$$

After clearing fractions there exist integral polynomials $a(x)$, $b(x)$ and a positive integer μ such that $a(x)c_d(x) + b(x)c_A(x) = \mu$. On taking

$$N = \min\{\mu \in \mathbf{Z}_+ : a(x)c_d(x) + b(x)c_A(x) = \mu; a(x), b(x) \in \mathbf{Z}[x]\}$$

we see that $\mathbf{Z}[x]/((c_d(x)) + (c_A(x))) \cong (\mathbf{Z}/N)[x]/((c_d(x)) + (c_A(x)))$. □

By a *cyclotomic ring* we mean one of the form $\mathbf{Z}[x]/(c_A(x))$ where A is a nonempty finite set of positive integers. There are two examples of importance for us. Firstly, if m is a positive integer we may represent the integral group ring $\mathbf{Z}[C_m]$ in the form $\mathbf{Z}[x]/(x^m - 1)$. Thus $x^m - 1 = c_A(x)$ and so $\mathbf{Z}[C_m] \cong \mathbf{Z}[x]/c_A(x)$ where A is the set of positive divisors of m; that is:

For each positive integer $m \geq 2$ the integral group ring $\mathbf{Z}[C_m]$ is a cyclotomic ring.

(10.16)

Let $I(G)$ denote the integral augmentation ideal of the finite group G; then the $\mathbf{Z}[G]$ dual $I^*(G)$ has a natural ring structure, namely the quotient $\mathbf{Z}[G]/(\Sigma)$ by the two-sided ideal generated by $\Sigma = \sum_g g$. Writing $B = \{b \in \mathbf{Z}_+ : b \text{ divides } m \text{ and } b \neq 1\}$ we see that then $I^*(C_m) \cong \mathbf{Z}[x]/(c_B(x))$; thus:

For each positive integer $m \geq 2$, $I^*[C_m]$ is a cyclotomic ring. (10.17)

Recall that in Sect. 9.6 we considered the class of iterated fibre products $\mathcal{J}(\mathbf{D}, \mathbf{L})$ where $\mathbf{D}$ is the class of Dedekind domains and $\mathbf{L}$ is the class of finite products of local rings each with nilpotent radical. Then we have:

Proposition 10.18 *If R is a cyclotomic ring then $R \in \mathcal{J}(\mathbf{D}, \mathbf{L})$.*

Proof We must show that $\mathbf{Z}[x]/(c_A(x)) \in \mathcal{J}(\mathbf{D}, \mathbf{L})$ whenever $A \subset \mathbf{Z}_+$ is a finite nonempty set of positive integers.

The proof is by induction on $|A|$. If $|A| = 1$ then $c_A(x) = c_a(x)$ where $A = \{a\}$. Then $\mathbf{Z}[x]/(c_a(x))$ is the ring of integers in the algebraic number field $\mathbf{Q}[x]/(c_a(x))$ ([10], p. 88). In particular, $\mathbf{Z}[x]/(c_A(x)) = \mathbf{Z}[x]/(c_a(x))$ is a Dedekind domain and so belongs to $\mathcal{J}_1(\mathbf{D}, \mathbf{L})$.

Now suppose that it is established that $\mathbf{Z}[x]/(c_B(x)) \in \mathcal{J}_{|B|}(\mathbf{D}, \mathbf{L})$ when $|B| \leq k - 1$ and suppose that $|A| = k$. Choose $d \in A$ and put $B = A - \{d\}$. Then $c_A(x) = c_d(x)c_B(x)$ and we have a fibre product

$$\mathbf{Z}[x]/(c_d(x)c_B(x)) \rightarrow \quad \mathbf{Z}[x]/(c_B(x))$$
$$\downarrow \qquad\qquad\qquad\qquad \downarrow$$
$$\mathbf{Z}[x]/(c_d(x)) \quad \rightarrow \mathbf{Z}[x]/(c_d(x)) + (c_B(x))$$

in which the arrows are the identification maps. In particular, all arrows are surjective and the square satisfies the Milnor condition. Again $\mathbf{Z}[x]/(c_d(x)) \in \mathcal{J}_1(\mathbf{D}, \mathbf{L})$

as in the induction base. Now $|B| = k - 1$ so, by induction, $\mathbf{Z}[x]/(c_B(x)) \in$ $\mathcal{J}_{k-1}(\mathbf{D}, \mathbf{L})$. Finally, if $N \geq 2$ then by Proposition 10.15 $\mathbf{Z}[x]/(c_A(x)) + (c_B(x))$ is a finite product of local rings, each with nilpotent radical and so

$$\mathbf{Z}[x]/(c_A(x)) \cong \mathbf{Z}[x]/(c_d(x)c_B(x)) \in \mathcal{J}_k(\mathbf{D}, \mathbf{L}).$$

In the case where $N = 1$ then $\mathbf{Z}[x]/(c_A(x)) \cong \mathbf{Z}[x]/(c_d(x) \times \mathbf{Z}[x]/(c_B(x))$ and again $\mathbf{Z}[x]/(c_A(x)) \in \mathcal{J}_k(\mathbf{D}, \mathbf{L})$. This completes the proof.   □

From (9.26) and the above it now follows that:

Corollary 10.19 *Let S be a cyclotomic ring; then $S[C_\infty]$ has property* SFC.

In particular, writing $\mathbf{Z}[C_\infty \times C_m] \cong R[C_\infty]$ where $R = \mathbf{Z}[C_m]$ we see that:

Theorem 10.20 *For any positive integer $m \geq 2$ the group ring $\mathbf{Z}[C_\infty \times C_m]$ has the* SFC *property.*

Theorem 10.20 is a result of Bass and Murthy [3]. The above proof is, however, more direct than the original.

10.4 Stably Free Modules over $\mathbf{Z}[F_n \times C_4]$

Evidently Theorem 10.20 provides a partial generalization of (10.14), so it is natural to ask whether (10.14) generalizes completely; that is:

Question Does $\mathbf{Z}[F_n \times C_m]$ have the SFC property when $n \geq 2$ and m is not prime?

O'Shea [78] has answered this question in the negative when m is divisible by p^2 for some prime p. He shows that $\mathbf{Z}[F_n \times C_m]$ then has infinitely many isomorphically distinct stably free modules of rank 1. Below we give the simplest case, $m = 4$; our account is a slight variation on O' Shea's original argument. We begin with the Milnor square

$$\mathcal{A} = \begin{cases} \mathbf{Z}[x]/(x^4 - 1) \to \mathbf{Z}[x]/(x^2 + 1) \\ \quad\quad\downarrow \quad\quad\quad\quad\quad\quad \downarrow \\ \mathbf{Z}[x]/(x^2 - 1) \to \mathbf{F}_2[x]/(x^2 - 1) \end{cases}$$

Writing $A[C_n] = A[x]/(x^n - 1)$ this becomes

$$\mathcal{A} = \begin{cases} \mathbf{Z}[C_4] \to \quad R \\ \quad\downarrow \quad\quad\quad\quad \downarrow \\ \mathbf{Z}[C_2] \to \mathbf{F}_2[C_2] \end{cases}$$

where $R = \mathbf{Z}[x]/(x^2 + 1)$. We write $R = \mathbf{Z}[i]$ where $i^2 = -1$.

Let G be a group with the property that $B[G]$ has only trivial units whenever B is an integral domain. Tensoring with $\mathbf{Z}[G]$ we obtain

$$\mathcal{A}[G] = \begin{cases} \mathbf{Z}[G \times C_4] \to & R[G] \\ \downarrow & \downarrow \\ \mathbf{Z}[G \times C_2] \to \mathbf{F}_2[G \times C_2] \end{cases}$$

Proposition 10.21 $\mathbf{Z}[G \times C_2]$ *has only trivial units.*

Proof Write $A = \mathbf{Z}[G]$ and $C_2 = \langle t | t^2 = 1 \rangle$ so that $\mathbf{Z}[G \times C_2] \cong A[C_2]$ and let $u = u_1 + u_2 t \in A[C_2]^*$. We claim that either $u_1 = 0$ or $u_2 = 0$. Thus consider the ring homomorphism

$$\varphi : A[C_2] \to A \times A; \quad \varphi(a + bt) = (a + b, a - b).$$

Then $\varphi(u) \in A^* \times A^*$ so that $u_1 + u_2 \in A^*$ and $u_1 - u_2 \in A^*$. As $A = \mathbf{Z}[G]$ has only trivial units then

$$u_1 + u_2 = \eta_1 g; \qquad u_1 - u_2 = \eta_2 h,$$

where $\eta_i = \pm 1$ and $g, h \in G$.

$$2u_1 = \eta_1 g + \eta_2 h; \qquad 2u_2 = \eta_1 g - \eta_2 h.$$

If $h \neq g$ then $u_1 \notin \mathbf{Z}[G]$. Thus $h = g$ and the four possibilities for the pair (η_1, η_2) give $(u_1 = g, u_2 = 0)$; $(u_1 = 0, u_2 = g)$; $(u_1 = -g, u_2 = 0)$; $(u_1 = 0, u_2 = -g)$ and so either $u_1 = 0$ or $u_2 = 0$ as claimed. In either case, u is a trivial unit in $\mathbf{Z}[G \times C_2]$. $\qquad\square$

We denote by $[u]$ the class of $u \in \mathbf{F}_2[G \times C_2]^*$ in $\mathbf{Z}[G \times C_2]^* \backslash \mathbf{F}_2[G \times C_2]^* / R[G]^*$. Putting $s = t + 1 \in \mathbf{F}_2[G]$ we have $s^2 = 0$. When $\alpha \in \mathbf{F}_2[G]$ we regard $1 + s\alpha$ as an element of $\mathbf{F}_2[G \times C_2]$ via the identifications

$$1 + s\alpha \sim 1 \otimes 1 + \alpha \otimes s \in \mathbf{F}_2[G] \otimes \mathbf{F}_2[C_2] \cong \mathbf{F}_2[G \times C_2].$$

Observe that $1 + s\alpha \in \mathbf{F}_2[G \times C_2]^*$ with $(1 + s\alpha)^{-1} = 1 + s\alpha$.

Write $\alpha \in \mathbf{F}_2[G]$ in the form $\alpha = \sum_{g \in \mathrm{supp}(\alpha)} \alpha_g g$. Then '$\alpha_1 = 0$' is equivalent to the statement that '$1 \notin \mathrm{supp}(\alpha)$'. Observe that if $1 \notin \mathrm{supp}(\alpha)$ then $1 \notin \mathrm{supp}(g\alpha g^{-1})$ for any $g \in G$.

Proposition 10.22 *Let* $\alpha, \beta \in \mathbf{F}_2[G]$ *be such that* $\alpha_1 = \beta_1 = 0$; *if* $[1 + s\alpha] = [1 + s\beta]$ *then* $|\mathrm{supp}(\alpha)| = |\mathrm{supp}(\beta)|$.

Proof The class $[1 + s\alpha]$ consists of all elements of the form $v_1(w_1)(1 + s\alpha)v_2(w_2)$ where $w_1 \in \mathbf{Z}[G \times C_2]^*$ and $w_2 \in R[G]^*$. As $\mathbf{Z}$ is an integral domain then $\mathbf{Z}[G]$ has only trivial units. By Proposition 10.21 the units of $\mathbf{Z}[G \times C_2]$ are also trivial so that

$$w_1 = \eta_1 g t^{e_1},$$

where $g \in G$ $\eta_1 = \pm 1$, $e_1 \in \{0, 1\}$. Likewise R is also an integral domain so that $w_2 = uh$ where $h \in G$ and $u \in R^*$. However $R^* = \{1, i, -1, -i\}$ so that

$$w_2 = \eta_2 h i^{e_2},$$

where $h \in G$ $\eta_2 = \pm 1$, $e_2 \in \{0, 1\}$. We note that $st^e = s$ in $F_2[G \times C_2]$ for any exponent e and that $v_2(i) = v_1(t) = t$. As $v_r(-1) = 1$ we have

$$v(w_1)(1 + s\alpha)v(w_2) = (ght^{e_1+e_2} + st^{e_1+e_2}(g\alpha h)) = (ght^{e_1+e_2} + s(g\alpha h)).$$

Let $\pi_* : F_2[G \times C_2] \to F_2[G]$ denote the homomorphism induced from the projection $\pi : G \times C_2 \to G$. Then $\pi_*(t) = 1$ and $\pi_*(s) = 0$ and so $\pi(ght^{e_1+e_2} + s(g\alpha h)) = gh$. Thus if $v(w_1)(1 + s\alpha)v(w_2) = 1 + s\beta$ then $h = g^{-1}$ and $v(w_1)(1 + s\alpha)v(w_2) = t^{e_1+e_2} + s(g\alpha g^{-1})$. Hence

$$1 + s\beta = t^{e_1+e_2} + s(g\alpha g^{-1}).$$

If $t^{e_1+e_2} = t$ then $(1 + \beta) + t\beta = g\alpha g^{-1} + t(1 + g\alpha g^{-1})$. This is a contradiction as $1 \notin \text{supp}(\beta)$ and $1 \notin \text{supp}(g\alpha g^{-1})$. Thus $t^{e_1+e_2} = 1$ so that $\beta = g\alpha g^{-1}$ and hence $|\text{supp}(\alpha)| = |\text{supp}(\beta)|$. $\qquad \square$

We now specialize to the case where $G = F_n = \langle x_1, \ldots, x_n | \emptyset \rangle$ is the free group on n generators where $n \geq 2$. Let $\mathbf{Z}_+$ denote the set of positive integers and let $A \subset \mathbf{Z}_+$ be a finite subset. For $g \in G$ we put

$$p_A(g) = \sum_{a \in A} g^a \in F_2[F_n]$$

and we define $\tilde{A} = p_A(x_1 x_2 x_1^{-1}) + p_A(x_2) \in F_2[F_n]$. Then with the above notation:

Proposition 10.23 $|\text{supp}(\tilde{A})| = 2|A|$ and $1 \notin \text{supp}(\tilde{A})$.

It follows from Propositions 10.22 and 10.23 that the image of the mapping

$$\{\text{Finite subsets of } \mathbf{Z}_+\} \longrightarrow Z[F_n \times C_2]^* \backslash F_2[F_n \times C_2]^* / R[F_n]^*,$$

$$A \qquad \mapsto \qquad [1 + s\tilde{A}]$$

is infinite. In particular, let $A(k) = \{r \in \mathbf{Z}_+ : 1 \leq r \leq k\}$; then:

Proposition 10.24 The classes $\{[1 + s\widetilde{A(k)}]\}_{1 \leq k}$ are pairwise distinct.

Proposition 10.25 $1 + s\tilde{A} \in [F_2[F_n \times C]^*, F_2[F_n \times C_2]^*]$.

Proof As $x_1 + s \in F_2[F_n \times C_2]^*$ with $(x_1 + s)^{-1} = x_1^{-1} + sx_1^{-2}$ one checks easily that

$$(x_1 + s)(1 + sp_A(x_2))(x_1 + s)^{-1}(1 + sp_A(x_2))^{-1}$$
$$= 1 + s(x_1 p_A(x_2)x_1^{-1} + p_A(x_2)).$$

However, $x_1 p_A(x_2)x_1^{-1} = p_A(x_1 x_2 x_1^{-1})$ so that

$$1 + s\tilde{A} = (x_1 + s)(1 + sp_A(x_2))(x_1 + s)^{-1}(1 + sp_A(x_2))^{-1}. \qquad \square$$

Now consider the Milnor square

$$\mathcal{A}[F_n] = \begin{cases} \mathbf{Z}[F_n \times C_4] \to & R[F_n] \\ \qquad\downarrow & \quad\downarrow \\ \mathbf{Z}[F_n \times C_2] \to & \mathbf{F}_2[F_n \times C_2] \end{cases}$$

and the stabilization maps $\sigma_{m,m+k} : \overline{GL_m}(\mathcal{A}[F_n]) \to \overline{GL_{m+k}}(\mathcal{A}[F_n])$.

Proposition 10.26 *For each finite subset $A \subset \mathbf{Z}_+$*

$$\sigma_{1,2}([1 + s\widetilde{A}]) = * \in \overline{GL_2}(\mathcal{A}[F_n]).$$

Proof By Proposition 2.62 it follows that $\sigma_{1,2}(\eta) \in E_2(\mathbf{F}_2[F_n \times C_2])$ if $\eta \in [\mathbf{F}_2[F_n \times C]^*, \mathbf{F}_2[F_n \times C_2]^*]$. By Proposition 10.25 it follows that

$$\sigma_{1,2}([1 + s\widetilde{A}]) = * \in \overline{E_2}(\mathcal{A}[F_n]).$$

However, $\overline{E_2}(\mathcal{A}[F_n]) = \{*\}$ as $\mathcal{A}[F_n]$ satisfies the Milnor condition. $\square$

It follows from (3.39) that the classes $\{[1 + s\widetilde{A(k)}]\}_{1 \leq k}$ represent pairwise non-isomorphic locally free modules of rank 1 which become free after a single stabilization; that is:

Corollary 10.27 *For each $n \geq 2$, $\mathbf{Z}[F_n \times C_4]$ admits infinitely many distinct isomorphism classes of stably free modules of rank 1.*

Chapter 11
Group Rings of Dihedral Groups

In this chapter we continue the study of stably free cancellation over the integral group rings $\mathbf{Z}[F_n \times \Phi]$ in the case where Φ is the dihedral group of order $2m$ defined by the presentation

$$D_{2m} = \langle x, y | x^m = y^2 = 1, yx = x^{m-1}y \rangle.$$

Our main result, first proved in [57], is that $\mathbf{Z}[F_n \times D_{2p}]$ has SFC when p is an odd prime. This breaks down for $p = 2$. Although $\mathbf{Z}[C_\infty \times D_4]$ still has SFC (the case $n = 1$) when $n \geq 2$ a result of O'Shea shows that $\mathbf{Z}[F_n \times D_4]$ has infinitely many isomorphically distinct stably free modules of rank 1.

11.1 Stably Free Cancellation for a Class of Cyclic Algebras

We recall the cyclic algebra construction: suppose that $\theta : B \to B$ is an involution on a commutative ring B and let $b \in B$ satisfy $\theta(b) = b$. We define the *cyclic ring* $\mathcal{C}(B, \theta, b)$ to be the (two-sided) B-module $\mathcal{C}(B, \theta, b) = B + By$ which is free of rank 2 over B with basis $\{1, y\}$ and with multiplication determined by the relations

$$y^2 = b; \qquad y\xi = \theta(\xi)y \quad (\xi \in B).$$

In the special case where $b = 1$ we simply write $\mathcal{C}(B, \theta)$ and when θ is clear from context we abbreviate this to $\mathcal{C}(B)$. In this section we take B to be the cyclotomic ring $R = R(p)$ where p is an odd prime. Then R has an involution θ defined by $\theta(\zeta) = \zeta^{-1}$. Thus $\mathcal{C}(R)$ is the free R-module of rank 2 with basis $\{1, y\}$ and multiplication given by:

$$y\zeta = \zeta^{-1}y; \qquad y^2 = 1.$$

So defined, $\mathcal{C}(R)$ becomes an algebra over the fixed ring $R_0 = \{x \in R : \theta(x) = x\}$.

We now take p to be an odd prime and, as in Sect. 10.2, take $R = \mathbf{Z}[\zeta]/c_p(\zeta)$. Let $\theta : R \to R$ be the involution corresponding to complex conjugation; then $R_0 = \mathbf{Z}[\mu]$ where $\mu = \zeta + \zeta^{-1}$ and it is known that:

$$R_0 \text{ is the ring of integers in } \mathbf{Q}[\mu]. \tag{11.1}$$

F.E.A. Johnson, *Syzygies and Homotopy Theory*, Algebra and Applications 17,
DOI 10.1007/978-1-4471-2294-4_11, © Springer-Verlag London Limited 2012

We noted in Sect. 10.2 that R is a **Z**-lattice of rank $p - 1$ in $\mathbf{Q}[\zeta_p]$ and that $(\zeta - 1)R$ has index p in R. In fact ([38], p. 525):

$$(\zeta - 1)^{p-1} = pu \quad \text{for some unit } u \in R^*. \tag{11.2}$$

It follows that R/p is a finite local ring and that $\text{rad}(R/p)$ is the kernel of the canonical surjection $R/p \to R/(\zeta - 1)R \cong \mathbf{F}_p$. Indeed, the correspondence $t \mapsto \zeta - 1$ induces an isomorphism

$$\mathbf{F}_p[t]/t^{p-1} \cong R/pR. \tag{11.3}$$

There are corresponding statements for R_0;

$$R_0 \text{ is a } \mathbf{Z}\text{-lattice of rank } (p - 1)/2. \tag{11.4}$$

$$(\mu - 2)R_0 \text{ has index } p \text{ in } R_0; \text{ moreover:} \tag{11.5}$$

$$(\mu - 2)^{(p-1)/2} = pw \text{ for some unit } w \in R_0^*. \tag{11.6}$$

Thus R_0/p is also a finite local ring in which $\text{rad}(R_0/p)$ is the kernel of the canonical surjection $R_0/p \to R_0/(\mu - 2)R_0 \cong \mathbf{F}_p$. Likewise the correspondence $s \mapsto \mu - 2$ induces an isomorphism

$$\mathbf{F}_p[s]/s^{(p-1)/2} \cong R_0/pR_0. \tag{11.7}$$

Let $\langle \zeta \rangle$ denote the subgroup of R^* generated by ζ. It is known (cf. [31], p. 212) that the mapping $\langle \zeta \rangle \times R_0^* \to R^*$; $(\zeta^k, u) \mapsto \zeta^k u$ is an isomorphism. We have already observed that ζ maps to 1 under the canonical mapping $R \to \mathbf{F}_p$. From (10.10) we see also that:

The canonical homomorphism $R_0 \to \mathbf{F}_p$ has the lifting property for units.   (11.8)

We construct a ring homomorphism $\varphi : C(R) \to M_2(R_0)$ to the ring $M_2(R_0)$ of 2×2 matrices over R_0 via the assignments

$$\varphi(\zeta) = \begin{pmatrix} 1 & 1 \\ \mu - 2 & \mu - 1 \end{pmatrix}; \qquad \varphi(y) = \begin{pmatrix} -1 & 1 \\ 0 & 1 \end{pmatrix}.$$

To verify that φ defines a ring homomorphism we must check that

(i) $\varphi(\zeta)^p = \text{Id};$ (ii) $\varphi(y)^2 = \text{Id};$ (iii) $\varphi(y)\varphi(\zeta) = \varphi(\zeta)^{-1}\varphi(y).$

The relations (ii), (iii) are straightforward. To see (i) put

$$C = \begin{pmatrix} 1 - \zeta & -\zeta \\ 1 - \zeta^{-1} & -\zeta^{-1} \end{pmatrix} \in GL_2(R \otimes \mathbf{Q}).$$

Then one may check easily that

$$C\varphi(\zeta)C^{-1} = \begin{pmatrix} \zeta & 0 \\ 0 & \zeta^{-1} \end{pmatrix}.$$

Evidently $(C\varphi(\zeta)C^{-1})^p = \text{Id}$ so that $\varphi(\zeta)^p = \text{Id}$ as required.

φ is injective and $\text{Im}(\varphi)$ is a subring of index p in $M_2(R_0)$,   (11.9)

$$\text{Im}(\varphi) = \left\{ \begin{pmatrix} a & b \\ c(\mu - 2) & d \end{pmatrix} : a, b, c, d \in R_0 \right\}. \tag{11.10}$$

We note also that

$$pM_2(R_0) \subset \mathrm{Im}(\varphi). \tag{11.11}$$

For any ring A we denote by $\Delta_2(A)$ the diagonal subring of $M_2(A)$; that is:

$$\Delta_2(A) = \left\{ \begin{pmatrix} \delta_1 & 0 \\ 0 & \delta_2 \end{pmatrix} : \delta_i \in A \right\}.$$

It is clear from (11.10) that:

$$\Delta_2(R_0) \subset \mathrm{Im}(\varphi). \tag{11.12}$$

We now repeat the construction $\bmod p$. The same formal assignments

$$\zeta \mapsto \begin{pmatrix} 1 & 1 \\ \mu - 2 & \mu - 1 \end{pmatrix}; \qquad y \mapsto \begin{pmatrix} -1 & 1 \\ 0 & 1 \end{pmatrix}$$

define a homomorphism $\varphi_* : C(R/p) \to M_2(R_0/p)$ and we see also that:

$$\Delta_2(R_0/p) \subset \mathrm{Im}(\varphi_*). \tag{11.13}$$

Now suppose given $u_1, u_2 \in \mathrm{rad}(R_0/p)$; then we may write $u_i = (\mu - 2)x_i$ for some $x_i \in R_0/p$. By (11.13) choose $\xi \in C(R/p)$ such that

$$\varphi_*(\xi) = \begin{pmatrix} x_1 & 0 \\ 0 & x_2 \end{pmatrix}$$

and put $\eta = (\mu - 2)\xi$. As $(\mu - 2)^{(p-1)/2} = p$ then $\eta^{(p-1)/2} = 0 \in C(R/p)$. Putting $\upsilon = 1 + \eta \in C(R/p)^*$ we obtain an addendum to (11.13).

If $u_1, u_2 \in \mathrm{rad}(R_0/p)$ then there exists $\upsilon \in C(R/p)^*$ such that

$$\varphi_*(\upsilon) = \begin{pmatrix} 1 + u_1 & 0 \\ 0 & 1 + u_2 \end{pmatrix}. \tag{11.14}$$

We note that φ_* fails to be injective. Instead, (11.11) gives a filtration

$$p\,\mathrm{Im}(\varphi) \subset pM_2(R_0) \subset \mathrm{Im}(\varphi) \subset M_2(R_0).$$

Identifying $C(R/p) \cong C(R)/pC(R)$ with $\mathrm{Im}(\varphi)/p\,\mathrm{Im}(\varphi)$ and φ_* with the projection $\mathrm{Im}(\varphi)/p\,\mathrm{Im}(\varphi) \to M_2(R_0)/pM_2(R_0) \cong M_2(R_0/p)$ we see there is an exact sequence:

$$0 \to M_2(R_0)/\mathrm{Im}(\varphi) \to C(R/p) \xrightarrow{\varphi_*} M_2(R_0/p) \to 0. \tag{11.15}$$

Now consider the following square:

$$
\begin{array}{ccc}
C(R) & \xrightarrow{\varphi} & M_2(R_0) \\
\Big\downarrow{\natural} & & \Big\downarrow{\natural} \\
C(R/p) & \xrightarrow{\varphi_*} & M_2(R_0/p)
\end{array}
\qquad (\mathrm{I})
$$

As it stands, (I) is not a fibre square. We may modify it in two ways so as to become so. The first modification is to replace $C(R/p)$ by $C(R)/pM_2(R_0)$ with the homomorphisms adjusted appropriately thus:

$$
\begin{array}{ccc}
C(R) & \xrightarrow{\;\;\varphi\;\;} & M_2(R_0) \\[2mm]
\Big\downarrow{\scriptstyle\natural} & & \Big\downarrow{\scriptstyle\natural} \\[2mm]
C(R)/pM_2(R_0) & \xrightarrow{\;\;\varphi_*\;\;} & M_2(R_0/p)
\end{array}
\qquad\qquad (\bar{\mathrm{I}})
$$

The second way of modifying it is to replace $C(R)$ by the formal pullback $\widehat{C}(R)$ of the corner associated with (I) thus:

$$
\begin{array}{ccc}
\widetilde{C}(R) & \xrightarrow{\;\;\pi_-\;\;} & M_2(R_0) \\[2mm]
\Big\downarrow{\scriptstyle\pi_+} & & \Big\downarrow{\scriptstyle\natural} \\[2mm]
C(R/p) & \xrightarrow{\;\;\varphi_*\;\;} & M_2(R_0/p)
\end{array}
\qquad\qquad (\widetilde{\mathrm{I}})
$$

Both $(\bar{\mathrm{I}})$ and $(\widetilde{\mathrm{I}})$ are now fibre squares. We note:

Proposition 11.16 $C(R)$ *is a retract of* $\widetilde{C}(R)$.

Proof The surjection $r : C(R/p) \to C(R)/pM_2(R_0)$ induces a homomorphism of fibre squares $r : (\widetilde{\mathrm{I}}) \to (\bar{\mathrm{I}})$

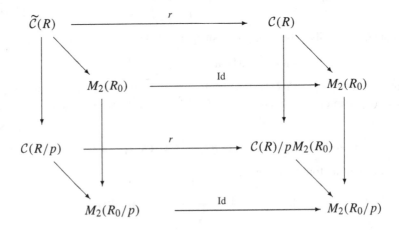

Furthermore, it follows from the universal property of pullbacks that there is a unique homomorphism $i : C(R) \to \widetilde{C}(R)$ making the following commute:

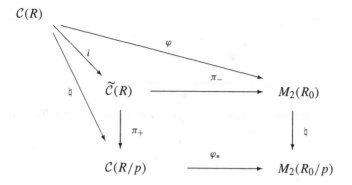

It is straightforward to show that $r \circ i = \mathrm{Id}$.   □

Lemma 11.17 *Let* $\Delta = \left(\begin{smallmatrix} \delta_1 & 0 \\ 0 & \delta_2 \end{smallmatrix}\right)$ *with* $\delta_i \in (R_0/p)^*$*; then* $\Delta = \varphi_*(\upsilon)\natural(\widetilde{\Delta})$ *for some* $\upsilon \in \mathcal{C}(R/p)^*, \widetilde{\Delta} \in GL_2(R_0)$.

Proof Consider the diagram

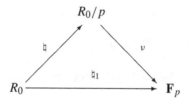

where $\natural$ is reduction mod p, $\natural_1$ is reduction mod $\mu - 2$ and ν is the unique ho-momorphism. Note that $\mathrm{Ker}(\nu) = \mathrm{rad}(R_0/p)$. Given $\Delta = \left(\begin{smallmatrix} \delta_1 & 0 \\ 0 & \delta_2 \end{smallmatrix}\right)$ with $\delta_i \in (R_0/p)^*$ then $\nu(\delta_i) \in \mathbf{F}_p^*$. By (11.8) there exist $\tilde{\delta}_i \in R_0^*$ such that $\natural_1(\tilde{\delta}_i) = \nu(\delta_i)$. Put $\gamma_i = \natural(\tilde{\delta}_i) \in (R_0/p)^*$ and put

$$\widetilde{\Delta} = \begin{pmatrix} \tilde{\delta}_1 & 0 \\ 0 & \tilde{\delta}_2 \end{pmatrix} \in GL_2(R_0); \qquad \Gamma = \begin{pmatrix} \gamma_1 & 0 \\ 0 & \gamma_2 \end{pmatrix} \in GL_2(R_0/p).$$

Put $u_i = \delta_i \gamma_i^{-1} - 1$ then $\Delta\Gamma^{-1} = \left(\begin{smallmatrix} 1+u_1 & 0 \\ 0 & 1+u_2 \end{smallmatrix}\right)$ and $u_i \in \mathrm{rad}(R_0/p)$ as $\nu(u_i) = 0$. By (11.14) there exists $\upsilon \in \mathcal{C}(R/p)^*$ such that $\varphi_*(\upsilon) = \Delta\Gamma^{-1}$. Hence $\Delta = \varphi_*(\upsilon)\Gamma = \varphi_*(\upsilon)\natural(\widetilde{\Delta})$ and this completes the proof.   □

There is a surjective ring homomorphism $\overline{\nu}_* : R/p \rightarrow \mathbf{F}_p$ making the following commute:

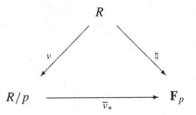

Applying the cyclic algebra construction we obtain a commutative diagram

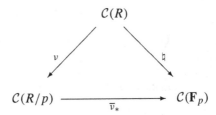

As $\bar{\nu}_* : R/p \to \mathbf{F}_p$ has nilpotent kernel, the kernel of the induced map $\bar{\nu}_* : \mathcal{C}(R/p) \to \mathcal{C}(\mathbf{F}_p)$ is also nilpotent. However, $\mathcal{C}(\mathbf{F}_p) \cong \mathbf{F}_p[C_2] \cong \mathbf{F}_p \times \mathbf{F}_p$ has the SFC property. By the Bourbaki-Nakayama Lemma Proposition 9.15 we have:

$$\mathcal{C}(R/p) \text{ has the SFC property.} \qquad (11.18)$$

We can now show:

Theorem 11.19 $\widetilde{\mathcal{C}}(R)$ *has the* SFC *property.*

Proof Note that, as $\natural : M_2(R_0) \to M_2(R_0/p)$ is surjective, the fibre square

$$
\begin{array}{ccc}
\widetilde{\mathcal{C}}(R) & \xrightarrow{\;\;\pi_-\;\;} & M_2(R_0) \\[4pt]
\Big\downarrow{\scriptstyle \pi_+} & & \Big\downarrow{\scriptstyle \natural} \qquad\qquad (\widetilde{\mathrm{I}}) \\[4pt]
\mathcal{C}(R/p) & \xrightarrow[\;\;\varphi_*\;\;]{} & M_2(R_0/p)
\end{array}
$$

satisfies the Milnor condition. Moreover as R_0 is a Dedekind domain, it has the SFC property so that:

(i) $M_2(R_0)$ has the SFC property.

Now R_0/p is a finite local ring and so is weakly Euclidean. It follows that:

(ii) $M_2(R_0/p)$ is weakly Euclidean.

We saw in (11.18) that:

(iii) $\mathcal{C}(R/p)$ has the SFC property.

To show that $\widetilde{C}(R)$ has the SFC property it suffices, by Corollary 3.48, to show that $(\widetilde{I})$ is pointlike in dimension one. Thus suppose that $X \in M_2(R_0/p)^* = GL_2(R_0/p)$. As R_0/p is weakly Euclidean we may decompose X as a product $X = \Delta E$ in which $E \in E_2(R_0/p)$ and

$$\Delta = \begin{pmatrix} \delta & 0 \\ 0 & 1 \end{pmatrix}; \quad \delta \in (R_0/p)^*.$$

As $\natural : R_0 \to R_0/p$ is surjective we may choose $\widetilde{E} \in E_2(R_0)$ such that $E = \natural(\widetilde{E})$. By Lemma 11.17 we may write $\Delta = \varphi_*(\upsilon)\natural(\widetilde{\Delta})$ so for some $\widetilde{\Delta} \in GL_2(R_0)$. Thus $X = \varphi_*(\upsilon)\natural(\widetilde{\Delta}\widetilde{E})$ where $\upsilon \in C(R/p)^*$ and $\widetilde{\Delta}\widetilde{E} \in GL_2(R_0)$. Thus $(\widetilde{I})$ is pointlike in dimension one and this completes the proof that $\widetilde{C}(R)$ has the SFC property. $\qquad \square$

As $C(R)$ is a retract of $\widetilde{C}(R)$ it follows from (9.1) that:

Corollary 11.20 $C(R)$ *has the* SFC *property.*

11.2 Extending over Free Group Rings

In this section we extend the conclusion of Sect. 11.1 from R to the group ring $R[F_n]$. Explicitly, identify $R[F_n]$ with $R \otimes \mathbf{Z}[F_n]$ and replace θ by $\theta \otimes 1 : R \otimes \mathbf{Z}[F_n] \to R \otimes \mathbf{Z}[F_n]$. Tensoring with $\mathbf{Z}[F_n]$, the square (I) of Sect. 11.1 now becomes

$$
\begin{array}{ccc}
C(R[F_n]) & \xrightarrow{\varphi} & M_2(R_0[F_n]) \\
\downarrow{\scriptstyle\natural} & & \downarrow{\scriptstyle\natural} \\
C(R/p[F_n]) & \xrightarrow{\varphi_*} & M_2(R_0/p[F_n])
\end{array} \qquad \text{(I)}
$$

As before, this fails to be a fibre square and we replace it by the formal pullback $\widehat{C}$ of the corner associated with (I) thus:

$$
\begin{array}{ccc}
\widehat{C} & \xrightarrow{\pi_-} & M_2(R_0[F_n]) \\
\downarrow{\scriptstyle\pi_+} & & \downarrow{\scriptstyle\natural} \\
C(R/p)[F_n] & \xrightarrow{\varphi_*} & M_2((R_0/p)[F_n])
\end{array} \qquad \widehat{(I)}
$$

From the formal properties of pullback and $- \otimes_{\mathbf{Z}} \mathbf{Z}[F_n]$ we see that:

$$\widehat{C} \cong \widetilde{C}(R)[F_n]. \qquad (11.21)$$

In particular, it follows directly from Proposition 11.16 that:

$$C(R)[F_n] \text{ is a retract of } \widehat{C}. \qquad (11.22)$$

Note also that it follows directly from (11.13) that:

$$\Delta_2(R_0/p)[F_n] = \Delta_2((R_0/p)[F_n]) \subset \operatorname{Im}(\varphi_*).\tag{11.23}$$

Now suppose given $u_1, u_2 \in \operatorname{rad}(R_0/p)[F_n]$; then we may write $u_i = (\mu - 2)x_i$ for some $x_i \in (R_0/p)[F_n]$. By (11.23) choose $\xi \in \mathcal{C}(R/p)[F_n]$ such that

$$\varphi_*(\xi) = \begin{pmatrix} x_1 & 0 \\ 0 & x_2 \end{pmatrix}$$

and put $\eta = (\mu - 2)\xi$. As $(\mu - 2)^{(p-1)/2} = p$ then $\eta^{(p-1)/2} = 0 \in \mathcal{C}(R/p)[F_n]$. It follows that $1 + \eta \in \mathcal{C}(R/p)^*$. Putting $\upsilon = 1 + \eta$ we obtain an addendum to (11.23):

If $u_1, u_2 \in \operatorname{rad}(R_0/p)[F_n]$ then there exists $\upsilon \in \mathcal{C}(R/p)[F_n]^*$ such that

$$\varphi_*(\upsilon) = \begin{pmatrix} 1+u_1 & 0 \\ 0 & 1+u_2 \end{pmatrix}.\tag{11.24}$$

Before showing that $\widehat{\mathcal{C}}$ possesses the SFC property we first establish the analogue of Lemma 11.17.

Lemma 11.25 *Let* $\Delta = \begin{pmatrix} \delta_1 & 0 \\ 0 & \delta_2 \end{pmatrix}$ *with* $\delta_i \in (R_0/p)^*[F_n]$; *then* $\Delta = \varphi_*(\upsilon)\natural(\widetilde{\Delta})$ *for some* $\upsilon \in \mathcal{C}(R/p)[F_n]^*$, $\widetilde{\Delta} \in \Delta_2(R_0[F_n])^*$.

Proof Consider the diagram

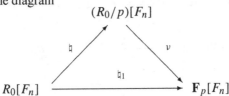

$$(R_0/p)[F_n]$$

$$R_0[F_n] \xrightarrow{\hspace{3cm}} \mathbf{F}_p[F_n]$$

where $\natural$ is reduction mod p, $\natural_1$ is reduction mod $\mu - 2$ and υ is the unique homomorphism and note that $\operatorname{Ker}(\upsilon) = \operatorname{rad}(R_0/p)[F_n]$.

Now suppose given $\Delta = \begin{pmatrix} \delta_1 & 0 \\ 0 & \delta_2 \end{pmatrix}$ with $\delta_i \in (R_0/p)[F_n]^*$. As F_n satisfies the $\mathcal{TUP}^1$ condition $\mathbf{F}_p[F_n]$ has only trivial units so we may write $\upsilon(\delta_i) = \alpha_i g_i$ with $\alpha_i \in \mathbf{F}_p^*$ and $g_i \in F_n$. By (11.8) there exist $\tilde{\alpha}_i \in R_0^*$ such that $\natural_1(\tilde{\alpha}_i) = \alpha_i$. Now define

$$\tilde{\delta}_i = \tilde{\alpha}_i g_i \in R_0[F_n]^*; \qquad \gamma_i = \natural(\tilde{\delta}_i) \in (R_0/p)[F_n]^*;$$

$$\widetilde{\Delta} = \begin{pmatrix} \tilde{\delta}_1 & 0 \\ 0 & \tilde{\delta}_2 \end{pmatrix} \in GL_2(R_0[F_n]);$$

$$\Gamma = \begin{pmatrix} \gamma_1 & 0 \\ 0 & \gamma_2 \end{pmatrix} \in GL_2((R_0/p)[F_n]).$$

---

[1] See Appendix C.

Then $\natural(\widetilde{\Delta}) = \Gamma$ and $\Delta\Gamma^{-1} = \begin{pmatrix} 1+u_1 & 0 \\ 0 & 1+u_2 \end{pmatrix}$ where $u_i = \delta_i \gamma_i^{-1} - 1$. Moreover $u_i \in \mathrm{rad}(R_0/p)[F_n]$ as $v(u_i) = 0$. By (11.24) choose $\upsilon \in \mathcal{C}(R/p)[F_n]^*$ such that $\varphi_*(\upsilon) = \Delta\Gamma^{-1}$. Then $\Delta = \varphi_*(\upsilon)\Gamma = \varphi_*(\upsilon)\natural(\widetilde{\Delta})$ and this completes the proof. $\square$

There is a surjective ring homomorphism $\overline{v}_* : (R/p)[F_n] \to \mathbf{F}_p[F_n]$ making the following commute:

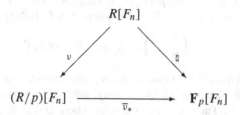

$$R[F_n]$$
$$v \qquad\qquad \natural$$
$$(R/p)[F_n] \xrightarrow{\ \overline{v}_*\ } \mathbf{F}_p[F_n]$$

Applying the cyclic algebra construction we obtain a commutative diagram

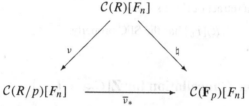

$$\mathcal{C}(R)[F_n]$$
$$v \qquad\qquad \natural$$
$$\mathcal{C}(R/p)[F_n] \xrightarrow{\ \overline{v}_*\ } \mathcal{C}(\mathbf{F}_p)[F_n]$$

One checks easily that the surjective homomorphism $\overline{v}_* : \mathcal{C}(R/p)[F_n] \to \mathcal{C}(\mathbf{F}_p)[F_n]$ has nilpotent kernel. However $\mathcal{C}(\mathbf{F}_p)[F_n] \cong \mathbf{F}_p[F_n \times C_2] \cong \mathbf{F}_p[F_n] \times \mathbf{F}_p[F_n]$ and from Bass' generalization of Sheshadri's Theorem [1, 2], or alternatively by the Theorem of Cohn [17], $\mathbf{F}_p[F_n]$ has the SFC property. It follows from the Bourbaki-Nakayama Lemma that:

$$\mathcal{C}(R/p)[F_n] \text{ has the SFC property.} \tag{11.26}$$

We can now show:

Theorem 11.27 $\widehat{C}$ *has the* SFC *property.*

Proof Note that, as $\natural : M_2(R_0[F_n]) \to M_2((R_0/p)[F_n])$ is surjective, the fibre square

$$\widehat{C} \xrightarrow{\ \pi_-\ } M_2(R_0[F_n])$$
$$\pi_+ \downarrow \qquad\qquad\qquad \downarrow \natural \qquad\qquad (\mathrm{I})$$
$$\mathcal{C}(R/p)[F_n] \xrightarrow{\ \varphi_*\ } M_2((R_0/p)[F_n])$$

satisfies the Milnor condition. Moreover, we saw in (11.26) that $C(R/p)[F_n]$ has the SFC property. As R_0 is a Dedekind domain then $R_0[F_n]$ has the SFC property by (9.5). Thus $M_2(R_0[F_n])$ also the SFC property by (9.20). In addition, by Theorem 2.51 of Chap. 2, $(R_0/p)[F_n]$ is weakly Euclidean and so also $M_2((R_0/p)[F_n])$ is weakly Euclidean. By Corollary 3.48, to show that $\widehat{C}$ has the SFC property it is enough to show that $\widehat{(I)}$ is pointlike in dimension one.

Suppose that $X \in M_2((R_0/p)[F_n])^* = GL_2((R_0/p)[F_n])$. As $(R_0/p)[F_n]$ is weakly Euclidean we may write $X = \Delta E$ where $E \in E_2((R_0/p)[F_n])$ and

$$\Delta = \begin{pmatrix} \delta & 0 \\ 0 & 1 \end{pmatrix}; \qquad \delta \in (R_0/p)[F_n]^*.$$

As $\natural : R_0[F_n] \to (R_0/p)[F_n]$ is surjective we may choose $\widetilde{E} \in E_2((R_0)[F_n])$ such that $E = \natural(\widetilde{E})$. By Lemma 11.25 we may write $\Delta = \varphi_*(\upsilon)\natural(\widetilde{\Delta})$ so for some $\widetilde{\Delta} \in GL_2((R_0)[F_n])$. Thus $X = \varphi_*(\upsilon)\natural(\widetilde{\Delta}\widetilde{E})$ where $\upsilon \in C(R/p)[F_n]^*$ and $\widetilde{\Delta}\widetilde{E} \in GL_2(R_0[F_n])$. Hence $\widehat{(I)}$ is pointlike in dimension one; this completes the proof that $\widehat{C}$ has the SFC property. $\qquad\square$

As $C(R)[F_n]$ is a retract of $\widehat{C}$ we see that:

$$C(R)[F_n] \text{ has the SFC property.} \tag{11.28}$$

11.3 Stably Free Cancellation for $\mathbf{Z}[C_\infty \times D_{2p}]$

We begin by considering again the fibre square

$$\mathcal{Z}(p) = \begin{cases} \mathbf{Z}[C_p] \xrightarrow{\pi_-} R \\ \downarrow\pi_+ \qquad\quad \downarrow\nu \\ \mathbf{Z} \xrightarrow{\quad\nu\quad} \mathbf{F}_p \end{cases}$$

familiar from Sect. 10.2. We note that $\mathcal{Z}(p)$ may also interpreted as a diagram of involuted rings. We denote all involutions by θ. On $\mathbf{Z}[C_p]$ we take θ to be the canonical group ring involution $\theta(x) = x^{-1}$. So defined, θ projects forward to the 'complex conjugation' on R, $\theta(\zeta) = \zeta^{-1}$. In turn this induces an involution on R/p. Finally we take θ to be the identity on both $\mathbf{Z}$ and $\mathbf{F}_p$. Applying the cyclic algebra construction functorially to $\mathcal{Z}(p)$ we obtain the commutative square

$$
\begin{array}{ccc}
\mathbf{Z}[D_{2p}] & \xrightarrow{\quad\eta\quad} & C(R) \\
\downarrow{\scriptstyle\epsilon} & & \downarrow{\scriptstyle\nu} \\
\mathbf{Z}[C_2] & \xrightarrow{\quad\natural\quad} & \mathbf{F}_p[C_2]
\end{array}
\tag{11.29}
$$

Theorem 11.30 $\mathbf{Z}[D_{2p}]$ *has the SFC property.*

Proof Note that in (11.29) both $\natural$ and ν are surjective. Moreover, as the cyclic alge-bra construction preserves pullbacks we see that (11.29) is both a fibre square and satisfies the Milnor condition. We also note that:

(i) $\mathbf{Z}[C_2]$ has the SFC property by (10.11);
(ii) $\mathcal{C}(R)$ has the SFC property by Corollary 11.20;
(iii) $\mathbf{F}_p[C_2] \cong \mathbf{F}_p \times \mathbf{F}_p$ is weakly Euclidean.

Thus to show that $\mathbf{Z}[D_{2p}]$ has the SFC property it suffices to show that (11.29) is pointlike in dimension one. To see this, observe that a unit $u \in \mathbf{F}_p[C_2]$ has one of two forms:

(i) $\alpha \cdot 1$ where $\alpha \in \mathbf{F}_p^*$;
(ii) $\alpha \cdot y$ where y is the nontrivial element of C_2.

By (10.10) the canonical map on units $R^* \to \mathbf{F}_p^*$ is surjective. If $u \in \mathbf{F}_p[C_2]^*$ is of type (i) then write $u = \nu(\widehat{\alpha} \cdot 1)$ for some $\widehat{\alpha} \in R^*$. Similarly if u is of type (ii) we may write $u = \nu(\widehat{\alpha} \cdot y)$ for some $\widehat{\alpha} \in R^*$. In any case the map on units $\nu : \mathcal{C}(R)^* \to \mathbf{F}_p[C_2]^*$ is surjective so that (III) is pointlike in dimension one. Hence $\mathbf{Z}[D_{2p}]$ has the SFC property. □

The verification of the SFC property for $\mathbf{Z}[F_n \times D_{2p}]$ is parallel to that for $\mathbf{Z}[D_{2p}]$. We indicate the changes involved. Applying $- \otimes_\mathbf{Z} \mathbf{Z}[F_n]$ to (11.29) yields a fibre square:

$$
\begin{array}{ccc}
\mathbf{Z}[F_n \times D_{2p}] & \xrightarrow{\eta} & \mathcal{C}(R)[F_n] \\
\downarrow{\scriptstyle \epsilon} & & \downarrow{\scriptstyle \nu} \\
\mathbf{Z}[F_n \times C_2] & \xrightarrow{\natural} & \mathbf{F}_p[F_n \times C_2]
\end{array}
\qquad (11.31)
$$

We note that (11.31) satisfies the Milnor condition as $\natural : \mathbf{Z}[F_n \times C_2] \to \mathbf{F}_p[F_n \times C_2]$ is surjective. Moreover, both $\mathbf{Z}[F_n \times C_2]$ and $\mathcal{C}(R)[F_n]$ have the SFC property by (10.14) and (11.28) respectively. Also $\mathbf{F}_p[F_n \times C_2] \cong \mathbf{F}_p[F_n] \times \mathbf{F}_p[F_n]$ is weakly Euclidean. Thus, again by Corollary 3.48, to show that $\mathbf{Z}[F_n \times D_{2p}]$ has the SFC property it suffices to show that the square (11.31) is pointlike in dimension one. We show that the map on units $\nu : \mathcal{C}(R)[F_n]^* \to \mathbf{F}_p[F_n \times C_2]^*$ is surjective.

To see this note that, by the $\mathcal{TUP}$ property of F_n, $\mathbf{F}_p[F_n]$ has only trivial units so that a unit $u \in \mathbf{F}_p[F_n \times C_2] \cong \mathbf{F}_p[F_n] \times \mathbf{F}_p[F_n]$ has one of two forms:

(a) $\alpha \cdot g$ where $\alpha \in \mathbf{F}_p^*$ and $g \in F_n$;
(b) $\alpha \cdot gy$ where $\alpha \in \mathbf{F}_p^*$, $g \in F_n$ and y is the nontrivial element of C_2.

By (10.10) the canonical map $R^* \to \mathbf{F}_p^*$ is surjective. Again by the $\mathcal{TUP}$ property for F_n the induced map $R[F_n]^* \to \mathbf{F}_p[F_n]^*$ is also surjective. As y is a unit in $\mathcal{C}(R)[F_n]$ then in either case (a) or (b) u is in the image of $\nu : \mathcal{C}(R)[F_n]^* \to \mathbf{F}_p[F_n \times C_2]^*$. Thus (11.31) is pointlike in dimension one and so by (3.37):

Theorem 11.32 $\mathbf{Z}[F_n \times D_{2p}]$ *has the* SFC *property.*[2]

11.4 Stably Free Modules over $\mathbf{Z}[F_n \times D_4]$

It is easier to handle D_4 via its alternative description as $C_2 \times C_2$. We start with the Milnor square

$$\mathbf{Z}[x, y]/(x^2 - 1)(y^2 - 1) \to \mathbf{Z}[y]/(y^2 - 1)$$
$$\downarrow \qquad\qquad\qquad\qquad \downarrow$$
$$\mathbf{Z}[x]/(x^2 - 1) \qquad \to \mathbf{F}_2[t]/(t^2 - 1)$$

which we rewrite as

$$\mathcal{C} = \begin{cases} \mathbf{Z}[C_2 \times C_2] \to \mathbf{Z}[C_2] \\ \quad\downarrow \qquad\qquad \downarrow \\ \mathbf{Z}[C_2] \quad \to \mathbf{F}_2[C_2]. \end{cases}$$

We first establish:

Proposition 11.33 $\mathbf{Z}[C_\infty \times C_2 \times C_2]$ *has property* SFC.

Proof The group ring $\mathbf{F}_2[C_2]$ is a local ring and $\mathbf{F}_2[C_2])/\mathrm{rad} \cong \mathbf{F}_2$. Moreover, the canonical homomorphism $\mathbf{F}_2[C_2] \to \mathbf{F}_2$ has the strong lifting property for units. As C_∞ has the $\mathcal{TUP}$ property then by Proposition 2.50 the induced homomorphism $\natural : (\mathbf{F}_2[C_2])[C_\infty] \to \mathbf{F}_2[C_\infty]$ also has the strong lifting property for units. As $\mathbf{F}_2[C_\infty]$ is weakly Euclidean by Theorem 2.49 it follows from Corollary 2.52 that $\mathbf{F}_2[C_\infty \times C_2] \cong (\mathbf{F}_2[C_2])[C_\infty]$ is also weakly Euclidean. Tensoring $\mathcal{C}$ above with $\mathbf{Z}[C_\infty]$ and making the obvious identifications gives us the Milnor square

$$\mathbf{Z}[C_\infty \times C_2 \times C_2] \xrightarrow{\pi_-} \mathbf{Z}[C_\infty \times C_2]$$
$$\downarrow{\pi_+} \qquad\qquad\qquad\qquad \downarrow{\nu}$$
$$\mathbf{Z}[C_\infty \times C_2] \xrightarrow{\nu} \mathbf{F}_2[C_\infty \times C_2]$$

Now $\mathbf{Z}[C_\infty \times C_2]$ has property SFC by (10.14) and we have observed above that $\mathbf{F}_2[C_\infty \times C_2]$ is weakly Euclidean. The hypotheses of Corollary 3.54 apply; the conclusion that $\mathbf{Z}[C_\infty \times C_2 \times C_2] \cong A[C_2 \times C_2]$ has property SFC now follows. $\qquad\square$

The conclusion of this theorem fails when C_∞ is replaced by a free group of higher rank. By slightly modifying the argument of O'Shea given in Sect. 10.4, we proceed to show that $\mathbf{Z}[F_n \times C_2 \times C_2]$ has infinitely many stably free modules of rank 1 whenever $n \geq 2$.

---

[2] When $n = 1$ this can be regarded as saying $R[D_{2p}]$ has stably free cancellation where $R = \mathbf{Z}[t, t^{-1}]$ is the ring of Laurent polynomials over $\mathbf{Z}$. The corresponding result over the ring $\mathbf{Z}[t]$ of genuine polynomials was established by Strouthos using Quillen patching [89].

Tensoring the Milnor square $\mathcal{C}$ above with $\mathbf{Z}[F_n]$ we obtain

$$\mathcal{C}[F_n] = \begin{cases} \mathbf{Z}[F_n \times C_2 \times C_2] \to \mathbf{Z}[F_n \times C_2] \\ \qquad\quad \downarrow \qquad\qquad\qquad\quad \downarrow \\ \mathbf{Z}[F_n \times C_2] \qquad \to \mathbf{F}_2[F_n \times C_2] \end{cases}$$

Here we denote by $\langle u \rangle$ the class of $u \in \mathbf{F}_2[F_n \times C_2]^*$ in $\mathbf{Z}[F_n \times C_2]^* \backslash \mathbf{F}_2[F_n \times C_2]^* / \mathbf{Z}[F_n \times C_2]^*$. With the same definition of $\widetilde{A(k)}$ a very similar proof to Proposition 10.24 shows that:

Proposition 11.34 *The classes $\{\langle 1 + s\widetilde{A(k)} \rangle\}_{1 \le k}$ are pairwise distinct.*

The remainder of the argument is formally identical with that of Sect. 10.4. It remains true that $1 + s\widetilde{A(k)} \in [\mathbf{F}_2[F_n \times C]^*, \mathbf{F}_2[F_n \times C_2]^*]$ and so the classes $\langle 1 + s\widetilde{A(k)} \rangle$ map to $*$ under the stabilization maps $\sigma_{1,2} : \overline{GL}_1(\mathcal{C}[F_n]) \to \overline{GL}_2(\mathcal{C}[F_n])$. We see that:

Corollary 11.35 *For each $n \ge 2$, $\mathbf{Z}[F_n \times C_2 \times C_2]$ admits infinitely many distinct isomorphism classes of stably free modules of rank 1.*

Chapter 12
Group Rings of Quaternion Groups

In this chapter we extend the study of stably free cancellation for $\mathbf{Z}[F_n \times \Phi]$ to the cases where Φ is the quaternion group $Q(4m)$ of order $4m$ defined by the presentation

$$Q(4m) = \langle x, y | x^m = y^2, xyx = y \rangle.$$

Here we find a marked contrast with the dihedral and cyclic cases. We first show by a delicate calculation that $\mathbf{Z}[C_\infty \times Q(8)]$ has infinitely many distinct stably free modules of rank 1. Whilst this result might seem unduly specific, it nevertheless implies a similar conclusion for $\mathbf{Z}[F_n \times Q(8m)]$ whenever $m, n \geq 1$. We conclude with a brief survey of what is known for the group rings $\mathbf{Z}[F_n \times Q(4m)]$ when m is odd.

12.1 An Elementary Corner Calculation

We adopt the following notation throughout:

$\mathbf{Z}_{(p)} =$ the local ring obtained from $\mathbf{Z}$ by inverting all primes $q \neq p$;

$\widehat{\mathbf{Z}}_{(p)} =$ the ring of p-adic integers; that is, the completion of $\mathbf{Z}_{(p)}$ at p;

$\widehat{\mathbf{Q}}_{(p)} =$ the field of p-adic numbers; that is, the field of fractions of $\widehat{\mathbf{Z}}_{(p)}$.

For a prime p we denote by $\mathcal{T}(p)$ the corner

$$\mathcal{T}(p) = \begin{cases} & \widehat{\mathbf{Z}}_{(p)}[t, t^{-1}] \\ & \downarrow \\ \widehat{\mathbf{Q}}_{(p)} \to & \widehat{\mathbf{Q}}_{(p)}[t, t^{-1}] \end{cases}$$

F.E.A. Johnson, *Syzygies and Homotopy Theory*, Algebra and Applications 17, DOI 10.1007/978-1-4471-2294-4_12, © Springer-Verlag London Limited 2012

so that $\overline{GL_2}(\mathcal{T}(p)) = GL_2(\widehat{\mathbf{Q}}_{(p)})\backslash GL_2(\widehat{\mathbf{Q}}_{(p)}[t, t^{-1}])/GL_2(\widehat{\mathbf{Z}}_{(p)}[t, t^{-1}])$. For $n \geq 1$ we put

$$Z(n) = \begin{pmatrix} 1 & \frac{t^n}{p} \\ 0 & 1 \end{pmatrix} \in GL_2(\widehat{\mathbf{Q}}_{(p)}[t, t^{-1}]).$$

Theorem 12.1 *The matrices $Z(n)$ represent pairwise distinct classes in $\overline{GL_2}(\mathcal{T}(p))$.*

Proof Suppose that the assertion is false; then for some integers m, n with $1 \leq m < n$ there exist $X \in GL_2(\widehat{\mathbf{Q}}_{(p)})$, $Y \in GL_2(\widehat{\mathbf{Z}}_{(p)}[t, t^{-1}])$ such that $Z(m) = XZ(n)Y$. For convenience we express this in the form

$$X^{-1}Z(m) = Z(n)Y. \tag{I}$$

Write

$$X^{-1} = \begin{pmatrix} a & b \\ c & d \end{pmatrix}; \qquad Y = \begin{pmatrix} A & B \\ C & D \end{pmatrix},$$

where $a, b, c, d \in \widehat{\mathbf{Q}}_{(p)}$ and $A, B, C, D \in \widehat{\mathbf{Z}}_{(p)}[t, t^{-1}]$. Expanding (I) gives

$$\begin{pmatrix} a & b + a\frac{t^m}{p} \\ c & d + c\frac{t^m}{p} \end{pmatrix} = \begin{pmatrix} A + C\frac{t^n}{p} & B + D\frac{t^n}{p} \\ C & D \end{pmatrix}. \tag{II}$$

Now write $A = \sum_r A_r t^r$; $B = \sum_r B_r t^r$; $C = \sum_r C_r t^r$; $D = \sum_r D_r t^r$ and equate constants and coefficients of t^r. Equating entries in the $(2, 1)$ position we see that C is a constant polynomial; that is:

$$C_0 = c \quad \text{and} \quad C_r = 0 \quad \text{for } r > 0. \tag{III}$$

Now substituting back and equating entries in the $(1, 1)$ position we get

$$A = a - \left(\frac{c}{p}\right)t^n$$

that is

$$A_0 = a; \qquad A_n = -\frac{c}{p}; \qquad A_r = 0 \quad \text{for } r \neq 0, n. \tag{IV}$$

A similar calculation for the $(2, 2)$ position gives

$$D_0 = d; \qquad D_m = \frac{c}{p}; \qquad D_r = 0 \quad \text{for } r \neq 0, m. \tag{V}$$

Substituting back and equating entries in the $(1, 2)$ position gives

$$B = b + \left(\frac{a}{p}\right)t^m - \left(\frac{d}{p}\right)t^n - \left(\frac{c}{p^2}\right)t^{m+n}.$$

Now since $1 \le m < n$ we see that

$$b = B_0 \in \widehat{\mathbf{Z}}_{(p)}; \qquad \left(\frac{a}{p}\right) = B_m \in \widehat{\mathbf{Z}}_{(p)},$$

$$\left(\frac{d}{p}\right) = -B_n \in \widehat{\mathbf{Z}}_{(p)}; \qquad \left(\frac{c}{p^2}\right) = -B_{m+n} \in \widehat{\mathbf{Z}}_{(p)}.$$

Write $a = p\alpha$; $b = \beta$; $c = p^2\gamma$; $d = p\delta$ where $\alpha, \beta, \gamma, \delta \in \widehat{\mathbf{Z}}_{(p)}$. Substitution gives

$$Y = \begin{pmatrix} p\alpha - p\gamma t^n & \beta + \alpha t^m - \delta t^n - \gamma t^{m+n} \\ p^2\gamma & p\delta + p\gamma t^m \end{pmatrix}$$

and $\det(Y) = p^2(\alpha\delta - \beta\gamma)$, and in particular, $\det(Y)$ is, by calculation, a constant polynomial. Now $Y \in GL_2(\widehat{\mathbf{Z}}_{(p)}[t, t^{-1}])$ so that $\det(Y) \in (\widehat{\mathbf{Z}}_{(p)}[t, t^{-1}])^*$. As the unit group of $\widehat{\mathbf{Z}}_{(p)}[t, t^{-1}]$ consists of polynomials of the form ut^m with $u \in \widehat{\mathbf{Z}}_{(p)}^*$ and $m \in \mathbf{Z}$ and, as we have already calculated, $\det(Y)$ is a constant polynomial then $\det(Y) \in \widehat{\mathbf{Z}}_{(p)}^*$. However $(\alpha\delta - \beta\gamma) \in \widehat{\mathbf{Z}}_{(p)}$ so that $\det(Y) = p^2(\alpha\delta - \beta\gamma) \notin \widehat{\mathbf{Z}}_{(p)}^*$. This is a contradiction. Thus $Z(m)$, $Z(n)$ represent distinct classes in $\overline{GL_2(\mathcal{T}(p))}$. □

12.2 Local Properties of Quaternions at Odd Primes

For any commutative ring R the quaternion algebra $(\frac{-1,-1}{R})$ is obtained by imposing on the free R-module of rank 4, with basis elements $\{1, i, j, k\}$ the (associative) multiplication determined by

$$i^2 = j^2 = -1; \qquad k = ij = -ji.$$

We put $\Omega = (\frac{-1,-1}{\mathbf{Z}})$. Evidently Ω is an order in $(\frac{-1,-1}{\mathbf{Q}})$. However it is not a maximal order. We denote by Γ the unique maximal order containing Ω. It may be described explicitly as $\Gamma = \mathrm{span}_{\mathbf{Z}}\{1, i, j, \omega\}$ where $\omega = \frac{1}{2}(1 + i + j + k)$. We note a classical result of Hurwitz [45]; (see also [83, p. 83]):

$$\Gamma \text{ is a (noncommutative) principal ideal domain.} \tag{12.2}$$

For each prime p put $\Omega_{(p)} = \Omega \otimes_{\mathbf{Z}} \mathbf{Z}_{(p)}$ and $\Gamma_{(p)} = \Gamma \otimes_{\mathbf{Z}} \mathbf{Z}_{(p)}$ It follows from (12.2) that $\Gamma_{(p)}$ is a principal ideal domain for every prime p. By contrast, neither Ω nor $\Omega_{(2)}$ is a principal ideal domain; both have infinite global dimension. However, when p is odd, $\Omega_{(p)} = \Gamma_{(p)}$ so that:

$$\Omega_{(p)} \text{ is a principal ideal domain for each odd prime } p. \tag{12.3}$$

Now put $\widehat{\Omega}_{(p)} = \Omega \otimes_{\mathbf{Z}} \widehat{\mathbf{Z}}_{(p)}$. When p is odd there is a ring isomorphism $\widehat{\Omega}_{(p)} \cong M_2(\widehat{\mathbf{Z}}_{(p)})$ with the ring of 2×2 matrices over $\widehat{\mathbf{Z}}_{(p)}$. For the sake of completeness we give a proof beginning with:

Proposition 12.4 *Let R be a commutative ring in which 2 is invertible and in which there exist $\xi, \eta \in R$ such that $\xi^2 + \eta^2 = -1$; then there is an isomorphism of R-algebras*

$$\left(\frac{-1, -1}{R} \right) \cong M_2(R).$$

Proof One shows easily that the R-linear map $\theta : (\frac{-1,-1}{R}) \longrightarrow M_2(R)$ defined by

$$\theta(1) = \begin{pmatrix} 1 & 0 \\ 0 & 1 \end{pmatrix}; \qquad \theta(i) = \begin{pmatrix} 0 & 1 \\ -1 & 0 \end{pmatrix};$$

$$\theta(j) = \begin{pmatrix} \xi & \eta \\ \eta & -\xi \end{pmatrix}; \qquad \theta(k) = \begin{pmatrix} \eta & -\xi \\ -\xi & -\eta \end{pmatrix}$$

is a ring homomorphism and is bijective when 2 is invertible in R. $\square$

To show that the equation $\xi^2 + \eta^2 = -1$ has a solution in $\widehat{\mathbf{Z}}_{(p)}$ we begin by showing it has a solution in $\mathbf{F}_p$, the field with p elements (compare [77, p. 162]). Note that in $\mathbf{F}_p$ the set $(\mathbf{F}_p^*)^2$ of nonzero squares is a subgroup of index two in $\mathbf{F}_p^*$. If $x, y \in \mathbf{F}_p^* - (\mathbf{F}_p^*)^2$ then

$$\frac{x}{y} \in (\mathbf{F}_p^*)^2.$$

As a preliminary observation note that the mapping $\psi : \mathbf{F}_p \to \mathbf{F}_p; \ \psi(x) = x + 1$ has the property that $\mathbf{F}_p = \{\psi(1), \psi^2(1), \ldots, \psi^p(1)\}$.

Proposition 12.5 *For any odd prime p there exist $\xi, \eta \in \mathbf{F}_p$ such that $\xi^2 + \eta^2 = -1$.*

Proof There are two cases according to whether or not -1 is a square in $\mathbf{F}_p$. First suppose that -1 is not a square and consider the restriction of ψ to $(\mathbf{F}_p^*)^2$. Assume that $\psi((\mathbf{F}_p^*)^2) \subset (\mathbf{F}_p^*)^2$; then for all $r \geq 1$, $\psi^r((\mathbf{F}_p^*)^2) \subset (\mathbf{F}_p^*)^2$. However $1 \in (\mathbf{F}_p^*)^2$ so that $\mathbf{F}_p = \{\psi(1), \psi^2(1), \ldots, \psi^p(1)\} \subset (\mathbf{F}_p^*)^2$. This is a contradiction as $0 \notin (\mathbf{F}_p^*)^2$. Thus there exists $\xi \in \mathbf{F}_p^*$ such that $\xi^2 + 1 \notin (\mathbf{F}_p^*)^2$. However, $-1 \notin (\mathbf{F}_p^*)^2$ so that

$$\frac{(\xi^2 + 1)}{-1} = -1 - \xi^2 \in (\mathbf{F}_p^*)^2;$$

that is, there exists $\eta \in \mathbf{F}_p^*$ such that $\eta^2 = -\xi^2 - 1$ so solving the equation $\xi^2 + \eta^2 = -1$. The case where -1 is a square is trivial; if $\eta \in \mathbf{F}_p$ satisfies $\eta^2 = -1$ then choosing $\xi = 0$ the equation $\xi^2 + \eta^2 = -1$ is again solved. $\qquad\square$

Corollary 12.6 *For any odd prime* p, $(\frac{-1,-1}{\mathbf{F}_p}) \cong M_2(\mathbf{F}_p)$.

We proceed to show that $\xi^2 + \eta^2 = -1$ has a solution over $\widehat{\mathbf{Z}}_{(p)}$. To avoid making a formal statement of Hensel's Lemma we proceed as follows; say that a ring homomorphism $\varphi : \widehat{L} \to L$ has *property* $\mathcal{L}$ when (i) φ is surjective and the induced map on units $\varphi : \widehat{L}^* \to L^*$ is also surjective; (ii) 2 is invertible in $\widehat{L}$; (iii) $\mathrm{Ker}(\varphi)^2 = 0$.

Proposition 12.7 *Let* $\varphi : \widehat{L} \to L$ *be a ring homomorphism with property* $\mathcal{L}$ *and suppose that* $\xi \in L$ *and* $\eta \in L^*$ *satisfy* $\xi^2 + \eta^2 = -1$. *Then there exist* $\widehat{\xi} \in \widehat{L}$ *and* $\widehat{\eta} \in \widehat{L}^*$ *such that* $\varphi(\widehat{\xi}) = \xi$ *and* $\varphi(\widehat{\eta}) = \eta$ *and such that* $\widehat{\xi}^2 + \widehat{\eta}^2 = -1$.

Proof As φ is surjective, choose $\widehat{\xi} \in \widehat{L}$ such that $\varphi(\widehat{\xi}) = \xi$. As φ is also surjective on units choose $\mu \in \widehat{L}^*$ such that $\varphi(\mu) = \eta$; then

$$\varphi(1 + \widehat{\xi}^2 + \mu^2) = 1 + \xi^2 + \eta^2 = 0.$$

Both 2 and μ are invertible in $\widehat{L}$ so put $k = \frac{1}{2\mu}(1 + \widehat{\xi}^2 + \mu^2) \in \mathrm{Ker}(\varphi)$ and $\widehat{\eta} = \mu - k$. Then $\varphi(\widehat{\eta}) = \varphi(\mu) = \eta$ as $\varphi(k) = 0$. Moreover, $k^2 = 0$ as $\mathrm{Ker}(\varphi)^2 = 0$. Hence

$$1 + \widehat{\xi}^2 + \widehat{\eta}^2 = 1 + \widehat{\xi}^2 + \mu^2 - 2\mu k = 0$$

and $\widehat{\xi}^2 + \widehat{\eta}^2 = -1$ as required. Furthermore, $\widehat{\eta} \in \widehat{L}^*$ as $\widehat{\eta}(\mu + k) = \mu^2$ and $\mu^2 \in \widehat{L}$. $\qquad\square$

The canonical homomorphism $\varphi_n : \mathbf{Z}/p^{n+1} \to \mathbf{Z}/p^n$ has property $\mathcal{L}$ when p is odd. Choose $\xi, \eta \in \mathbf{F}_p$ with $\eta \neq 0$ so that $\xi^2 + \eta^2 = -1$ and apply Proposition 12.7 iteratively to the homomorphisms φ_n to construct a sequence $\{(\xi_n, \eta_n)\}_{1 \leq n}$ with $\xi_n, \eta_n \in \mathbf{Z}/p^n$ such that (i) $\xi_1 = \xi$ and $\eta_1 = \eta$; (ii) for each $n \geq 1$, $\varphi_n(\xi_{n+1}) = \xi_n$, $\varphi_n(\eta_{n+1}) = \eta_n$ and $\eta_{n+1} \in (\mathbf{Z}/p^{n+1})^*$; (iii) $\xi_n^2 + \eta_n^2 = -1$. Identifying $\widehat{\mathbf{Z}}_{(p)}$ with $\underleftarrow{\lim}(\varphi_n)$ it follows that:

Corollary 12.8 *For each odd prime* p *there exist* $\widehat{\xi}, \widehat{\eta} \in \widehat{\mathbf{Z}}_{(p)}$ *such that* $\widehat{\xi}^2 + \widehat{\eta}^2 = -1$.

Corollary 12.9 *For any odd prime* p, $(\frac{-1,-1}{\widehat{\mathbf{Z}}_{(p)}}) \cong M_2(\widehat{\mathbf{Z}}_{(p)})$.

12.3 A Quaternionic Corner Calculation

In this section p will denote an odd prime. Start from the corner

$$\begin{cases} & \widehat{\mathbf{Z}}_{(p)}[t,t^{-1}] \\ & \qquad\downarrow \\ \mathbf{Q}[t,t^{-1}] \to & \widehat{\mathbf{Q}}_{(p)}[t,t^{-1}] \end{cases}$$

Then applying the functor $(\frac{-1,-1}{\underline{\quad}})$ we obtain the corner

$$\mathcal{Q}(p) = \begin{cases} & (\frac{-1,-1}{\widehat{\mathbf{Z}}_{(p)}[t,t^{-1}]}) \\ & \qquad\downarrow \\ (\frac{-1,-1}{\mathbf{Q}[t,t^{-1}]}) \to & (\frac{-1,-1}{\widehat{\mathbf{Q}}_{(p)}[t,t^{-1}]}) \end{cases}$$

We will show that the unit set $\overline{GL}_1(\mathcal{Q}(p))$ is infinite. To see this, first modify the above by replacing $(\frac{-1,-1}{\mathbf{Q}[t,t^{-1}]})$ by $(\frac{-1,-1}{\mathbf{Q}})$ thus;

$$\widetilde{\mathcal{Q}}(p) = \begin{cases} & (\frac{-1,-1}{\widehat{\mathbf{Z}}_{(p)}[t,t^{-1}]}) \\ & \qquad\downarrow \\ (\frac{-1,-1}{\mathbf{Q}}) \to & (\frac{-1,-1}{\widehat{\mathbf{Q}}_{(p)}[t,t^{-1}]}) \end{cases}$$

Observe that $(\frac{-1,-1}{\mathbf{Q}[t,t^{-1}]}) \cong (\frac{-1,-1}{\mathbf{Q}})[t,t^{-1}]$. However, $(\frac{-1,-1}{\mathbf{Q}})$ is an integral domain. It follows from the 'two unique products' criterion (Appendix C) that $(\frac{-1,-1}{\mathbf{Q}})[t,t^{-1}]$ has only trivial units; $((\frac{-1,-1}{\mathbf{Q}})[t,t^{-1}])^* \cong (\frac{-1,-1}{\mathbf{Q}})^* \times \{t^k : k \in \mathbf{Z}\}$. The powers $t^k \in (\frac{-1,-1}{\mathbf{Q}[t,t^{-1}]})^*$ are central in $(\frac{-1,-1}{\widehat{\mathbf{Q}}_{(p)}[t,t^{-1}]})$ and can equally well be regarded as originating in $(\frac{-1,-1}{\widehat{\mathbf{Z}}_{(p)}[t,t^{-1}]})^*$. On taking double cosets we see that:

Proposition 12.10 $\overline{GL}_1(\mathcal{Q}(p)) = \overline{GL}_1(\widetilde{\mathcal{Q}}(p))$.

We repeat, for emphasis, that the sets $\overline{GL}_1(\mathcal{Q}(p))$ and $\overline{GL}_1(\widetilde{\mathcal{Q}}(p))$ are *identical*. Now as in Sect. 12.2 choose $\widehat{\xi}, \widehat{\eta} \in \widehat{\mathbf{Z}}_{(p)}$ such that $\widehat{\xi}^2 + \widehat{\eta}^2 = -1$; the assignments

$$1 \mapsto \begin{pmatrix} 1 & 0 \\ 0 & 1 \end{pmatrix}; \qquad i \mapsto \begin{pmatrix} 0 & 1 \\ -1 & 0 \end{pmatrix};$$

$$j \mapsto \begin{pmatrix} \widehat{\xi} & \widehat{\eta} \\ \widehat{\eta} & -\widehat{\xi} \end{pmatrix}; \qquad k \mapsto \begin{pmatrix} \widehat{\eta} & -\widehat{\xi} \\ -\widehat{\xi} & -\widehat{\eta} \end{pmatrix}$$

define an isomorphism $\theta_0 : (\frac{-1,-1}{\widehat{\mathbf{Q}}_{(p)}[t,t^{-1}]}) \to M_2(\widehat{\mathbf{Q}}_{(p)}[t,t^{-1}])$ which in turn induces a homomorphism of corners $\theta : \widetilde{\mathcal{Q}}(p) \to M_2(\mathcal{T}(p))$;

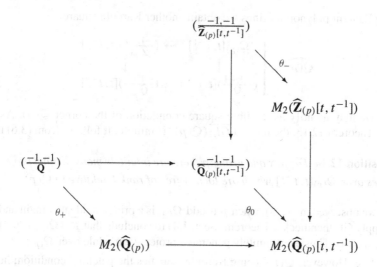

As θ_0 is an isomorphism we see that:

The induced map $\theta_* : \overline{GL_1}(\widetilde{\mathcal{Q}}(p)) \to \overline{GL_1}(M_2(\mathcal{T}(p))(= \overline{GL_2}(\mathcal{T}(p))$ is surjective.
$$\tag{12.11}$$

It now follows that $\overline{GL_1}(\mathcal{Q}(p))$ is infinite. To be precise put

$$\zeta(n) = 1 + \frac{1}{2p}\{i - \widehat{\eta}j + \widehat{\xi}k\}t^n \in \left(\frac{-1,-1}{\widehat{\mathbf{Q}}_{(p)}[t,t^{-1}]}\right);$$

then:

$$\theta_0(\zeta(n)) = Z(n) = \begin{pmatrix} 1 & \frac{t^n}{p} \\ 0 & 1 \end{pmatrix} \in GL_2(\widehat{\mathbf{Q}}_{(p)}[t,t^{-1}]). \tag{12.12}$$

Then from Theorem 12.1, Proposition 12.10, (12.11) and (12.12) it follows that:

Theorem 12.13 *The elements $\{\zeta(n)\}_{1 \le n}$ represent pairwise distinct classes in* $\overline{GL_1}(\mathcal{Q}(p))$.

12.4 Stably Free Modules over $\Omega[t, t^{-1}]$

Taking as regular submonoid $S = \{p^r : r \ge 0\}$ it is straightforward to see that the following is a Karoubi square:

$$\begin{cases} \Omega_{(p)} & \to (\frac{-1,-1}{\mathbf{Z}_{(p)}}) \\ \downarrow & \downarrow \\ (\frac{-1,-1}{\mathbf{Q}}) & \to (\frac{-1,-1}{\widehat{\mathbf{Q}}_{(p)}}) \end{cases}$$

Taking Laurent polynomial rings we obtain another Karoubi square

$$\widehat{\mathcal{Q}(p)} = \begin{cases} \Omega_{(p)}[t,t^{-1}] & \to & (\frac{-1,-1}{\mathbf{Z}_{(p)}})[t,t^{-1}] \\ \downarrow & & \downarrow \\ (\frac{-1,-1}{\mathbf{Q}})[t,t^{-1}] & \to & (\frac{-1,-1}{\mathbf{Q}_{(p)}})[t,t^{-1}] \end{cases}$$

which we may identify as the fibre square completion of the corner $\mathcal{Q}(p)$. As we saw in Theorem 12.13, the unit set $\overline{GL}_1(\mathcal{Q}(p))$ is infinite. It follows from (3.6) that:

Proposition 12.14 *For any odd prime p there are infinitely many distinct projective modules over $\Omega_{(p)}[t,t^{-1}]$ which are locally free of rank 1 relative to $\widehat{\mathcal{Q}(p)}$.*

As we observed in (12.3), when p is odd $\Omega_{(p)}$ is a principal ideal domain and we may apply Grothendieck's Theorem (Sect. 1.4) to conclude that $\widetilde{K}_0(\Omega_{(p)}[t,t^{-1}]) \cong \widetilde{K}_0(\Omega_{(p)}) = 0$. Thus every finitely generated projective module over $\Omega_{(p)}[t,t^{-1}]$ is stably free. However, $\widehat{\mathcal{Q}(p)}$, being Karoubi, satisfies the patching condition; hence by Theorem 3.8 and Proposition 12.14 we have:

Theorem 12.15 *For any odd prime p, there are infinitely many distinct stably free modules of rank 1 over $\Omega_{(p)}[t,t^{-1}]$.*

Now consider the following fibre square where $\Omega_{[p]} = \Omega \otimes \mathbf{Z}[\frac{1}{p}]$:

$$\widehat{\mathcal{K}} = \begin{cases} \Omega[t,t^{-1}] & \xrightarrow{\ i\ } & \Omega_{(p)}[t,t^{-1}] \\ \downarrow \varphi & & \downarrow \nu \\ \Omega_{[p]}[t,t^{-1}] & \xrightarrow{\ j\ } & (\frac{-1,-1}{\mathbf{Q}})[t,t^{-1}] \end{cases}$$

Then $\widehat{\mathcal{K}}$ is Karoubi and so is E-surjective (Sect. 3.7). As $(\frac{-1,-1}{\mathbf{Q}})$ is a division ring then $(\frac{-1,-1}{\mathbf{Q}})[t,t^{-1}]$ is weakly Euclidean by Theorem 2.49 and has property SFC by (9.13); as $\widehat{\mathcal{K}}$ satisfies the hypothesis of Corollary 3.36 there is a surjection $i : \mathcal{SF}_1(\Omega[t,t^{-1}]) \to \mathcal{SF}_1(\Omega_{(p)}[t,t^{-1}])$. From Theorem 12.15 we now conclude that:

Theorem 12.16 *$\Omega[t,t^{-1}]$ has infinitely many distinct isomorphism classes of stably free modules of rank 1.*

It is paradoxical that in the context of quaternion algebras it is the odd primes which behave badly. Though the details need not concern us here it can be shown that $\Omega_{(2)}[t,t^{-1}]$ has no nontrivial stably free modules.

Observe that the ring $\widehat{\mathbf{Z}}_{(p)}[t,t^{-1}]$ has Krull dimension equal to 2. It follows from Suslin's stability theorem [67, p. 111] that $\widehat{\mathbf{Z}}_{(p)}[t,t^{-1}]$ is 4-weakly Euclidean; thus $(\frac{-1,-1}{\widehat{\mathbf{Z}}_{(p)}})[t,t^{-1}]$, being isomorphic to the ring $M_2(\widehat{\mathbf{Z}}_{(p)}[t,t^{-1}])$, is 2-weakly Euclidean. Moreover, $(\frac{-1,-1}{\widehat{\mathbf{Z}}_{(p)}})[t,t^{-1}]$ also has property SFC and both $(\frac{-1,-1}{\mathbf{Q}})[t,t^{-1}]$

and $(\frac{-1,-1}{\mathbf{Q}_{(p)}})[t, t^{-1}]$ are weakly Euclidean with property SFC. Thus, in contrast to Theorem 12.15, it follows from Corollary 3.45 that:

When p is odd, $\Omega_{(p)}[t, t^{-1}]$ has no nontrivial stably free module of rank > 1.

$$(12.17)$$

12.5 Stably Free Modules over $\mathbf{Z}[C_\infty \times Q(8)]$

We again recall the cyclic algebra construction for involuted rings; given an involution θ on a ring A and an element $\mathbf{a} \in A$ such that $\theta(\mathbf{a}) = \mathbf{a}$, the cyclic algebra $\mathcal{C}(A, \theta, \mathbf{a})$ is defined to be the free A-module of rank two with basis $\{1, y\}$ and multiplication determined by $y\lambda = \theta(\lambda)y$; $y^2 = \mathbf{a}$.

Next note that the factorization $x^4 - 1 = (x^2 - 1)(x^2 + 1)$ gives rise to a fibre square

$$\begin{cases} \mathbf{Z}[x]/(x^4 - 1) \xrightarrow{\pi_-} \mathbf{Z}[x]/(x^2 + 1) \\ \quad\downarrow \pi_+ \qquad\qquad\qquad \downarrow \varphi_- \\ \mathbf{Z}[x]/(x^2 - 1) \xrightarrow{\varphi_+} \mathbf{F}_2[x]/(x^2 - 1) \end{cases}$$

We write the cyclic group of order n in the form $C_n = \langle x \mid x^n = 1 \rangle$ and, for any ring A, make the identification $A[C_n] = A[x]/(x^n - 1)$. The above fibre square then becomes

$$(*) = \begin{cases} \mathbf{Z}[C_4] \longrightarrow \mathbf{Z}[i] \\ \quad\downarrow \qquad\qquad \downarrow \\ \mathbf{Z}[C_2] \longrightarrow \mathbf{F}_2[C_2] \end{cases}$$

where $\mathbf{Z}[i] = \mathbf{Z}[x]/(x^2 + 1)$ is the ring of Gaussian integers. Now write

$$Q(8) = \langle x, y \mid x^2 = y^2, \ x^4 = 1, \ yxy^{-1} = x^{-1} \rangle.$$

C_4 is contained with index 2 in $Q(8)$ and conjugation by y induces on $\mathbf{Z}[C_4]$ the canonical involution $\widehat{} : \mathbf{Z}[C_4] \to \mathbf{Z}[C_4], \widehat{g} = g^{-1}$. Applying the cyclic algebra construction to the involuted ring $(\mathbf{Z}[C_4], \widehat{})$ with fixed element x^2 we see that

$$\mathbf{Z}[Q(8)] \cong \mathcal{C}(\mathbf{Z}[C_4], \widehat{}, x^2).$$

The canonical involution $\widehat{}$ induces involutions on the remaining corners of the above square; in turn, it induces complex conjugation (denoted below by θ) on $\mathbf{Z}[i]$ and the identity on both $\mathbf{Z}[C_2]$ and $\mathbf{F}_2[C_2]$. Moreover, the fixed elements corresponding to x^2 are, again in turn, -1 in $\mathbf{Z}[i]$ and 1 in both $\mathbf{Z}[C_2]$ and $\mathbf{F}_2[C_2]$. Observe that $\mathcal{C}(\mathbf{Z}[i], \theta, -1)$ is isomorphic to $\Omega = (\frac{-1,-1}{\mathbf{Z}})$. Applying the cyclic algebra construction to $(*)$ we obtain the fibre square:

$$\widehat{\mathcal{M}} = \begin{cases} \mathbf{Z}[Q(8)] \longrightarrow \Omega \\ \quad\downarrow \qquad\qquad \downarrow \\ \mathbf{Z}[C_2 \times C_2] \longrightarrow \mathbf{F}_2[C_2 \times C_2] \end{cases}$$

Taking tensor product $- \otimes \mathbf{Z}[t, t^{-1}]$ we obtain another fibre square

$$\widehat{\mathcal{M}}[t, t^{-1}] = \begin{cases} \mathbf{Z}[C_\infty \times Q(8)] & \longrightarrow & \Omega[t, t^{-1}] \\ \downarrow & & \downarrow \\ \mathbf{Z}[C_\infty \times C_2 \times C_2] & \longrightarrow & \mathcal{T}[t, t^{-1}] \end{cases}$$

where $\mathcal{T} = \mathbf{F}_2[C_2 \times C_2]$. If $\epsilon : \mathcal{T} = \mathbf{F}_2[C_2 \times C_2] \to \mathbf{F}_2$ is the augmentation and $\mathbf{m} = \mathrm{Ker}(\epsilon)$ then $\mathbf{m}^2 = 0$. Thus $\mathcal{T}$ is a local ring with nilpotent radical and hence $\mathcal{T}[t, t^{-1}]$ is weakly Euclidean by Corollary 2.52. Furthermore, $\mathcal{T}[t, t^{-1}]$ has property SFC by (9.18); that is:

$$\mathcal{T}[t, t^{-1}] \text{ is weakly Euclidean and has property SFC.} \tag{12.18}$$

Clearly $\widehat{\mathcal{M}}[t, t^{-1}]$ is a Milnor square and so has the patching property. Thus $\widehat{\mathcal{M}}[t, t^{-1}]$ satisfies the hypotheses of Corollary 3.36 so that $\mathcal{SF}_1(\mathbf{Z}[C_\infty \times Q(8)]) \to \mathcal{SF}_1(\Omega[t, t^{-1}])$ is surjective. From Theorem 12.16 we now see:

Theorem 12.19 $\mathbf{Z}[C_\infty \times Q(8)]$ *admits infinitely many isomorphically distinct stably free modules of rank* 1.

There is a corollary of Theorem 12.19 which, though obvious, is still perhaps worth pointing out. Suppose that G is a finitely generated group such that $H_1(G; \mathbf{Q}) \neq 0$. This condition is equivalent to the existence of a surjective group homomorphism $\rho : G \to C_\infty$. Put $r = \rho \times \mathrm{Id} : G \times Q(8) \to C_\infty \times Q(8)$. Choosing $g \in G$ so that $\rho(g) = t$ the correspondence $(t, q) \mapsto (g, q)$ determines a homomorphism $i : C_\infty \times Q(8) \to G \times Q(8)$ such that $r \circ i = \mathrm{Id}_{C_\infty \times Q(8)}$; that is, $C_\infty \times Q(8)$ is a retract of $C_\infty \times Q(8)$. If $\{S_\xi\}_{\xi \in X}$ is an infinite collection of pairwise nonisomorphic stably free modules of rank 1 over $\mathbf{Z}[C_\infty \times Q(8)]$ then $\{i_*(S_\xi)\}_{\xi \in X}$ is likewise an infinite collection of pairwise nonisomorphic stably free modules of rank 1 over $\mathbf{Z}[G \times Q(8)]$; hence we obtain:

Corollary 12.20 *If* G *is a finitely generated group with* $H_1(G; \mathbf{Q}) \neq 0$ *then* $\mathbf{Z}[G \times Q(8)]$ *admits infinitely many isomorphically distinct stably free modules of rank* 1.

12.6 Extension to Generalized Quaternion Groups

It is convenient to treat the generalized quaternion groups as two different families:

(I) $Q(8m) = \langle x, y \mid x^{2m} = y^2, xyx = y \rangle$ where m is arbitrary.
(II) $Q(4m) = \langle x, y \mid x^m = y^2, xyx = y \rangle$ where m is odd.

We begin with (I); we reduce the problem for the groups $Q(8m)$ with $m \geq 2$ to that of $Q(8)$ considered already in Sect. 12.5. Thus supppose $m \geq 2$ and take the following Milnor square where $q_m(x) = x^{m-1} + x^{m-2} + \cdots + x + 1$:

$$\begin{cases} \mathbf{Z}[x]/(x^{4m} - 1) \overset{\eta}{\longrightarrow} \mathbf{Z}[x]/q_m(x^4) \\ \quad \psi \downarrow \qquad\qquad\qquad \downarrow \\ \mathbf{Z}[x]/(x^4 - 1) \longrightarrow (\mathbf{Z}/m)[x]/(x^4 - 1) \end{cases} \qquad (12.21)$$

Now take θ to be the canonical involution on $\mathbf{Z}[x]/(x^{4m} - 1) = \mathbf{Z}[C(4m)]$; $\theta(x^r) = x^{4m-r}$. Observe that θ induces involutions on the other corners. As the above square is now a commutative diagram of involuted rings, we may apply the cyclic algebra construction; that is we introduce a new variable y subject to the relations $yx^r = x^{4m-r}y$ and $y^2 = x^{2m} = \theta(x^{2m})$. We get a Milnor square

$$\begin{cases} \mathcal{C}(\mathbf{Z}[C(4m)], \theta, x^{2m}) \overset{\eta}{\longrightarrow} \mathcal{C}(\mathbf{Z}[x]/q_m(x^4), \theta, x^{2m}) \\ \quad\quad \psi \downarrow \qquad\qquad\qquad\qquad \downarrow \\ \mathcal{C}(C(4), \theta, x^2) \quad\longrightarrow\quad \mathcal{C}(\mathbf{Z}/m, \theta, x^2) \end{cases}$$

which, on putting $\mathcal{C}(m) = \mathcal{C}(\mathbf{Z}[x]/q_m(x^4), \theta, x^{2m})$, we may write as

$$\begin{cases} \mathbf{Z}[Q(8m)] \overset{\eta}{\longrightarrow} \mathcal{C}(m) \\ \quad\; \psi \downarrow \qquad\qquad\quad \downarrow \\ \mathbf{Z}[C_\infty \times Q(8)] \longrightarrow (\mathbf{Z}/m)[Q(8)]. \end{cases} \qquad (12.22)$$

Now put $\mathcal{E}(m) = (\mathbf{Z}/m)[C_\infty \times Q(8)]$; tensoring (12.22) with $\mathbf{Z}[C_\infty]$ and we obtain:

$$\begin{cases} \mathbf{Z}[C_\infty \times Q(8m)] \overset{\eta}{\longrightarrow} \mathcal{C}(m)[C_\infty] \\ \quad\quad \psi \downarrow \qquad\qquad\qquad \downarrow \\ \mathbf{Z}[C_\infty \times Q(8)] \quad\longrightarrow\quad \mathcal{E}(m) \end{cases} \qquad (12.23)$$

To proceed we must examine the structure of $\mathcal{E}(m) = (\mathbf{Z}/m)[C_\infty \times Q(8)]$. We first deal with the case $m = 2$; then $\mathbf{Z}/m$ is the field $\mathbf{F}_2$ and we have:

Proposition 12.24 $\mathbf{F}_2[Q(8)]$ *is a local ring with nilpotent radical.*

Proof Observe that $Q(8)^{ab} \cong C_2 \times C_2$ and that $\mathbf{F}_2[C_2 \times C_2]$ is a local ring; indeed, the augmentation homomorphism $\epsilon : \mathbf{F}_2[C_2 \times C_2] \to \mathbf{F}_2$ is surjective with nilpotent kernel. Now the projection $Q(8) \to Q(8)^{ab}$ induces a surjective ring homomorphism

$$\natural : \mathbf{F}_2[Q(8)] \to \mathbf{F}_2[C_2 \times C_2].$$

In terms of the presentation $Q(8) = \langle x, y \mid x^2 = y^2, xyx = y \rangle$ it is straightforward to see that $\mathrm{Ker}(\natural)$ is spanned by the elements $\{1 + z, x + xz, y + yz, xy + xyz\}$ where $z = x^2 = y^2$. Moreover, $1 + z$ is central and $(1 + z)^2 = 0$. Thus $\mathrm{Ker}(\natural)$ is nilpotent. The result now follows as $\epsilon \circ \natural : \mathbf{F}_2[Q(8)] \to \mathbf{F}_2$ is surjective with nilpotent kernel. $\qquad\square$

Proposition 12.25 *For each* $e \geq 1$, $(\mathbf{Z}/2^e)[Q(8)]$ *is a local ring with nilpotent radical.*

Proof For $e = 1$ this is simply Proposition 12.24. When $e \geq 2$, the natural projection $\mathbf{Z}/2^e \to \mathbf{F}_2$ induces a surjective ring homomorphism $\natural : (\mathbf{Z}/2^e)[Q(8] \to \mathbf{F}_2[Q(8)]$. Composition with $\epsilon \circ \natural$ as per Proposition 12.24 gives a surjective ring homomorphism $(\mathbf{Z}/2^e)[Q(8] \to \mathbf{F}_2$ with nilpotent kernel.   □

Proposition 12.26 *For each $e \geq 1$, $\mathcal{E}(2^e)$ is weakly Euclidean and has property* SFC.

Proof $\mathcal{E}(2^e) \cong ((\mathbf{Z}/2^e)[Q(8])[C_\infty]$ is weakly Euclidean by Corollary 2.52 and Proposition 12.25. Likewise, from (9.18) and Proposition 12.25 it follows that $\mathcal{E}(2^e)$ has SFC.   □

Turning to odd primes, we have:

Proposition 12.27 *$\mathcal{E}(p^e)$ is weakly Euclidean and has property SFC for any odd prime p and any $e \geq 1$.*

Proof For any commutative ring A in which 2 is invertible it is easy to see that

$$A[Q(8)] \cong A \times A \times A \times A \times \left(\frac{-1, -1}{A}\right):$$

Taking $A = \mathbf{Z}/p^e$ then, as in Corollary 12.6, Proposition 12.7, $(\frac{-1,-1}{\mathbf{Z}/p^e}) \cong M_2(\mathbf{Z}/p^e)$ and so

$$\mathcal{E}(p^m) \cong (\mathbf{Z}/p^e)[C_\infty]^{(4)} \times M_2((\mathbf{Z}/p^e)[C_\infty]).$$

Now $(\mathbf{Z}/p^e)[C_\infty]$ is weakly Euclidean, by Corollary 2.52 and has SFC , by (9.18). Thus $M_2((\mathbf{Z}/p^e)[C_\infty])$ is weakly Euclidean, by Theorem 2.48 and has SFC, by (9.20). The result now follows from Corollary 2.46 and (9.21).   □

In general, write m as a product of powers of distinct primes $m = 2^e p_1^{e_1} \cdots p_k^{e_k}$ so that $\mathbf{Z}/m \cong \mathbf{Z}/p^{e_1} \times \cdots \times \mathbf{Z}/p^{e_k}$ and hence

$$\mathcal{E}(m) = \mathcal{E}(p_1^{e_1}) \times \cdots \times \mathcal{E}(p_k^{e_k}).$$

It follows from Propositions 12.26, 12.27, Corollary 2.46 and (9.21) that:

For any integer $m \geq 2$, $\mathcal{E}(m)$ is weakly Euclidean and has property SFC. (12.28)

The square in (12.23) satisfies the hypotheses of Corollary 3.36. It follows that

$$\psi \times \eta : \mathcal{SF}_1(\mathbf{Z}[C_\infty \times Q(8m]) \to \mathcal{SF}_1(\mathbf{Z}[C_\infty \times Q(8)]) \times \mathcal{SF}_1(\mathcal{C}(m)[C_\infty])$$

is surjective. In general the task of giving a complete description of $\mathcal{SF}_1(\mathcal{C}(m)[C_\infty])$ is complicated; ignoring this factor we nevertheless see that:

For each $m \geq 2$, $\mathcal{SF}_1(\mathbf{Z}[C_\infty \times Q(8m]) \overset{\psi}{\to} \mathcal{SF}_1(\mathbf{Z}[C_\infty \times Q(8)]$ is surjective.
(12.29)

From Theorem 12.19 we see that:[1]

$$\mathcal{SF}_1(\mathbf{Z}[C_\infty \times Q(8m)] \text{ is infinite when } m \geq 1. \tag{12.30}$$

If G is a finitely generated group with $H_1(G; \mathbf{Q}) \neq 0$ then $C_\infty \times Q(8m)$ is a retract of $G \times Q(8m)$ so we also obtain:

Let G be a finitely generated group with $H_1(G; \mathbf{Q}) \neq 0$; then

$$\mathcal{SF}_1(\mathbf{Z}[G \times Q(8m)]) \text{ is infinite for each } m \geq 1. \tag{12.31}$$

In particular, we could take $G = F_n$ to be a finitely generated free group.

We now suppose that $m > 1$ is odd and turn to the groups $Q(4m)$; consider the following Milnor square where $q_m(x) = x^{m-1} + x^{m-2} + \cdots + x + 1$:

$$
\begin{cases}
\mathbf{Z}[x]/(x^{2m} - 1) \xrightarrow{\eta} \mathbf{Z}[x]/q_m(x^2) \\
\quad \psi \downarrow \qquad\qquad\qquad \downarrow \\
\mathbf{Z}[x]/(x^2 - 1) \longrightarrow (\mathbf{Z}/m)[x]/(x^2 - 1)
\end{cases}
\tag{12.32}
$$

In applying the cyclic algebra construction, as m is odd, the relations $y^2 = x^m$ and $x^2 = 1$ at the bottom lefthand corner force $y^2 = x$ and so allow us to eliminate x at this position. We then obtain a Milnor square

$$
\begin{cases}
\mathcal{C}(\mathbf{Z}[C(2m)], \theta, x^m) \xrightarrow{\eta} \mathcal{C}(\mathbf{Z}[x]/q_m(x^2), \theta, x^m) \\
\quad \psi \downarrow \qquad\qquad\qquad\qquad \downarrow \\
\mathbf{Z}[y]/(y^4 - 1) \longrightarrow (\mathbf{Z}/m)[y]/(y^4 - 1)
\end{cases}
$$

which we may write as:

$$
\begin{cases}
\mathbf{Z}[Q(4m)] \xrightarrow{\eta} \mathcal{C}(\mathbf{Z}[x]/q_m(x^2), \theta, x^m) \\
\quad \psi \downarrow \qquad\qquad\qquad \downarrow \\
\mathbf{Z}[C(4)] \longrightarrow (\mathbf{Z}/m)[C(4)]
\end{cases}
\tag{12.33}
$$

Now let F_n be the free group of rank n. We shall write $\mathcal{K}(m, n) = \mathcal{C}(\mathbf{Z}[x]/q_m(x^2), \theta, x^m)[F_n]$ and $\mathcal{F}(m, n) = (\mathbf{Z}/m)[F_n \times C_4] \cong ((\mathbf{Z}/m)[C_4])[F_n]$ so that we have a Milnor square:

$$
\begin{cases}
\mathbf{Z}[F_n \times Q(4m)] \xrightarrow{\eta} \mathcal{K}(m, n) \\
\quad \psi \downarrow \qquad\qquad\qquad \downarrow \\
\mathbf{Z}[F_n \times C_4] \longrightarrow \mathcal{F}(m, n)
\end{cases}
\tag{12.34}
$$

Given a commutative ring A in which 2 is invertible one sees easily that

$$A[C(4)] \cong A \times A \times A[t]/(t^2 + 1). \tag{12.35}$$

---

[1]This generalizes a result of Pouya Kamali; in his thesis [61], using a different system of fibre squares, Kamali was able to show that $\mathcal{SF}_1(\mathbf{Z}[C_\infty \times Q(8m)]$ is infinite when $m > 1$ is *not a power of 2*.

When p is an odd prime $(\mathbf{Z}/p^a)[t]/(t^2+1) = \mathbf{Z}/(p^a) \times \mathbf{Z}/(p^a)$ if $p \equiv 1 \bmod 4$ whilst $(\mathbf{Z}/p^a)[t]/(t^2+1)$ is a local ring with nilpotent radical if $p \equiv 3 \bmod 4$. If m is not a prime power we write it as a product of powers of distinct odd primes $m = p_1^{e_1} \cdots p_k^{e_k}$ so that $\mathbf{Z}/m \cong \mathbf{Z}/p^{e_1} \times \cdots \times \mathbf{Z}/p^{e_k}$. Collecting our results we see that in each case $(\mathbf{Z}/m)[C(4)]$ is a finite product of local rings each with nilpotent radical. In consequence of Corollary 2.46 and Proposition 2.47 we now have:

$$\mathcal{F}(m,n) \text{ is weakly Euclidean.} \tag{12.36}$$

Likewise from (9.18) and (9.21) it follows that:

$$\mathcal{F}(m,n) \text{ has property SFC.} \tag{12.37}$$

Thus the square in (12.34) satisfies the hypotheses of Corollary 3.36. It follows that

$$\psi \times \eta : \mathcal{SF}_1(\mathbf{Z}[F_n \times Q(4m]) \to \mathcal{SF}_1(\mathbf{Z}[F_n \times C(2) \times C(2)]) \times \mathcal{SF}_1(\mathcal{K}(m,n))$$

is surjective. Again the task of giving a complete description of $\mathcal{SF}_1(\mathcal{K}(m,n))$ is complicated. If we ignore $\mathcal{SF}_1(\mathcal{K}(m,n))$ we see that:

$$\mathcal{SF}_1(\mathbf{Z}[F_n \times Q(4m]) \overset{\psi}{\to} \mathcal{SF}_1(\mathbf{Z}[F_n \times C_4]) \text{ is surjective for } m \text{ odd.} \tag{12.38}$$

From Corollary 10.27 $\mathcal{SF}_1(\mathbf{Z}[F_n \times C_4])$ is infinite when $n \geq 2$; thus:

$$\mathcal{SF}_1(\mathbf{Z}[F_n \times Q(4m)]) \text{ is infinite when } m \text{ is odd and } n \geq 2. \tag{12.39}$$

When m is odd the problem for $n = 1$, that is for $C_\infty \times Q(4m)$, is rather more delicate. We pose the following question for any odd integer $m > 1$:

$$\text{Is } \mathcal{SF}_1(\mathbf{Z}[C_\infty \times Q(4m)]) \text{ infinite?} \tag{12.40}$$

Chapter 13
Parametrizing $\Omega_1(\mathbf{Z})$: Generic Case

In this chapter we will take G to be a finitely generated group and we denote by $\mathcal{SF}_+$ the isomorphism classes of finitely generated nonzero stably free modules over $\mathbf{Z}[G]$. As before we denote by $\Omega_1(\mathbf{Z})$ the first syzygy of $\mathbf{Z}$ over $\mathbf{Z}[G]$; that is, the stable class $[J]$ of any module J which occurs in an exact sequence of Λ-modules

$$0 \to J \to \Lambda^m \to \mathbf{Z} \to 0$$

where $\Lambda = \mathbf{Z}[G]$. We have previously seen that both $\Omega_1(\mathbf{Z})$ and $\mathcal{SF}_+$ have the structure of trees in which the roots do not extend infinitely downwards and, for certain G at least, we considered the structure of $\mathcal{SF}_+$ in some detail. Here we seek to parametrize $\Omega_1(\mathbf{Z})$ by $\mathcal{SF}_+$; that is, the 'unknown' by the 'known'. We will show that, under suitable conditions, there is a height preserving mapping of trees $\kappa : \mathcal{SF}_+ \to \Omega_1(\mathbf{Z})$; compare also [54]. We then proceed to establish conditions under which κ is injective and/or surjective. The problem divides naturally into two cases, according to whether $\mathrm{Ext}^1(\mathbf{Z}, \Lambda)$ is zero or not.

Generic Case: $\mathrm{Ext}^1(\mathbf{Z}, \Lambda) = 0$
Singular Case: $\mathrm{Ext}^1(\mathbf{Z}, \Lambda) \neq 0$

The Generic Case admits of a reasonably complete conclusion, which we detail below. The Singular Case, however, is more intricate and is considered in Chaps. 14 and 15.

13.1 Minimality of the Augmentation Ideal

Evidently a mapping κ with the properties considered above must transform the minimal level of $\mathcal{SF}_+$ to that of $\Omega_1(\mathbf{Z})$. Therein lies our first difficulty. The minimal level of $\mathcal{SF}_+$ consists simply of the isomorphism classes of stably free modules of rank 1. By contrast, the minimal level of $\Omega_1(\mathbf{Z})$ is less easy to characterize. To explain this, let $\epsilon : \mathbf{Z}[G] \to \mathbf{Z}$ denote the augmentation homomorphism; then the augmentation ideal $\mathcal{I} = \text{Ker}(\epsilon)$ represents an element of $\Omega_1(\mathbf{Z})$, occuring as it does in an exact sequence

$$0 \to \mathcal{I} \to \Lambda \to \mathbf{Z} \to 0.$$

As the middle term Λ has minimal rank we may expect thereby that $\mathcal{I}$ lies at the minimal level of $\Omega_1(\mathbf{Z})$. Unfortunately, this need not be the case. As we note below, $\mathcal{I}$ fails to be minimal when G is a free group of rank ≥ 2. Our first task is therefore to establish reasonable conditions which guarantee that $\mathcal{I}$ does indeed lie at the minimal level of $\Omega_1(\mathbf{Z})$.

In what follows we assume, without further comment, that G is a finitely generated group with integral group ring $\Lambda = \mathbf{Z}[G]$; we establish:

First minimality criterion $\mathcal{I}$ is minimal in $\Omega_1(\mathbf{Z})$ if $\text{Ext}^1(\mathbf{Z}, \Lambda) = 0$.   (13.1)

Proof Let $h : J \oplus \Lambda^\beta \xrightarrow{\approx} \mathcal{I} \oplus \Lambda^\alpha$ be an isomorphism of Λ-modules. From the extension $0 \to \mathcal{I} \to \Lambda \xrightarrow{\epsilon} \mathbf{Z} \to 0$ defining $\mathcal{I}$ we may construct a succession of exact sequences thus:

$$0 \to \mathcal{I} \oplus \Lambda^\alpha \xrightarrow{i} \Lambda^{\alpha+1} \xrightarrow{\epsilon} \mathbf{Z} \to 0,$$

$$0 \to J \oplus \Lambda^\beta \xrightarrow{j} \Lambda^{\alpha+1} \xrightarrow{\epsilon} \mathbf{Z} \to 0,$$

$$0 \to J \to S \to \mathbf{Z} \to 0,$$

where $S = \Lambda^{\alpha+1}/j(\Lambda^\beta)$ and $j = i \circ h$. Evidently $\text{Hom}_\Lambda(S, \mathbf{Z}) \neq 0$. Moreover, since $\text{Ext}^1(\mathbf{Z}, \Lambda) = 0$ then S is projective by Proposition 5.17. In particular, the exact sequence

$$0 \to \Lambda^\beta \to \Lambda^{\alpha+1} \to S \to 0$$

defining S splits and there is an isomorphism $\Lambda^{\alpha+1} \cong \Lambda^\beta \oplus S$. Applying $\text{Hom}_\Lambda(-, \mathbf{Z})$ we see that

$$\mathbf{Z}^{\alpha+1} \cong \mathbf{Z}^\beta \oplus \text{Hom}_\Lambda(S, \mathbf{Z})$$

and since $\text{Hom}_\Lambda(S, \mathbf{Z}) \neq 0$ it follows that $\beta < \alpha + 1$ so that $\beta \leq \alpha$ as required. $\square$

The necessity for some minimality criterion is shown by the following:

Proposition 13.2 *Let $G = \Gamma * C_\infty$; then $\mathcal{I}$ fails to be minimal in $\Omega_1(\mathbf{Z})$.*

Proof Write $\mathcal{I}_G$ for the integral augmentation ideal of G; when $G = \Gamma * \Delta$ we see that ([34], p. 140)

$$\mathcal{I}_G \cong (\mathcal{I}_\Gamma \otimes_{\mathbf{Z}[\Gamma]} \mathbf{Z}[G]) \oplus (\mathcal{I}_\Delta \otimes_{\mathbf{Z}[\Delta]} \mathbf{Z}[G]).$$

On taking Δ to be the infinite cyclic group $C_\infty = \langle t | \emptyset \rangle$ the following exact sequence

$$0 \to \mathbf{Z}[C_\infty] \xrightarrow{t-1} \mathbf{Z}[C_\infty] \xrightarrow{\epsilon} \mathbf{Z} \to 0$$

shows that $\mathcal{I}_{C_\infty} \cong \mathbf{Z}[C_\infty]$ and hence $\mathcal{I}_{C_\infty} \otimes_{\mathbf{Z}[C_\infty]} \mathbf{Z}[G] \cong \mathbf{Z}[G]$. On substituting $\Delta = C_\infty$ in the above we see that

$$\mathcal{I}_G \cong (\mathcal{I}_\Gamma \otimes_{\mathbf{Z}[\Gamma]} \mathbf{Z}[G]) \oplus \mathbf{Z}[G];$$

hence $\mathcal{I}_\Gamma \otimes_{\mathbf{Z}[\Gamma]} \mathbf{Z}[G])$ lies below $\mathcal{I}_G$ in $\Omega_1^G(\mathbf{Z})$. □

Taking $\Gamma = F_{n-1}$ one sees iteratively that $\mathcal{I}_{F_n} \cong \mathbf{Z}[F_n]^n$ so that $\mathcal{I}_{F_n}$ departs progressively from minimality as n increases. Moreover, even when Γ is the trivial group, Proposition 13.2 still shows that 0 lies below $\mathcal{I}_{C_\infty}$ in $\Omega_1^{C_\infty}(\mathbf{Z})$.

13.2 Parametrizing $\Omega_1(\mathbf{Z})$ in the Generic Case

In this section we will continue to take G to be a finitely generated group and put $\Lambda = \mathbf{Z}[G]$. Moreover, we assume that $\mathrm{Ext}^1(\mathbf{Z}, \Lambda) = 0$ so that, by (13.1), we know that:

Min: $\mathcal{I}$ is a minimal element in $\Omega_1(\mathbf{Z})$.

As before we denote by $\mathcal{SF}_+$ the isomorphism classes of finitely generated stably free modules over $\mathbf{Z}[G]$. A consequence of the weak finite ($= \mathbf{WF}$) property for Λ is that a finitely generated *nonzero* stably free module S over $\Lambda = \mathbf{Z}[G]$ has a well defined rank $\mathrm{rk}(S) \geq 1$ defined by $\mathrm{rk}(S) = m \iff S \oplus \Lambda^n \cong \Lambda^{m+n}$. In what follows S will denote a finitely generated *nonzero* stably free module over $\Lambda = \mathbf{Z}[G]$. When $\mathrm{rk}(S) = m$ we have $\mathrm{Hom}_\Lambda(S, \mathbf{Z}) \oplus \mathrm{Hom}_\Lambda(\Lambda^m, \mathbf{Z}) \cong \mathrm{Hom}_\Lambda(\Lambda^{m+n}, \mathbf{Z})$. However $\mathrm{Hom}_\Lambda(\Lambda^k, \mathbf{Z}) \cong \mathbf{Z}^k$ so that $\mathrm{Hom}_\Lambda(S, \mathbf{Z}) \oplus \mathbf{Z}^m \cong \mathbf{Z}^{m+n}$. From the classification of finitely generated abelian groups it follows that:

$$\text{If} \quad \mathrm{rk}(S) = m \quad \text{then} \quad \mathrm{Hom}_\Lambda(S, \mathbf{Z}) \cong \mathbf{Z}^m. \tag{13.3}$$

In particular we see that there exists a nonzero Λ-homomorphism $\eta : S \to \mathbf{Z}$. The possibility of comparing $\Omega_1(\mathbf{Z})$ with $\mathbf{SF}_+$ arises from:

Theorem 13.4 *Let $\eta : S \to \mathbf{Z}$ be a nonzero Λ-homomorphism; then*

(i) $\mathrm{Ker}(\eta) \in \Omega_1(\mathbf{Z})$;
(ii) *if* $\mathrm{rk}(S) = 1$ *and* $\eta' : S \to \mathbf{Z}$ *is also a nonzero Λ-homomorphism then* $\mathrm{Ker}(\eta') = \mathrm{Ker}(\eta)$;
(iii) *if* $S \cong \Lambda^m$ *then* $\mathrm{Ker}(\eta) \cong \mathcal{I} \oplus \Lambda^{m-1}$.

Proof (i) First consider the special case where η is surjective. On applying Schanuel's Lemma to the exact sequences

$$0 \to \mathrm{Ker}(\eta) \to S \xrightarrow{\eta} \mathbf{Z} \to 0; \qquad 0 \to \mathcal{I} \to \Lambda \xrightarrow{\epsilon} \mathbf{Z} \to 0$$

we see that $\mathcal{I} \oplus S \cong \mathrm{Ker}(\eta) \oplus \Lambda$ and for any positive integer k, $\mathcal{I} \oplus S \oplus \Lambda^k \cong \mathrm{Ker}(\eta) \oplus \Lambda^{k+1}$. As S is stably free $S \oplus \Lambda^k \cong \Lambda^{m+k}$ for some k. Hence for some k, $\mathcal{I} \oplus \Lambda^{m+k} \cong \mathrm{Ker}(\eta) \oplus \Lambda^{k+1}$ and so $\mathrm{Ker}(\eta) \in \Omega_1(\mathbf{Z})$. If η is only assumed to be nonzero then $\operatorname{Im}(\eta) = (d)$ for some nonzero integer d. Putting $\xi = \frac{1}{d}\eta$ we see that ξ is surjective. The conclusion now follows from the above special case as $\operatorname{Ker}(\eta) = \operatorname{Ker}(\xi)$.

(ii) As $\operatorname{rk}(S) = 1$ then $\operatorname{Hom}_\Lambda(S, \mathbf{Z}) \cong \mathbf{Z}$. Let $\xi \in \operatorname{Hom}_\Lambda(S, \mathbf{Z})$ be a generator. Then for some nonzero integers n, n' we have $\eta = n\xi$ and $\eta' = n'\xi$ and so $\operatorname{Ker}(\eta) = \operatorname{Ker}(\xi) = \operatorname{Ker}(\eta')$ as required.

(iii) For $1 \le i \le m$ put $\epsilon_i = \epsilon \circ \pi_i$ where $\epsilon : \Lambda \to \mathbf{Z}$ is the augmentation homomorphism and $\pi_i : \Lambda^m \to \Lambda$ is the ith projection. As $\{\epsilon_i\}_{1 \le i \le m}$ is a $\mathbf{Z}$-basis for $\operatorname{Hom}_\Lambda(\Lambda^m, \mathbf{Z}) \cong \mathbf{Z}^m$ a given Λ-homomorphism $\eta : \Lambda^m \to \mathbf{Z}$ may be expressed uniquely in the form

$$\eta = x_1\epsilon_1 + \cdots + x_m\epsilon_m$$

for some $x_i \in \mathbf{Z}$. Regarding $\mathbf{x} = (x_1, \ldots, x_m)$ as the $\mathbf{Z}$-linear mapping $\mathbf{x} : \mathbf{Z}^m \to \mathbf{Z}$

$$\mathbf{x} \begin{pmatrix} y_1 \\ \vdots \\ y_m \end{pmatrix} = x_1 y_1 + \cdots + x_m y_m$$

then $\eta = \mathbf{x} \circ \epsilon_*$ where $\epsilon_* : \Lambda^m \to \mathbf{Z}^m$ is induced coordinate-wise by ϵ. The Smith normal form applied to $\mathbf{x}$ gives $\alpha \in E_n(\mathbf{Z})$ making the following commute

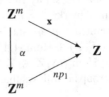

where p_1 is projection onto the first factor and where the integer $n(= n_\mathbf{x})$ generates the (nonzero) ideal $(x_1, \ldots, x_m)$ of $\mathbf{Z}$. As $\epsilon : \Lambda \to \mathbf{Z}$ is surjective then $\epsilon_* : E_m(\Lambda) \to E_m(\mathbf{Z})$ is surjective so that there exists $\tilde{\alpha} \in E_m(\Lambda)$ making the following commute:

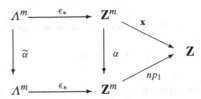

Thus $\operatorname{Ker}(\eta) = \operatorname{Ker}(\mathbf{x} \circ \epsilon_*) \cong \operatorname{Ker}(np_1 \circ \epsilon_*)$. However, $\operatorname{Ker}(np_1 \circ \epsilon_*) = \operatorname{Ker}(p_1 \circ \epsilon_*)$ as $n \ne 0$. The conclusion follows as $\operatorname{Ker}(p_1 \circ \epsilon_*) \cong \mathcal{I} \oplus \Lambda^{m-1}$. $\square$

For $k \ge 0$ we define $\Omega_1\langle k \rangle$ to be the set of modules in $\Omega_1(\mathbf{Z})$ at height k; that is:

$$J \in \Omega_1\langle k \rangle \iff J \oplus \Lambda^n \cong \mathcal{I} \oplus \Lambda^{k+n} \quad \text{for some } n \ge 1.$$

Likewise we define $\mathcal{SF}_+\langle k\rangle$ to be the set of modules in $\mathbf{SF}_+$ at height k; that is

$$S \in \mathcal{SF}_+\langle k\rangle \quad \Longleftrightarrow \quad S \oplus \Lambda^n \cong \Lambda^{k+n}.$$

When S is a nonzero finitely generated stably free Λ-module we denote by $[\kappa(S)]$ the set of isomorphism classes of all Λ-modules which occur in an exact sequence

$$0 \to J \to S \to \mathbf{Z} \to 0.$$

It follows from (13.3) and Theorem 13.4 that:

$$[\kappa(S)] \text{ is a non-empty subset of } \Omega_1(\mathbf{Z}) \text{ for each } S \in \mathbf{SF}_+. \tag{13.5}$$

We regard κ as a relation (that is, a multi-valued function) $\kappa : \mathcal{SF}_+ \rightsquigarrow \Omega_1(\mathbf{Z})$; that is, $\kappa(S)$ is a well defined element of $\Omega_1(\mathbf{Z})$ only when $[\kappa(S)]$ consists of a single element. Again from Theorem 13.4 it follows that:

$$\kappa(S) \text{ is well defined provided that } \textit{either } \text{rk}(S) = 1 \textit{ or } S \text{ is free.} \tag{13.6}$$

We also note that:

$$\text{If } J \in \kappa(S) \text{ then } h(J) = \text{rk}(S) - 1. \tag{13.7}$$

The height and rank functions on $\mathcal{SF}_+$ are related by $h(S) = \text{rk}(S) - 1$. Thus (13.7) can be rephrased to say that the relation κ is height preserving; that is:

$$\text{If } J \in \kappa(S) \text{ then } h(J) = h(S). \tag{13.8}$$

Proposition 13.9 *If* $\text{Ext}^1(\mathbf{Z}, \Lambda) = 0$ *then given any* $J \in \Omega_1(\mathbf{Z})$ *there exists* $S \in \mathcal{SF}_+$ *such that* $J \in [\kappa(S)]$; *that is, the relation* κ *is surjective.*

Proof Let $0 \to \mathcal{I} \xrightarrow{i} \Lambda \xrightarrow{\epsilon} \mathbf{Z} \to 0$ denote the defining exact sequence for $\mathcal{I}$ and suppose that $J \oplus \Lambda^m \cong \mathcal{I} \oplus \Lambda^n$. Choose an isomorphism $h : J \oplus \Lambda^m \xrightarrow{h} \mathcal{I} \oplus \Lambda^n$ and consider the exact sequence

$$0 \to J \oplus \Lambda^m \xrightarrow{j} \Lambda \oplus \Lambda^n \xrightarrow{\epsilon \circ \pi} \mathbf{Z} \to 0,$$

where $j = (i \oplus \text{Id}) \circ h$ and $\pi : \Lambda \oplus \Lambda^n \to \Lambda$ is the projection. Putting $S = (\Lambda \oplus \Lambda^n)/j(\Lambda^m)$ we have an exact sequence $0 \to J \to S \to \mathbf{Z} \to 0$. By hypothesis, $\mathbf{Z}$ is coprojective, and so, by the 'Desuspension Lemma' Proposition 5.17 it follows that S is projective. On splitting the exact sequence $0 \to \Lambda^m \to \Lambda^{n+1} \to S \to 0$ it follows that $S \oplus \Lambda^m \cong \Lambda^{n+1}$ so that S is stably free and, from the exact sequence $0 \to J \to S \to \mathbf{Z} \to 0$, it follows tautologically that $J \in [\kappa(S)]$. $\square$

The discussion now splits into two cases, according to whether G is finite or infinite. We note that both cases do, in fact, occur. When G is finite the condition $\text{Ext}^1(\mathbf{Z}, \Lambda) = 0$ holds automatically in consequence of the Eckmann-Shapiro Lemma. As an example with G infinite we may take any virtual duality group of (virtual) dimension ≥ 2.

13.3 Case I: G Finite

Let Γ be an infinite tree with a height function $h : \Gamma \to \mathbf{N}$ whose minimal level $h^{-1}(0)$ describes the 'roots' of the tree. We say that Γ is a fork when there is no branching except at the minimal level; that is, when $|h^{-1}(n)|$ for $n \geq 1$. The diagram (13.10) below illustrates the notion, with (**A**) showing proper branching and (**B**) representing the trivial case where there is no branching at all.

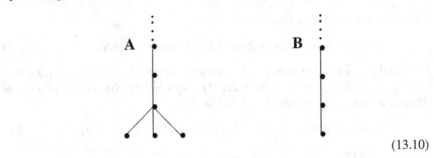

$$(13.10)$$

We note that when Γ, Δ are forks any level preserving mapping $\mu : \Gamma \to \Delta$ is completely described by its restriction to minimal levels $\mu : \Gamma^{\min} \to \Delta^{\min}$. It is a consequence of the theorem of Swan-Jacobinski [18, 46, 93] that when the group G is finite then both $S\mathcal{F}_+$ and $\Omega_1(\mathbf{Z})$ are forks. As in Sect. 8.4, for $J \in \Omega_1(\mathbf{Z})$ we denote by $\sigma(J)$ the 'Swan multiplicity'

$$\sigma(J) = |\mathrm{Im}(\nu_J)\backslash \mathrm{Ker}(S_J)|,$$

where $S_J : \mathrm{Aut}_{\mathcal{D}\mathrm{er}}(J) \to \widetilde{K}_0(\Lambda)$ is the Swan mapping and $\nu_J : \mathrm{Aut}_\Lambda(J) \to \mathrm{Aut}_{\mathcal{D}\mathrm{er}}(J)$ is the natural mapping. The situation for finite G may be stated thus:

Theorem 13.11 *Let G be finite; then*

(i) $\kappa : S\mathcal{F}_+ \to \Omega_1(\mathbf{Z})$ *is a surjective level preserving mapping; in addition*
(ii) *if $J \in \Omega_1(\mathbf{Z})$ lies above the minimal level then $|\kappa^{-1}(J)| = 1$; furthermore*
(iii) *if $J \in \Omega_1^{\min}(\mathbf{Z})$ then $|\kappa^{-1}(J)| = \sigma(J)$.*

Proof Let $S \in S\mathcal{F}_+$. If $h(S) = 0$, that is, if $\mathrm{rk}(S) = 1$, then $\kappa(S)$ is well defined by (13.6) above. If $h(S) \geq 1$, that is, if $\mathrm{rk}(S) \geq 2$, then S is free and so $\kappa(S)$ is well defined, again by (13.6). Thus κ is a mapping, and by (13.8), is level preserving. Finally, as G is finite then from the Eckmann-Shapiro Lemma it follows that $\mathrm{Ext}^1(\mathbf{Z}, \Lambda) = 0$. Thus κ is surjective by Proposition 13.9 and this proves (i). Moreover (ii) is simply the statement that neither $\Omega_1(\mathbf{Z})$ nor $S\mathcal{F}_+$ branch above the minimal level which follows, as we have already noted, from the Swan-Jacobinski Theorem. Finally (iii) follows from Theorem 8.15.   □

It follows from Theorem 13.11 that if $S\mathcal{F}_+$ is a trivial fork then so also is $\Omega_1(\mathbf{Z})$. As we indicated in Sect. 10.1, this is the case for very many but not all finite groups G. For example, in the case of the quaternion group $Q(4n) = \langle x, y | x^n = y^2,$ $xyx = y \rangle$ Swan shows that $S\mathcal{F}_+$ is a nontrivial fork whenever $n \geq 6$. In some (and probably all) such cases $\Omega_1(\mathbf{Z})$ is also nontrivial [50].

13.4 Case II: G Infinite and $\text{Ext}^1_\Lambda(\mathbf{Z}, \Lambda) = 0$

This case differs from the Case I in relation to behaviour with respect to duality. When $M = \Lambda$ we note that the evaluation map $\text{Hom}_\Lambda(\Lambda, \Lambda) \to \Lambda; f \mapsto f(1)$ is an isomorphism; that is:

$$\Lambda^* \cong \Lambda. \tag{13.12}$$

Evidently $(M \oplus N)^* \cong M^* \oplus N^*$ so we see easily that:

If P is projective then P^* is also projective and $P^{**} \cong P$. (13.13)

The converse to (13.13) is false in general. The simplest example is to take the trivial module $\mathbf{Z}$ with G infinite; then as we observed in Proposition 6.31, $\mathbf{Z}^* = \text{Hom}_\Lambda(\mathbf{Z}, \Lambda) = 0$. Thus $\mathbf{Z}^*$ is (trivially) projective whilst $\mathbf{Z}$ is not. This example has the following consequence:

Proposition 13.14 *If G is infinite and $\text{Ext}^1_\Lambda(\mathbf{Z}, \Lambda) = 0$ then $\mathcal{I}^* \cong \Lambda$.*

Proof Applying $\text{Hom}_\Lambda(-, \Lambda)$ to the augmentation sequence

$$0 \to \mathcal{I} \xrightarrow{i} \Lambda \xrightarrow{\epsilon} \mathbf{Z} \to 0$$

gives a long exact sequence in cohomology from which we extract the following portion: $\text{Hom}_\Lambda(\mathbf{Z}, \Lambda) \to \Lambda^* \xrightarrow{i^*} \mathcal{I}^* \to \text{Ext}^1(\mathbf{Z}, \Lambda)$. Now $\text{Hom}_\Lambda(\mathbf{Z}, \Lambda) = 0$ by Proposition 6.31 whilst $\text{Ext}^1_\Lambda(\mathbf{Z}, \Lambda) = 0$ by hypothesis. Thus $\mathcal{I}^* \cong \Lambda^* \cong \Lambda$. $\square$

More generally, suppose $J \in \Omega_1(\mathbf{Z})$ so that, for some m, n, we have $J \oplus \Lambda^m \cong \mathcal{I} \oplus \Lambda^n$; dualization then gives $J^* \oplus (\Lambda^*)^m \cong \mathcal{I}^* \oplus (\Lambda^*)^n$ so that, by (13.12) and Proposition 13.14

$$J^* \oplus \Lambda^m \cong \Lambda^{n+1}.$$

Thus $J^* \in \mathcal{SF}_+$ when $J \in \Omega_1(\mathbf{Z})$ and dualization gives a mapping $\delta : \Omega_1(\mathbf{Z}) \to \mathcal{SF}_+$; $\delta(J) = J^*$. There is also a duality involution $\theta : \mathcal{SF}_+ \to \mathcal{SF}_+$, $\theta(S) = S^*$. $\theta \circ \theta = \text{Id}$, which we are careful to distinguish notationally from δ. We make this discussion precise:

Theorem 13.15 *If G is infinite and $\text{Ext}^1_\Lambda(\mathbf{Z}, \Lambda) = 0$ then*

(i) *the correspondence $J \mapsto J^* = \text{Hom}_\Lambda(J, \Lambda)$ defines a surjective height-preserving mapping of trees $\delta : \Omega_1(\mathbf{Z}) \to \mathcal{SF}_+$;*
(ii) *the inverse relation δ^{-1} satisfies $\delta^{-1} = \kappa \circ \theta$; moreover*
(iii) *the induced mapping on minimum levels $\delta : \Omega_1^{\min}(\mathbf{Z}) \to \mathcal{SF}_+^{\min}$ is bijective.*

Proof By the preceding remarks the correspondence $J \mapsto J^*$ clearly defines a function $\delta : \Omega_1(\mathbf{Z}) \to \mathcal{SF}_+$. To show that δ preserves height first observe that $\mathcal{I}$ is at the minimal level by (13.1). If $J \in \Omega_1(\mathbf{Z})$ then for some $k \geq 0$ $J \oplus \Lambda^m \cong \mathcal{I} \oplus \Lambda^{m+k}$ and $h(J) = k$. The preceding calculation now gives $J^* \oplus \Lambda^m \cong \Lambda^{m+k+1}$ so that $\text{rk}(J^*) = k + 1$ and so $h(J^*) = k$ and δ preserves levels.

To show that δ is surjective, suppose $S \in \mathcal{SF}_+$, so that, for some $a \geq 1$, $S \oplus \Lambda^a \cong \Lambda^{m+a}$. On taking duals we see that $S^* \oplus \Lambda^a \cong \Lambda^{m+a}$, so that we also have $S^* \in \mathcal{SF}_+$. By (13.3) choose a nonzero homomorphism $\eta : S^* \to \mathbf{Z}$. Replacing $\mathbf{Z}$ by $\mathrm{Im}(\eta) \cong \mathbf{Z}$ if necessary we may suppose that $\eta : S^* \to \mathbf{Z}$ is surjective. It follows from Theorem 13.4 that $\mathrm{Ker}(\eta) \in \Omega_1(\mathbf{Z})$. We proceed to show that $\delta(\mathrm{Ker}(\eta)) \cong S$.

As in Proposition 13.14, applying $\mathrm{Hom}_\Lambda(-, \Lambda)$ to the sequence

$$0 \to \mathrm{Ker}(\eta) \xrightarrow{j} S^* \xrightarrow{\eta} \mathbf{Z} \to 0$$

gives an exact sequence: $\mathrm{Hom}_\Lambda(\mathbf{Z}, \Lambda) \to S^{**} \xrightarrow{j^*} \mathrm{Ker}(\eta)^* \to \mathrm{Ext}^1(\mathbf{Z}, \Lambda)$. As before, the two end terms are zero leaving an isomorphism $j^* : S^{**} \xrightarrow{\cong} \mathrm{Ker}(\eta)^*$ so that $\delta(\mathrm{Ker}(\eta)) \cong S^{**} \cong S$ as claimed. Hence δ is surjective, completing the proof of (i).

That there is equality *of relations* $\delta^{-1} = \kappa \circ \theta$ is the statement that

$$J \in [\kappa(S^*)] \quad \Longleftrightarrow \quad \delta(J) \cong S.$$

This is implicit in the proof of (i). To make it explicit, first suppose that $J \in [\kappa(S*)]$. Then there exists an exact sequence $0 \to J \to S^* \to \mathbf{Z} \to 0$. Repeating the argument of (i) gives $J^* \cong S^{**}$ so that, as $S^{**} \cong S$ we have $\delta(J) \cong S$.

Conversely, suppose that $J \in \Omega_1(\mathbf{Z})$ has the property that $\delta(J) \cong S$. In Proposition 13.9 we showed:

($\dagger$): $J \in [\kappa(T)]$ for some $T \in \mathcal{SF}_+$.

That is, there is an exact sequence $0 \to J \to T \to \mathbf{Z} \to 0$. Dualizing as in (i) above we see that $\delta(J) = J^* \cong T^*$. However, $\delta(J) \cong S$. Thus $S \cong T^*$ and so $T \cong S^*$. By ($\dagger$) it follows that $J \in [\kappa(S^*)]$, and this proves (ii).

To complete the proof we show that $\delta : \Omega_1^{\min}(\mathbf{Z}) \to \mathcal{SF}_+^{\min}$ is bijective. We showed in (13.6) that the relation κ actually defines a function $\kappa : \mathcal{SF}_+^{\min} \to \Omega_1^{\min}(\mathbf{Z})$. Moreover, as θ and δ are both level preserving we have mappings

$$\theta : \mathcal{SF}_+^{\min} \to \mathcal{SF}_+^{\min}; \qquad \delta : \Omega_1^{\min}(\mathbf{Z}) \to \mathcal{SF}_+^{\min}$$

in which θ, being self inverse, is bijective and δ is surjective. In what follows, where it is not explicit we assume that all mappings are restricted to the minimum levels in both $\Omega_1(\mathbf{Z})$ and $\mathcal{SF}_+$. From (ii) above we see that $\delta\kappa\theta = \mathrm{Id}$. Conjugation by $\theta = \theta^{-1}$ now gives

($\dagger\dagger$) $\theta\delta\kappa = \mathrm{Id}$.

Thus κ is injective. However, we showed in Proposition 13.9 that κ is surjective. Explicitly, given $J \in \Omega_1^{\min}(\mathbf{Z})$ there exists $S \in \mathcal{SF}_+$ such that $J \in [\kappa(S)]$. As κ preserves levels then $S \in \mathcal{SF}_+^{\min}$ and, in this case, by (13.6) $\kappa(S)$ is a single well defined element. Thus $\kappa : \mathcal{SF}_+^{\min} \to \Omega_1^{\min}(\mathbf{Z})$ is invertible. From ($\dagger\dagger$) we now see that $\delta = \theta^{-1} \circ \kappa : \Omega_1^{\min}(\mathbf{Z}) \to \mathcal{SF}_+^{\min}$ is bijective as claimed, so completing the proof. $\qquad\square$

Chapter 14
Parametrizing $\Omega_1(\mathbf{Z})$: Singular Case

In this chapter we work under the blanket assumption that G is a finitely generated group with abelianization G^{ab} and integral group ring $\Lambda = \mathbf{Z}[G]$, and that $\mathrm{Ext}^1_\Lambda(\mathbf{Z}, \Lambda) \neq 0$. Then G is necessarily infinite. We investigate minimality of $\mathcal{I}$ in $\Omega_1(\mathbf{Z})$ and first establish:

Second minimality criterion: $\mathcal{I}$ lies at the minimal level of $\Omega_1(\mathbf{Z})$ if G^{ab} is finite.

This second criterion also applies to many cases where $\mathrm{Ext}^1_\Lambda(\mathbf{Z}, \Lambda) = 0$ although we do not need to use it there. We employ it in Sect. 14.3 to give examples of groups G with infinite splitting in $\Omega_1(\mathbf{Z})$.

14.1 The Second Minimality Criterion

Let R be a (not necessarily commutative) ring R which is free as an algebra over $\mathbf{Z}$; the augmentation homomorphism $\epsilon_R : R[G] \to R$ is again defined by $\epsilon(g) = 1$ for all $g \in G$ and we denote by $I_R(G) = \mathrm{Ker}(\epsilon_R)$ the augmentation ideal over R. Let N be a Λ-module on which G acts trivially. Observing that $\mathrm{Ext}^1_{R[G]}(\Lambda, N) = 0$ then from the augmentation exact sequence $0 \to I_R(G) \xrightarrow{i} \Lambda \xrightarrow{\epsilon_R} R \to 0$ we get an exact sequence in cohomology

$$\xrightarrow{\epsilon_R^*} \mathrm{Hom}_{R[G]}(\Lambda, N) \xrightarrow{i^*} \mathrm{Hom}_{R[G]}(I_R(G), N) \xrightarrow{\delta} \mathrm{Ext}^1_{R[G]}(R, N) \to 0.$$

For $\alpha \in \mathrm{Hom}_{R[G]}(\Lambda, N)$ and $g \in G$, $i^*(\alpha)(g-1) = \alpha(i(g-1)) = \alpha(g) - \alpha(1)$. However G acts trivially on N so that $\alpha(g) = \alpha(1)g = \alpha(1)$ and $i^*(\alpha)(g-1) = 0$ for all $g \in G$. Hence $i^*(\alpha) = 0$ since I_R is generated over R by elements of the form $g - 1$. The right hand end of the above exact sequence simplifies to an isomorphism

$$\delta : \mathrm{Hom}_{R[G]}(I_R(G), N) \xrightarrow{\approx} \mathrm{Ext}^1_{R[G]}(R, N).$$

In the special case where $N = R$ we obtain:

$$\mathrm{Hom}_{R[G]}(I_R(G), R) \cong \mathrm{Ext}^1_{R[G]}(R, R). \tag{14.1}$$

F.E.A. Johnson, *Syzygies and Homotopy Theory*, Algebra and Applications 17,
DOI 10.1007/978-1-4471-2294-4_14, © Springer-Verlag London Limited 2012

By regarding R as a trivial module over $\mathbf{Z}[G]$ the same argument shows:

$$\mathrm{Hom}_{\mathbf{Z}[G]}(I_{\mathbf{Z}}(G), R) \cong \mathrm{Ext}^1_{\mathbf{Z}[G]}(\mathbf{Z}, R). \tag{14.2}$$

However $I_R(G) \cong I_{\mathbf{Z}}(G) \otimes_{\mathbf{Z}} R$ and we have by 'change of rings' that

$$\mathrm{Hom}_{\mathbf{Z}[G]}(I_{\mathbf{Z}}(G), R) \cong \mathrm{Hom}_{R[G]}(I_R(G), R). \tag{14.3}$$

We obtain isomorphisms $\mathrm{Hom}_{R[G]}(I_R(G), R) \cong \mathrm{Ext}^1_{R[G]}(R, R) \cong \mathrm{Ext}^1_{\mathbf{Z}[G]}(\mathbf{Z}, R)$. Finally, $\mathrm{Ext}^1_{\mathbf{Z}[G]}(\mathbf{Z}, R) \cong H^1(G, R)$ whilst by the Universal Coefficient Theorem

$$H^1(G, R) \cong \mathrm{Hom}_{\mathbf{Z}}(H_1(G; Z), R) \cong \mathrm{Hom}_{\mathbf{Z}}(G^{ab}, R).$$

Hence we see that there are isomorphisms

$$\mathrm{Hom}_{R[G]}(I_R(G), R) \cong \mathrm{Ext}^1_{R[G]}(R, R) \cong \mathrm{Hom}_{\mathbf{Z}}(G^{ab}, R). \tag{14.4}$$

Note that if Ψ is a finitely generated abelian group then $\mathrm{Hom}(\Psi, \mathbf{Z}) \neq 0$ except in the case where Ψ is finite. Reverting to the special case where $R = \mathbf{Z}$ and $\Lambda = \mathbf{Z}[G]$ we see that:

$$G^{ab} \text{ is finite} \iff \mathrm{Ext}^1_\Lambda(\mathbf{Z}, \mathbf{Z}) = 0 \iff \mathrm{Hom}_\Lambda(\mathcal{I}, \mathbf{Z}) = 0. \tag{14.5}$$

Second minimality criterion $\mathcal{I}$ is minimal in $\Omega_1(\mathbf{Z})$ if G^{ab} is finite. $\qquad$ (14.6)

Proof Suppose that $\mathcal{I} \oplus \Lambda^a \cong J \oplus \Lambda^b$; we must show that $b \leq a$. Clearly

$$\mathrm{Hom}_\Lambda(\mathcal{I}, \mathbf{Z}) \oplus \mathrm{Hom}_\Lambda(\Lambda^a, \mathbf{Z}) \cong \mathrm{Hom}_\Lambda(J, \mathbf{Z}) \oplus \mathrm{Hom}_\Lambda(\Lambda^b, \mathbf{Z}).$$

Under the assumption that G^{ab} is finite we see that $\mathrm{Hom}_\Lambda(\mathcal{I}, \mathbf{Z}) = 0$ by (14.4) and so $\mathbf{Z}^a \cong \mathrm{Hom}_\Lambda(J, \mathbf{Z}) \oplus \mathbf{Z}^b$. Thus $\mathbf{Z}^b$ imbeds as a subgroup of $\mathbf{Z}^a$ and so $b \leq a$. $\qquad \square$

We shall say that the ring R is *torsion free* when its additive group is torsion free; we note the following useful deduction from (14.4):

If G^{ab} is finite and R is torsion free then $\mathrm{Hom}_{R[G]}(I_R(G), R) = 0$. $\qquad$ (14.7)

We note that the conditions of the two minimality criteria are independent. Let $\mathcal{M}(1)$ be the condition that $\mathrm{Ext}^1_\Lambda(\mathbf{Z}, \Lambda) = 0$ and $\mathcal{M}(2)$ that G^{ab} is finite. When G is a free abelian group of finite rank $N \geq 2$, G satisfies Poincaré Duality in dimension N, and so $\mathrm{Ext}^r(\mathbf{Z}, \Lambda) = 0$ for $r \neq N$ [60]. In particular, G satisfies condition $\mathcal{M}(1)$. However, $G^{ab} \cong G$ is infinite and so G fails the condition $\mathcal{M}(2)$. Conversely, take $G = H_1 * H_2$ to be the free product of nontrivial finite groups H_1, H_2. If F denotes the kernel of the natural mapping $G \to H_1 \times H_2$ then by the Kurosh subgroup theorem (for example in the form given in [41, p. 118]) F is a free group of rank $(|H_1| - 1)(|H_2| - 1) \geq 2$. Put $\Omega = \mathbf{Z}[F]$; F has finite index in G so applying the

Eckmann-Shapiro Lemma we conclude that $\operatorname{Ext}_\Lambda^1(\mathbf{Z}, \Lambda) \cong \operatorname{Ext}_\Omega^1(\mathbf{Z}, \Omega)$. Since F is a (generalized) duality group of dimension 1 it follows that $\operatorname{Ext}_\Omega^1(\mathbf{Z}, \Omega) \neq 0$; thus $\operatorname{Ext}_\Lambda^1(\mathbf{Z}, \Lambda) \neq 0$ and so G fails condition $\mathcal{M}(1)$. However, $G^{ab} \cong H_1^{ab} \times H_2^{ab}$ so that G satisfies $\mathcal{M}(2)$.

14.2 Parametrizing Ω_1

Again we seek to parametrize $\Omega_1(\mathbf{Z})$ by $\mathcal{SF}_+$. A difficulty arises according to whether the map induced from the augmentation $\epsilon_* : \operatorname{Ext}_\Lambda^1(\mathbf{Z}, \Lambda) \to \operatorname{Ext}_\Lambda^1(\mathbf{Z}, \mathbf{Z})$ is injective or not. In the injective case we obtain a conclusion which though weaker and more fragmentary than that of Chap. 13 is nevertheless of interest; the case where ϵ^* fails to be injective is much more obscure.[1] Thus we assume throughout this section that:

G **Infinite and** $\epsilon_* : \operatorname{Ext}_\Lambda^1(\mathbf{Z}, \Lambda) \to \operatorname{Ext}_\Lambda^1(\mathbf{Z}, \mathbf{Z})$ **Is Injective**

Note that as G is finitely generated then $\operatorname{Ext}_\Lambda^1(\mathbf{Z}, \mathbf{Z}) \cong H^1(G, \mathbf{Z}) \cong G^{ab}/\text{Torsion}$ is a finitely generated free abelian group. Since $\epsilon_* : \operatorname{Ext}_\Lambda^1(\mathbf{Z}, \Lambda) \to \operatorname{Ext}_\Lambda^1(\mathbf{Z}, \mathbf{Z})$ is injective we see that:

$$\operatorname{Ext}_\Lambda^1(\mathbf{Z}, \Lambda) \text{ is a finitely generated free abelian group.} \quad (14.8)$$

Next suppose S is a stably free module of rank 1 and that $S \oplus \Lambda^n \cong \Lambda^{n+1}$. It follows that $\operatorname{Ext}_\Lambda^1(\mathbf{Z}, S) \oplus \operatorname{Ext}_\Lambda^1(\mathbf{Z}, \Lambda)^n \cong \operatorname{Ext}_\Lambda^1(\mathbf{Z}, \Lambda) \oplus \operatorname{Ext}_\Lambda^1(\mathbf{Z}, \Lambda)^n$. As $\operatorname{Ext}_\Lambda^1(\mathbf{Z}, \Lambda)$ is a finitely generated free abelian group we conclude that:

$$\text{If } S \text{ is a stably free module of rank 1 then } \operatorname{Ext}_\Lambda^1(\mathbf{Z}, S) \cong \operatorname{Ext}_\Lambda^1(\mathbf{Z}, \Lambda). \quad (14.9)$$

Suppose we are now given a stably free module S of rank 1. It follows from (13.3) that $\operatorname{Hom}_\Lambda(S, \mathbf{Z}) \cong \mathbf{Z}$ so that there is a surjective homomorphism $\eta : S \to \mathbf{Z}$ which is unique up to sign. Thus there is an extension

$$S = (0 \to J \xrightarrow{i} S \xrightarrow{\eta} \mathbf{Z} \to 0).$$

We denote by $[S]$ the congruence class of S in $\operatorname{Ext}^1(\mathbf{Z}, J)$. With this notation and the ambient hypothesis we have:

Theorem 14.10 $\operatorname{Ext}_\Lambda^1(\mathbf{Z}, J) \cong \mathbf{Z}$ *and* $[S]$ *is a generator.*

---

[1] This case arises for groups which contain a nonabelian free group of finite index; for example, $SL_2(\mathbf{Z})$.

Proof First consider the augmentation sequence $(0 \to \mathcal{I} \xrightarrow{i} \mathbf{Z}[G] \xrightarrow{\epsilon} \mathbf{Z} \to 0)$. Since $\epsilon_* : \mathrm{Ext}^1_\Lambda(\mathbf{Z}, \Lambda) \to \mathrm{Ext}^1_\Lambda(\mathbf{Z}, \mathbf{Z})$ is injective the exact sequence

$$\mathrm{Hom}_\Lambda(\mathbf{Z}, \Lambda) \to \mathrm{Hom}_\Lambda(\mathbf{Z}, \mathbf{Z}) \to \mathrm{Ext}^1_\Lambda(\mathbf{Z}, \mathcal{I}) \to \mathrm{Ext}^1_\Lambda(\mathbf{Z}, \Lambda) \xrightarrow{\epsilon_*} \mathrm{Ext}^1_\Lambda(\mathbf{Z}, \mathbf{Z})$$

reduces to $\mathrm{Hom}_\Lambda(\mathbf{Z}, \Lambda) \to \mathrm{Hom}_\Lambda(\mathbf{Z}, \mathbf{Z}) \to \mathrm{Ext}^1_\Lambda(\mathbf{Z}, \mathcal{I}) \to 0$. However, as G is infinite, $\mathrm{Hom}_\Lambda(\mathbf{Z}, \Lambda) = 0$ so that $\mathrm{Ext}^1_\Lambda(\mathbf{Z}, \mathcal{I}) \cong \mathrm{Hom}_\Lambda(\mathbf{Z}, \mathbf{Z}) \cong \mathbf{Z}$. In the general case where $\mathcal{S} = (0 \to J \xrightarrow{i} S \xrightarrow{\eta} \mathbf{Z} \to 0)$ then $J \oplus \Lambda \cong \mathcal{I} \oplus S$ by Schanuel's Lemma. Hence

$$\mathrm{Ext}^1_\Lambda(\mathbf{Z}, J) \oplus \mathrm{Ext}^1_\Lambda(\mathbf{Z}, \Lambda) \cong \mathrm{Ext}^1_\Lambda(\mathbf{Z}, \mathcal{I}) \oplus \mathrm{Ext}^1_\Lambda(\mathbf{Z}, S) \cong \mathbf{Z} \oplus \mathrm{Ext}^1_\Lambda(\mathbf{Z}, \Lambda).$$

It follows from (14.8) that $\mathrm{Ext}^1_\Lambda(\mathbf{Z}, J) \cong \mathbf{Z}$. Take $\mathcal{X} = (0 \to J \to X \to \mathbf{Z} \to 0)$ to represent a generator of $\mathrm{Ext}^1_\Lambda(\mathbf{Z}, J) \cong \mathbf{Z}$. We will show that $[\mathcal{S}] = \pm[\mathcal{X}]$. Then $\mathrm{Ext}^1_\Lambda(S, J) = 0$ as S is projective so that from the exact sequence $\mathrm{Hom}_\Lambda(S, J) \xrightarrow{i^*} \mathrm{Hom}_\Lambda(J, J) \xrightarrow{\delta} \mathrm{Ext}^1_\Lambda(\mathbf{Z}, J) \to 0$ we see that the mapping $\delta : \mathrm{Hom}_\Lambda(J, J) \to \mathrm{Ext}^1_\Lambda(\mathbf{Z}, J); \delta(\alpha) = \alpha_*(\mathcal{S})$ is surjective.

Hence we may write $[\mathcal{X}] = [\alpha_*(\mathcal{S})]$ for some $\alpha \in \mathrm{Hom}_\Lambda(J, J)$. However, $[\mathcal{X}]$ generates $\mathrm{Ext}^1_\Lambda(\mathbf{Z}, J) \cong \mathbf{Z}$ so that for some $n \in \mathbf{Z}$ we may write $[\mathcal{S}] = n[\mathcal{X}]$. Thus $[\mathcal{X}] = n[\alpha_*(\mathcal{X})]$. Writing $[\alpha_*(\mathcal{X})] = m[\mathcal{X}]$ for some integer m we obtain $[\mathcal{X}] = mn[\mathcal{X}]$. As $mn \in \mathbf{Z}$ and $[\mathcal{X}]$ is a generator of $\mathrm{Ext}^1(\mathbf{Z}, J) \cong \mathbf{Z}$ then $mn = 1$ so that $n = \pm 1$ □

We have seen, (13.6), that $\kappa(S)$ is well defined when S is a stably free module of rank 1 and that, when G is infinite and $\mathrm{Ext}^1(\mathbf{Z}, \Lambda) = 0$, κ gives a bijection

$$\kappa : \mathcal{SF}^{\min}_+ \to \Omega^{\min}_1(\mathbf{Z}).$$

As we now see, κ continues to be injective on $\mathcal{SF}^{\min}_+$ under our present hypothesis:

Theorem 14.11 $\epsilon_* : \mathrm{Ext}^1_\Lambda(\mathbf{Z}, \Lambda) \to \mathrm{Ext}^1_\Lambda(\mathbf{Z}, \mathbf{Z})$ *is injective and G is finitely generated infinite then* $\kappa : SF^{\min}_+ \to \Omega_1(\mathbf{Z})$ *is injective.*

Proof Let S, $S' \in SF_1$ and suppose that $\kappa(S) = \kappa(S') = J$. We must show that $S \cong S'$. There are exact sequences $\mathcal{S} = (0 \to J \xrightarrow{i} S \xrightarrow{\epsilon} \mathbf{Z} \to 0)$; $\mathcal{S}' = (0 \to J \xrightarrow{i'} S' \xrightarrow{\epsilon'} \mathbf{Z} \to 0)$ and, by Theorem 14.10, both $[\mathcal{S}], [\mathcal{S}']$ generate $\mathrm{Ext}^1_\Lambda(\mathbf{Z}, J) \cong \mathbf{Z}$ so that $[\mathcal{S}'] = \pm[\mathcal{S}]$. Replacing ϵ' by $-\epsilon'$ if necessary we may suppose that $[\mathcal{S}'] = [\mathcal{S}]$. Thus there is a congruence

$$
\begin{matrix}
\mathcal{S} \\
c \downarrow \\
\mathcal{S}'
\end{matrix}
=
\begin{pmatrix}
0 \to J \xrightarrow{i} S \xrightarrow{\epsilon} \mathbf{Z} \to 0 \\
\mathrm{Id} \downarrow \quad c \downarrow \quad \mathrm{Id} \downarrow \\
0 \to J \xrightarrow{i'} S' \xrightarrow{\epsilon'} \mathbf{Z} \to 0
\end{pmatrix}
$$

and $c : S \to S'$ is the required isomorphism.   □

14.3 Infinite Branching in $\Omega_1(\mathbf{Z})$

We are careful not to over-interpret the statement of Theorem 14.11. Although it remains true that the image of $SF_+^{\min}$ under κ lies at the same height as the class $[\mathcal{I}]$ of the augmentation ideal, the minimality criteria of Sects. 13.1, 14.1 no longer apply and, in the absence of further information, we are careful not to claim that image of $SF_+^{\min}$ lies within $\Omega_1^{\min}(\mathbf{Z})$. Indeed, ϵ^* is injective when $G = C_\infty$ yet the minimal level of $\Omega_1(\mathbf{Z})$ is represented by 0. However, as we shall see in Sect. 16.3, in the case where $G = C_\infty \times \Phi$ and Φ is nontrivial and finite then $\Omega_1^{\min}(\mathbf{Z})$ *is* represented by $[\mathcal{I}]$.

In earlier chapters we showed the existence of groups G which admit an infinite number of stably free modules of rank 1 over $\mathbf{Z}[G]$; then by Theorem 14.11, $\Omega_1(\mathbf{Z})$ displays an infinite amount of branching. We illustrate this with some examples. In what follows we take $Q(8m)$ to be the generalized quaternion group of order $8m$

$$Q(8m) = \langle x, y \mid x^{2m} = y^2, xyx = y \rangle.$$

Begin by taking $G = C_\infty^N \times \Phi$ where Φ is a nontrivial finite group; put $\Lambda = \mathbf{Z}[G]$ and $\Lambda_0 = \mathbf{Z}[C_\infty^N]$. Then $\text{Ext}_\Lambda^1(\mathbf{Z}, \Lambda) \cong \text{Ext}_{\Lambda_0}^1(\mathbf{Z}, \Lambda_0)$ by the Eckmann-Shapiro Lemma. As C_∞^N is a Poincaré Duality group of dimension N it follows [60] that

$$\text{Ext}_\Lambda^1(\mathbf{Z}, \Lambda) = \begin{cases} \mathbf{Z} & N = 1, \\ 0 & N \geq 2. \end{cases} \tag{14.12}$$

By direct calculation one may first show:

$$\epsilon_* : \text{Ext}_{\mathbf{Q}[C_\infty]}^1(\mathbf{Q}, \mathbf{Q}[C_\infty]) \to \text{Ext}_{\mathbf{Q}[C_\infty]}^1(\mathbf{Q}, \mathbf{Q}) \text{ is an isomorphism.} \tag{14.13}$$

We now prove:

Proposition 14.14 $\epsilon_* : \text{Ext}_\Lambda^1(\mathbf{Z}, \Lambda) \to \text{Ext}_\Lambda^1(\mathbf{Z}, \mathbf{Z})$ *is injective.*

Proof The statement for $N \geq 2$ is trivial by (14.12) so that it suffices to consider the case $N = 1$. In this case, again by (14.12), $\text{Ext}_\Lambda^1(\mathbf{Z}, \Lambda) \cong \mathbf{Z}$ so that it suffices to prove that, taking rational coefficients, the corresponding map $\epsilon_* : \text{Ext}_{\mathbf{Q}[G]}^1(\mathbf{Q}, \mathbf{Q}[G]) \to \text{Ext}_{\mathbf{Q}[G]}^1(\mathbf{Q}, \mathbf{Q})$ is nonzero. This follows from the isomorphism already noted in (14.13) by applying the Künneth Theorem with rational coefficients to $G = C_\infty \times \Phi$ above. $\qquad\square$

We obtain:

Theorem 14.15 *Let* $G = C_\infty^N \times Q(8m)$ *where* $N \geq 1$ *and* $m \geq 1$; *then* $\Omega_1(\mathbf{Z})$ *has an infinite amount of branching at the level of* $\mathcal{I}$.

Proof It follows from Theorems 13.15 and 14.11 that $\kappa : SF_+^{\min} \to \Omega_1(\mathbf{Z})$ is injective. However, by (12.31) $\mathbf{Z}[G]$ admits infinitely many isomorphism types of stably free modules of rank 1 and this completes proof. $\qquad\square$

In the above, by (13.1), $\mathcal{I}$ is minimal in $\Omega_1(\mathbf{Z})$ provided $N \geq 2$. We shall see, in Sect. 16.3, that $\mathcal{I}$ is still minimal when $N = 1$. Anticipating this result, the above conclusion may be strengthened to conclude that, for $N, m \geq 1$:

$$\Omega_1(\mathbf{Z})^{\min} \text{ is infinite for } G = C_\infty^N \times Q(8m). \tag{14.16}$$

Chapter 15
Generalized Swan Modules

Let $G = C_\infty \times \Phi$ where Φ is a nontrivial finite group. In this chapter, subject to conditions on Φ, we begin the task of parametrizing the first syzygy $\Omega_1(\mathbf{Z})$ over $\mathbf{Z}[G]$. We do so via a generalization, introduced by Edwards in his thesis [25], of a type of module first studied by Swan. The original Swan modules arise as follows; let Φ be a finite group and let $\epsilon : \mathbf{Z}[\Phi] \to \mathbf{Z}$ be the augmentation homomorphism. For $n \in \mathbf{Z}$ we take $\mathbf{n} : \mathbf{Z} \to \mathbf{Z}$ to be the homomorphism $x \mapsto nx$. The eponymous Swan module $(\mathcal{I}, n)$ is defined as $(\mathcal{I}, n) = \varprojlim(\epsilon, \mathbf{n})$. From the commutative diagram

$$
\begin{array}{ccccccccc}
0 \to & \mathcal{I} & \longrightarrow & (\mathcal{I}, n) & \overset{\epsilon}{\longrightarrow} & \mathbf{Z} & \to 0 \\
& \downarrow\mathrm{Id} & & \cap & & \downarrow\mathbf{n} \\
0 \to & \mathcal{I} & \longrightarrow & \mathbf{Z}[\Phi] & \overset{\epsilon}{\longrightarrow} & \mathbf{Z} & \to 0
\end{array}
$$

we see that $(\mathcal{I}, n)$ imbeds in $\mathbf{Z}[\Phi]$ as a $\mathbf{Z}$-sublattice of index n. The main properties of the $(\mathcal{I}, n)$ were established by Swan [91]. Indeed, this was the original context for the projectivity criterion of Chap. 5; Swan showed that $(\mathcal{I}, n)$ is projective over $\mathbf{Z}[\Phi]$ if and only if n is a unit mod $|\Phi|$. This and other aspects generalize in a manner which we now proceed to describe.

15.1 Quasi-Augmentation Sequences and Swan Modules

A module S over a ring Λ is said to be *strongly Hopfian* when for all integers $n \geq 1$ any surjective Λ-homomorphism $\varphi : S^{(n)} \to S^{(n)}$ is necessarily an isomorphism. It is straightforward to see this entails an apparently stronger property, namely, that if $n_1 \leq n_2$ and $\varphi : S^{(n_1)} \to S^{(n_2)}$ is surjective homomorphism then φ is an isomorphism and $n_1 = n_2$. An exact sequence of finitely generated Λ-modules $S = (0 \to S_- \overset{i}{\to} S_0 \overset{p}{\to} S_+ \to 0)$ is called a *quasi augmentation sequence* when S_0 is stably free and when $S_+ \, S_-$ satisfy the following conditions;

F.E.A. Johnson, *Syzygies and Homotopy Theory*, Algebra and Applications 17,
DOI 10.1007/978-1-4471-2294-4_15, © Springer-Verlag London Limited 2012

(i) $\mathrm{Ext}^1_\Lambda(S_+, \Lambda) = 0$;
(ii) S_+, S_- are strongly Hopfian;
(iii) $\mathrm{Hom}_\Lambda(S_-, S_+) = 0$.

By a *Swan module* relative to S we mean a Λ-module M which occurs in an exact sequence of the form

$$\mathcal{E}_M = (0 \to S_- \to M \to S_+ \to 0).$$

Until further notice, Λ will denote a group ring $R[\Phi]$ where Φ is a finite group and R is a ring subject at least to the following restriction (†):

(†) R is free over $\mathbf{Z}$ and satisfies the *weak finiteness* condition of Sect. 1.1.

At the outset we do not assume that R is commutative. Whilst further restrictions will be imposed on R as we progess, the condition (†) will be assumed to hold without further mention. We first consider:

(15.1) The Standard Augmentation Sequence Let Φ a finite group and take Λ to be the group ring $R[\Phi]$. The augmentation map $\epsilon_R : R[\Phi] \to R$, $\epsilon_R(\sum_\varphi a_\varphi \varphi) = \sum_\varphi a_\varphi$ is a ring homomorphism and gives an exact sequence

$$\mathcal{R} = (0 \to I_R \overset{i}{\to} R[\Phi] \overset{\epsilon_R}{\to} R \to 0)$$

where $I_R = \mathrm{Ker}(\epsilon_R)$ and i is the inclusion; with this notation we have:

Proposition 15.2 $\mathcal{R}$ *is a quasi-augmentation sequence over* $R[\Phi]$.

Proof As in (14.4), $\mathrm{Hom}_{R[\Phi]}(I_R, R) \cong \mathrm{Ext}^1_{R[\Phi]}(R, R) \cong \mathrm{Hom}_{\mathbf{Z}}(\Phi^{ab}, R)$. As Φ is finite and R is free over $\mathbf{Z}$ then $\mathrm{Hom}_{R[\Phi]}(I_R(\Phi), R) = 0$. The weak finiteness assumption on R also implies that R is strongly Hopfian as a module over $R[\Phi]$. As I_R is free of rank $|\Phi| - 1$ over R then I_R is also strongly Hopfian. Finally, by the Eckmann-Shapiro Lemma, $\mathrm{Ext}^1_{R[\Phi]}(R, R[\Phi]) = \mathrm{Ext}^1_R(R, R) = 0$. □

(15.3) The Augmentation Ideal of $\mathbf{Z}[C_\infty \times \Phi]$ as a Swan Module In the original context of finite groups the integral augmentation ideal is *not* a Swan module. However, it is so in the case of $C_\infty \times \Phi$. What follows is a special case of Proposition 15.2 but we repeat the details for emphasis. Take $R = \mathbf{Z}[C_\infty] = \mathbf{Z}[t, t^{-1}]$ and, as in (15.1), take

$$\mathcal{R} = (0 \to I_R \overset{i}{\to} R[\Phi] \overset{\epsilon_R}{\to} R \to 0).$$

In addition to the R-augmentation ϵ_R, we also have the $\mathbf{Z}$-augmentation for C_∞,

$$\epsilon_\infty : R \to \mathbf{Z}; \quad \epsilon_\infty(t) = 1;$$

then the exact sequence $0 \to R \overset{t-1}{\longrightarrow} R \overset{\epsilon_\infty}{\longrightarrow} \mathbf{Z} \to 0$ is a complete resolution for $\mathbf{Z}$ over R. Finally, we have the $\mathbf{Z}$-augmentation of $C_\infty \times \Phi$, $\epsilon_{\mathbf{Z}}(t \otimes \varphi) = 1$. Observe that $\epsilon_Z = \epsilon_\infty \circ \epsilon_R$ so that $\mathcal{I} = \mathrm{Ker}(\epsilon_Z)$ occurs in the following diagram in which both rows are exact:

$$
\begin{array}{ccccccccc}
& & & & 0 & & & & \\
& & & & \downarrow & & & & \\
0 & \to & I_R & \to & \mathcal{I} & \longrightarrow & R & \to & 0 \\
& & \| & & \cap & & \downarrow{\scriptstyle t-1} & & \\
0 & \to & I_R & \to & \Lambda & \overset{\epsilon_R}{\longrightarrow} & R & \to & 0 \\
& & & & \downarrow{\scriptstyle \epsilon_\infty} & & & & \\
& & & & Z & & & & \\
& & & & \downarrow & & & & \\
& & & & 0 & & & &
\end{array}
$$

We see from the exact sequence $0 \to I_R \to \mathcal{I} \longrightarrow R \to 0$ that $\mathcal{I}$ is a Swan module over $\mathcal{R}$; in Swan's notation one would describe it as $\mathcal{I} = (I_R, t-1)$.

(15.4) The Dual Augmentation Sequence For this example, begin by taking R to be a commutative ring satisfying (†) and again take $\Lambda = R[\Phi]$ where Φ is a finite group. Now take the $R[\Phi]$-dual $\mathcal{R}^*$ of the above sequence which, by the 0-dimensional case of Eckmann–Shapiro is isomorphic to the R-dual.

$$\mathcal{R}^* = (0 \to R \overset{\epsilon^*}{\to} R[\Phi] \overset{i^*}{\to} I_R^* \to 0).$$

Although R and $R[\Phi]$ are always self-dual, I_R^* is not isomorphic to I_R unless Φ is cyclic.

The Swan modules obtained here are simply the $R[G]$-duals of those obtained from the standard sequence and it is largely a matter of taste as to which we use. In the case $R = Z$, Swan explicitly uses both [91, 94]. The dual sequence does, however, have the practical advantage that I_R^* is naturally a ring, being isomorphic to the quotient $R[G]/(\Sigma)$ where Σ is the principal two-sided ideal in $R[G]$ with generator the sum of the elements in G.

Our discussion throughout has been framed in terms of right modules. The requirement here that R be commutative is only made to avoid having to consider left R-modules upon dualizing. However, provided R possesses an (anti)-involution (if, for example, R is itself a group ring over a commutative ring) then one may relax this condition in the conventional way, replacing the R- dual by the conjugate dual with respect to the involution on R.

(15.5) The Quasi-Augmentation of a Stably Free Module of Rank 1 As a variation on (15.1), again take $\Lambda = R[\Phi]$ where Φ is finite and take a stably free module S of rank 1 over Λ satisfying $S \oplus \Lambda^m \cong \Lambda^{m+1}$. Assume, in addition, that R has the SFC-property. Then $\mathrm{Hom}_\Lambda(\Lambda, R) \cong R$ so that $\mathrm{Hom}_\Lambda(\Lambda^m, R) \cong R^m$ and so $\mathrm{Hom}_\Lambda(S, R) \oplus R^m \cong R^{m+1}$. From the assumption that R has SFC we see that $\mathrm{Hom}_\Lambda(S, R) \cong R$. In particular, there is a surjective Λ-homomorphism $\eta : S \to R$ which is unique up to multiplication by a unit in R. We claim that we have a quasi-augmentation sequence

$$\mathcal{S} = (0 \to \mathrm{Ker}(\eta) \xrightarrow{j} S \xrightarrow{\eta} R \to 0).$$

We must show that $\mathrm{Ker}(\eta)$ is strongly Hopfian and $\mathrm{Hom}_\Lambda(\mathrm{Ker}(\eta), R) = 0$.

For the first, note that for some m, $S \oplus \Lambda^m \cong \Lambda^{m+1}$. As Λ is free of rank $|\Phi|$ over R, and as R has the SFC property then S is also free of rank $|\Phi|$ over R. Splitting $\mathcal{S}$ over R we see that $\mathrm{Ker}(\eta)$ is free of rank $|\Phi| - 1$ over R. Hence $\mathrm{Ker}(\eta)$ is strongly Hopfian. Finally, comparing $\mathcal{S}$ with $\mathcal{R}$ and using Schanuel's Lemma we see that

$$\mathrm{Ker}(\eta) \oplus \Lambda \cong I_R \oplus S.$$

Thus

$$\mathrm{Hom}_\Lambda(\mathrm{Ker}(\eta), R) \oplus \mathrm{Hom}_\Lambda(\Lambda, R) \cong \mathrm{Hom}_\Lambda(S, R) \cong R.$$

As $\mathrm{Hom}_\Lambda(\Lambda, R) \cong R$ it follows that $\mathrm{Hom}_\Lambda(\mathrm{Ker}(\eta), R) = 0$ and $\mathcal{S}$ is a quasi augmentation as claimed.

As an example, we may take $R = \mathbf{Z}[F_m]$ to be the integral group ring of the free group of rank m. Then R has the SFC property and $R[\Phi]$ is the integral group ring $\mathbf{Z}[F_m \times \Phi]$ of the direct product. Taking $\Phi = Q(8p)$, as in Chap. 12, there are infinitely many stably free modules of rank 1 over $R[\Phi]$. Of course, if S is free then we simply retrieve the first example. However, when S is not free then $\mathrm{Ker}(\eta)$ need not be isomorphic to I_R; then the Swan modules obtained are distinct from, though stably equivalent to, those of (15.1).

15.2 Generalized Swan Modules and Rigidity

Fix a quasi augmentation $\mathcal{S} = (0 \to S_- \xrightarrow{i} S_0 \xrightarrow{p} S_+ \to 0)$ and consider Swan modules relative to $\mathcal{S}$. They have the property of being *rigid* in the sense that the exact sequence which defines them is essentially unique; that is:

Proposition 15.6 *Let M, N be Swan modules defined by exact sequences*

$$\mathcal{E}_M = (0 \to S_- \xrightarrow{i} M \xrightarrow{\varphi} S_+ \to 0); \qquad \mathcal{E}_N = (0 \to S_- \xrightarrow{j} N \xrightarrow{\psi} S_+ \to 0).$$

Then $M \cong N$ if and only if $\mathcal{E}_M \cong \mathcal{E}_N$.

Proof The implication ($\Longleftarrow$) is trivial. To prove ($\Longrightarrow$), suppose that $f : M \to N$ is an isomorphism of Λ-modules. Then the homomorphism $\psi \circ f \circ i : S_- \to S_+$ is necessarily zero, and so f induces homomorphisms on both kernels and cokernels:

$$
\begin{array}{ccccccccc}
0 \to & S_- & \xrightarrow{i} & M & \xrightarrow{\varphi} & S_+ & \to 0 \\
 & \downarrow f_- & & \downarrow f & & \downarrow f_+ & \\
0 \to & S_- & \xrightarrow{j} & N & \xrightarrow{\psi} & S_+ & \to 0
\end{array}
$$

that is, f induces a homomorphism of exact sequences $f : \mathcal{E}_M \to \mathcal{E}_N$. We must show that f_+ and f_- are isomorphisms.

To show that f_+ is an isomorphism observe that f_+ is evidently surjective and so is an isomorphism by the strong Hopfian property. It now follows easily that f_- is an isomorphism. Hence $f : \mathcal{E}_M \to \mathcal{E}_N$ is an isomorphism of exact sequences as claimed. □

A module M over Λ is said to be a *generalized Swan module* (relative to $\mathcal{S}$) when it occurs in an exact sequence

$$\mathcal{E}_M = (0 \to S_-^{(m)} \to M \to S_+^{(n)} \to 0)$$

for some positive integers m, n. As in the case of Swan modules the defining sequence $\mathcal{E}_M$ of M is then essentially unique. The proof of this requires a slight refinement of Proposition 15.6.

Theorem 15.7 (Rigidity Theorem) *Let M_1, M_2 be generalized Swan modules relative to $\mathcal{S}$:*

$$\mathcal{E}_1 = (0 \to S_-^{(\mu)} \xrightarrow{i} M_1 \xrightarrow{\varphi} S_+^{(\nu)} \to 0); \qquad \mathcal{E}_2 = (0 \to S_-^{(m)} \xrightarrow{j} M_2 \xrightarrow{\psi} S_+^{(n)} \to 0);$$

then $M_1 \cong M_2$ if and only if $\mathcal{E}_1 \cong \mathcal{E}_2$.

Proof Suppose that $f : M_1 \to M_2$ is an isomorphism of Λ-modules. By interchanging M_1 and M_2 and replacing f by f^{-1} we may, without loss of generality, assume that $\nu \le n$. The homomorphism $\psi \circ f \circ i : S_-^{(\mu)} \to S_+^{(n)}$ is necessarily zero, and so f induces homomorphisms on both kernels and cokernels and hence a homomorphism of exact sequences $f : \mathcal{E}_1 \to \mathcal{E}_2$;

$$
\begin{array}{ccccccccc}
0 & \to & S_-^{(\mu)} & \xrightarrow{i} & M_1 & \xrightarrow{\varphi} & S_+^{(\nu)} & \to & 0 \\
& & \downarrow f_- & & \downarrow f & & \downarrow f_+ & & \\
0 & \to & S_-^{(m)} & \xrightarrow{j} & M_2 & \xrightarrow{\psi} & S_+^{(n)} & \to & 0
\end{array}
$$

As f is an isomorphism it is surjective and so f_+ is also surjective. As S_+ is strongly Hopfian and $\nu \le n$ then f_+ is an isomorphism and $\nu = n$. By extending the sequence to the left by zeroes and using the Five Lemma it follows that f_- is also an isomorphism and so f induces an isomorphism of exact sequences $f : \mathcal{E}_M \to \mathcal{E}_N$. Also $\mu = m$ □

Observe that in the course of proving Theorem 15.7 we also proved:

Proposition 15.8 *Let M, N be generalized Swan modules relative to $\mathcal{S}$ thus:*

$$\mathcal{E}_1 = (0 \to S_-^{(\mu)} \xrightarrow{i} M_1 \xrightarrow{\varphi} S_+^{(\nu)} \to 0); \qquad \mathcal{E}_2 = (0 \to S_-^{(m)} \xrightarrow{j} M_2 \xrightarrow{\psi} S_+^{(n)} \to 0);$$

if $M_1 \cong M_2$ then $\mu = m$ and $\nu = n$.

A generalized Swan module M given by an exact sequence

$$(0 \to S_-^{(m)} \xrightarrow{j} M \xrightarrow{\psi} S_+^{(n)} \to 0)$$

is said to be of type (m, n). If $m = n$ we simply say that M is of rank m. If $m \neq n$ we say that M is of *mixed type*. As an nontrivial example of mixed type we may consider:

(15.9) The Integral Augmentation Ideal of $\mathbf{Z}[F_m \times \Phi]$ as a Swan Module Let $x_1, \ldots, x_m$ be a free generating set for F_m and let Φ be a nontrivial finite group. Taking $R = \mathbf{Z}[F_m]$ we have a complete resolution of $\mathbf{Z}$ over R

$$0 \to R^m \xrightarrow{X} R \longrightarrow \mathbf{Z} \to 0,$$

where $X = (x_1 - 1, \ldots, x_m - 1)$. We obtain a commutative diagram with exact rows

$$
\begin{array}{ccccccc}
0 \to & I_R & \to & \mathcal{I} & \longrightarrow & R^m & \to 0 \\
      & \| & & \cap & & \downarrow X & \\
0 \to & I_R & \to & R[\Phi] & \xrightarrow{\epsilon_R} & R & \to 0
\end{array}
$$

showing that $\mathcal{I}$ is a generalized Swan module of type $(1, m)$.

15.3 Classification of Generalized Swan Modules

Until further notice all generalized Swan modules will be taken relative to a fixed quasi augmentation sequence $\mathcal{S} = (0 \to S_- \to S \to S_+ \to 0)$. Relative to $\mathcal{S}$ the Rigidity Theorem (15.7) reduces the isomorphism classification of generalized Swan modules of type (m, n) to the isomorphism classification of exact sequences of the form $(0 \to S_-^{(m)} \xrightarrow{i} ? \xrightarrow{p} S_+^{(n)} \to 0)$. Up to *congruence*, that is, requiring identity maps on each end,

$$
\begin{array}{ccccccc}
0 \to & S_-^{(m)} & \to & ? & \to & S_+^{(n)} & \to 0 \\
      & \downarrow \text{Id} & & \downarrow & & \downarrow \text{Id} & \\
0 \to & S_-^{(m)} & \to & ?? & \to & S_+^{(n)} & \to 0
\end{array}
$$

such exact sequences are classified by $\mathrm{Ext}^1(S_+^{(n)}, S_-^{(m)})$. To allow for classification up to isomorphism, however, we must allow arbitrary automorphisms on the ends. There is a natural right action of $\mathrm{Aut}_\Lambda(S_+^{(n)})$ on $\mathrm{Ext}^1(S_+^{(m)}, S_-^{(m)})$

$$\mathrm{Ext}^1(S_+^{(n)}, S_-^{(m)}) \times \mathrm{Aut}_\Lambda(S_+^{(n)}) \to \mathrm{Ext}^1(S_+^{(n)}, S_-^{(m)})$$

$$(\mathcal{E}, \beta) \qquad\qquad\qquad \mapsto \qquad \beta^*(\mathcal{E})$$

Likewise there is a natural left action of $\text{Aut}_\Lambda(S_-^{(m)})$ on $\text{Ext}^1(S_+^{(n)}, S_-^{(m)})$

$$\text{Aut}_\Lambda(S_-^{(m)}) \times \text{Ext}^1(S_+^{(n)}, S_-^{(m)}) \to \text{Ext}^1(S_+^{(n)}, S_-^{(m)})$$

$$(\alpha, \mathcal{E}) \qquad\qquad \mapsto \qquad \alpha_*(\mathcal{E})$$

As is well known (see, for example, [68], p. 67, Lemma 7.6), the two actions commute in the sense that $\alpha_* \beta^*(\mathcal{E})$ is congruent to $\beta^* \alpha_*(\mathcal{E})$ so that we get a two sided action

$$\text{Aut}_\Lambda(S_-^{(m)}) \times \text{Ext}^1(S_+^{(n)}, S_-^{(m)}) \times \text{Aut}_\Lambda(S_+^{(n)}) \to \text{Ext}^1(S_+^{(n)}, S_-^{(m)})$$

$$(\alpha, \mathcal{E}, \beta) \qquad\qquad\qquad \mapsto \qquad \alpha_* \beta^*(\mathcal{E})$$

We get the following:

Theorem 15.10 *Let S be a quasi augmentation; then there is a $1-1$ correspondence*

$$\left\{ \begin{array}{c} \text{Isomorphism classes of} \\ \text{generalized Swan modules} \\ \text{of type } (m, n) \text{ over } S \end{array} \right\} \longleftrightarrow \text{Aut}(S_-^{(m)}) \backslash \text{Ext}^1(S_+^{(n)}, S_-^{(m)}) / \text{Aut}(S_+^{(n)}).$$

We can describe this in terms of 'coordinates': if σ denotes a sign $\sigma = \pm$ put $T_\sigma = \text{End}_\Lambda(S_\sigma)$ and make the identifications

$$\text{End}_\Lambda(S_-^{(m)}) \cong M_m(T_-); \qquad \text{Aut}_\Lambda(S_-^{(m)}) \cong GL_m(T_-),$$

$$\text{End}_\Lambda(S_+^{(n)}) \cong M_n(T_+); \qquad \text{Aut}_\Lambda(S_\sigma^{(n)}) \cong GL_n(T_+).$$

Let $M_{m,n}(A)$ denote the set of $m \times n$ matrices with entries in *a set A*. The additivity properties of Ext^1 allow us to identify

$$\text{Ext}^1(S_+^{(n)}, S_-^{(m)}) \longleftrightarrow M_{m,n}(\text{Ext}^1(S_+, S_-))$$

via the correspondence $\mathcal{E} \mapsto (\pi_*^r i_t^*(\mathcal{E}))_{1 \le r \le m, 1 \le t \le n}$ where $\pi^r : S_-^{(m)} \to S_-$ is projection to the rth-factor and $i_t : S_- \to S_-^{(m)}$ is inclusion of the tth-summand. Since S is stably free then S_- is a representative of the first syzygy $\Omega_1(S_+)$. Moreover, as $\text{Ext}^1(S_+, \Lambda) = 0$ the corepresentation formula (5.22) for cohomology holds to give

$$\text{Ext}^1(S_+, S_-) \cong \text{Hom}_{\mathcal{D}\text{er}}(S_-, S_-) \cong \text{End}_{\mathcal{D}\text{er}}(S_-).$$

Write $\Theta = \text{End}_{\mathcal{D}\text{er}}(S_-)$ then again since $\text{Ext}^1(S_+, \Lambda) = 0$ there are ring isomorphisms $\text{End}_{\mathcal{D}\text{er}}(S_+) \cong \text{End}_{\mathcal{D}\text{er}}(S_-) \cong \Theta$. In coordinate terms the natural surjective ring homomorphisms $\text{End}_\Lambda(S_-^{(m)}) \to \text{End}_{\mathcal{D}\text{er}}(S_-^{(m)}), \text{End}_\Lambda(S_+^{(n)}) \to \text{End}_{\mathcal{D}\text{er}}(S_+^{(n)})$ become surjective ring homomorphisms $M_m(T_-) \to M_m(\Theta), M_n(T_+) \to M_n(\Theta)$ which we write in the form $X_\sigma \mapsto [X_\sigma]$ and the above two sided action, in coordinate terms, becomes

$$GL_m(T_-) \times M_{m,n}(\Theta) \times GL_n(T_+) \to M_{m,n}(\Theta)$$

$$(X_-, \alpha, X_+) \qquad\qquad \mapsto [X_-]\alpha[X_+]$$

We get the following coordinatised version of the Classification Theorem:

Theorem 15.11 (Classification Theorem) *Let S be a quasi augmentation; then there is a $1 - 1$ correspondence*

$$\left\{ \begin{array}{c} \textit{Isomorphism classes of} \\ \textit{generalized Swan modules} \\ \textit{of type } (m, n) \textit{ over } S \end{array} \right\} \longleftrightarrow GL_m(T_-)\backslash M_{m,n}(\Theta)/GL_n(T_+).$$

When $m = n$ it is convenient to record the details of this correspondence explicitly. Given $f \in \mathrm{End}_{\mathcal{D}\mathrm{er}}(S_-^{(m)})$ we may form the pushout extension

$$\begin{array}{ccccccc} 0 \to & S_-^{(m)} & \stackrel{i}{\to} & S^{(m)} & \to & S_+^{(m)} & \to 0 \\ & \downarrow f & & \downarrow \hat{f} & & \downarrow \mathrm{Id} & \\ 0 \to & S_-^{(m)} & \stackrel{j}{\to} & \varinjlim(f, i) & \to & S_+^{(m)} & \to 0 \end{array}$$

Denote by $\mathrm{Iso}_m(S)$ the set of isomorphism classes of generalized Swan modules of type (m, m) over S. By making the identification $M_m(\Theta) \leftrightarrow \mathrm{End}_{\mathcal{D}\mathrm{er}}(S_-^{(m)})$ the bijection of Theorem 15.11 in the direction $GL_m(T_-)\backslash M_m(\Theta)/GL_m(T_+) \to \mathrm{Iso}_m(S)$ is given by the mapping

$$\varinjlim : GL_m(T_-)\backslash M_m(\Theta)/GL_m(T_+) \to \mathrm{Iso}_m(S); \quad [f] \mapsto \varinjlim(f, i).$$

This allows a parametrization of $\mathrm{Iso}_m(S)$ expressed entirely in terms of Θ. To see this, observe that we have a commutative diagram:

$$\begin{array}{ccc} E_m(T_-)\backslash M_m(\Theta)/E_m(T_+) & \stackrel{\nu_2}{\longrightarrow} & GL_m(T_-)\backslash M_m(\Theta)/GL_m(T_+) \\ \downarrow \nu_1 & & \downarrow \varinjlim \\ E_m(\Theta)\backslash M_m(\Theta)/E_m(\Theta) & \stackrel{\pi}{\longrightarrow} & \mathrm{Iso}_m(S) \end{array} \quad (15.12)$$

Here ν_1, ν_2 are the obvious maps. Note that the fibres of ν_2 are quotients of

$$GL_m(T_-)/E_m(T_-) \times E_m(T_+)\backslash GL_m(T_+).$$

To describe π, note that as $E_m(T_\sigma) \to E_m(\Theta)$ is surjective then ν_1 is a bijection. Defining $\pi = \varinjlim \circ \nu_2 \circ (\nu_1)^{-1}$ the diagram commutes as required. Since ν_1 and $\varinjlim$ are bijective and ν_2 is surjective then π is also surjective. Moreover ν_1 maps the fibres of ν_2 bijectively to the fibres of π so that we obtain:

Corollary 15.13 (Parametrization) *Let S be a quasi augmentation; then there is a surjective mapping*

$$\pi : E_m(\Theta)\backslash M_m(\Theta)/E_m(\Theta) \to \mathrm{Iso}_m(S)$$

each of whose fibres is a quotient of $GL_m(T_-)/E_m(T_-) \times E_m(T_+)\backslash GL_m(T_+).$

Suppose that $X \in M_m(\Theta)$ is a matrix representing the generalized Swan module J. When, as above, Θ is commutative it follows from Corollary 15.13 that the ideal $(\det(X)) \lhd \Theta$ depends only on J; thus when J is a generalized Swan module we may define an ideal $\langle \det(J) \rangle \lhd \Theta$ by the rule:

$$\langle \det(J) \rangle = (\det(X)) \quad \text{when } X \in M_m(\Theta) \text{ is a matrix representing } J. \quad (15.14)$$

15.4 Completely Decomposable Swan Modules

The Classification Theorem 15.11 shows that Swan modules of rank 1 are of the form $M(\theta)$ where, for $\theta \in \Theta$, $M(\theta) = \varinjlim(\theta, i)$ is obtained from the pushout diagram

$$
\begin{array}{ccc}
S_- & \xrightarrow{i} & S \\
\downarrow{\theta} & & \downarrow \\
S_- & \rightarrow & \varinjlim(\theta, i)
\end{array}
$$

This description requires the identification $\text{Ext}^1(S_+, S_-) \cong \text{End}_{\mathcal{D}er}(S_-) = \Theta$. Using the alternative identification $\text{Ext}^1(S_+, S_-) \cong \text{End}_{\mathcal{D}er}(S_+)$ under the canonical isomorphism $\rho : \text{End}_{\mathcal{D}er}(S_-) \to \text{End}_{\mathcal{D}er}(S_+)$ there is also a pullback description $M(\theta) = \varprojlim(\eta, \rho(\theta))$

$$
\begin{array}{ccc}
\varprojlim(\eta, \rho(\theta)) & \rightarrow & S_+ \\
\downarrow & & \downarrow{\rho(\theta)} \\
\Lambda & \xrightarrow{\eta} & S_+.
\end{array}
$$

As a special case of the parametrization theorem Corollary 15.13, the isomorphism class of $M(\theta)$ is determined entirely by the equivalence class $\langle \theta \rangle$ of θ in $T_-^* \backslash \Theta / T_+^*$; that is:

$$M(\theta) \cong M(\mu) \quad \Longleftrightarrow \quad \theta = [\alpha_-]\mu[\alpha_+] \quad \text{for some } \alpha_- \in T_-^*, \ \alpha_+ \in T_+^*. \quad (15.15)$$

From Theorem 5.41 we obtain the original form of Swan's projectivity criterion [91]; that is:

$$M(\theta) \text{ is projective} \quad \Longleftrightarrow \quad \theta \in \Theta^*. \quad (15.16)$$

Relative to $\mathcal{S}$, S itself is described as $S = M(1)$. We similarly obtain the original form of Swan's isomorphism criterion [91]:

$$M(\theta) \cong S \quad \Longleftrightarrow \quad \theta = [\alpha_-][\alpha_+] \quad \text{for some } \alpha_- \in T_-^*, \alpha_+ \in T_+^*. \quad (15.17)$$

More generally, for $\theta_1, \ldots, \theta_m \in \Theta$ we put $M(\theta_1, \ldots, \theta_m) = M(\theta_1) \oplus \cdots \oplus M(\theta_m)$. A generalized Swan module of this form is said to be *completely decomposable*. There is an obvious generalization of (15.16):

$$M(\theta_1, \ldots, \theta_m) \text{ is projective if and only if each } \theta_i \in \Theta^*. \quad (15.18)$$

Denote by $\mathcal{D}_m(\Theta)$ multiplicative submonoid of $M_m(\Theta)$ consisting of diagonal matrices thus

$$\Delta(\theta_1, \ldots, \theta_m) = \begin{pmatrix} \theta_1 & & 0 \\ & \ddots & \\ 0 & & \theta_m \end{pmatrix};$$

$\mathcal{D}_m(\Theta)^*$ will denote the group of *invertible* diagonal $m \times m$ matrices over Θ. We note that whilst a *surjective* ring homomorphism $\varphi : \Theta_1 \to \Theta_2$ induces a surjective monoid homomorphism $\varphi_* : \mathcal{D}_m(\Theta_1) \to \mathcal{D}_m(\Theta_2)$ the corresponding induced map on unit groups $\varphi_* : \mathcal{D}_m(\Theta_1)^* \to \mathcal{D}_m(\Theta_2)^*$ is, in general, *not surjective*. Up to isomorphism $M(\theta_1, \ldots, \theta_m)$ is classified by the image of $\Delta(\theta_1, \ldots, \theta_m)$ in $GL_m(T_-)\backslash M_m(\Theta)/GL_m(T_+)$ thus:

Proposition 15.19 *Relative to S every generalized Swan module is completely decomposable if and only if the natural mapping*

$$\natural : \mathcal{D}_m(\Theta) \to GL_m(T_-)\backslash M_m(\Theta)/GL_m(T_+)$$

is surjective.

The Parametrization Corollary 15.13 can be applied, in particular, to completely decomposable modules. In the notation of Sect. 2.3, for arbitrary $\theta_i, \mu_j \in \Theta$:

$$\Delta(\theta_1, \ldots, \theta_m) \sim \Delta(\mu_1, \ldots, \mu_m) \implies M(\theta_1, \ldots, \theta_m) \cong M(\mu_1, \ldots, \mu_m). \quad (15.20)$$

The considerations of Sect. 2.3 now give isomorphism relations between completely decomposable modules. In particular, from (2.25) we obtain:

Proposition 15.21 *Let $\theta_i \in \Theta^*$ for $1 \leq i \leq m$; then*

$$M(\theta_1, \ldots, \theta_m) \cong M\left(\prod_i \theta_i, 1, \ldots, 1\right).$$

Note that, by (2.27), the factors in the product $\prod_i \theta_i$ may be taken in any order.

To ensure simplicity of discussion, for the remainder of this section *we will suppose that T_+, T_- and hence Θ are all commutative*. These restrictions allow a useful generalization of (15.15):

Proposition 15.22 $\underbrace{M(\theta, 1, \ldots, 1)}_{m} \cong \underbrace{M(\mu, 1, \ldots, 1)}_{m} \iff \langle \theta \rangle = \langle \mu \rangle.$

Proof For ($\implies$), if $M(\theta, 1, \ldots, 1) \cong M(\mu, 1, \ldots, 1)$ then by Theorem 15.11 there exist $X_\sigma \in GL_m(T_\sigma)$ such that

$$[X_-]\Delta(\theta, 1, \ldots, 1)[X_+] = \Delta(\mu, 1, \ldots, 1).$$

Thus $[u_-]\theta[u_+] = \mu$ where $u_\sigma = \det(X_\sigma) \in T_\sigma^*$; that is, $\langle\theta\rangle = \langle\mu\rangle$. The implication $(\Longleftarrow)$ is trivial. $\qquad\square$

Corollary 15.23 *If* $M(\theta_1, \ldots, \theta_m)$, $M(\delta_1, \ldots, \delta_m)$ *are projective then*

$$M(\theta_1, \ldots, \theta_m) \cong M(\delta_1, \ldots, \delta_m) \quad \Longleftrightarrow \quad \left\langle \prod_i \theta_i \right\rangle = \left\langle \prod_i \delta_i \right\rangle.$$

Proof Since $M(\theta_1, \ldots, \theta_m)$ is projective then $\theta_i \in \Theta^*$ for each i, and so, by Proposition 15.21, $M(\theta_1, \ldots, \theta_m) \cong M(\prod_i \theta_i, 1, \ldots, 1)$. Likewise, $M(\delta_1, \ldots, \delta_m) \cong M(\prod_i \delta_i, 1, \ldots, 1)$. The conclusion now follows from Proposition 15.22. $\qquad\square$

Essentially the same argument as (15.18) now shows:

Corollary 15.24 *If* $M(\theta_1, \ldots, \theta_m)$ *is projective then*

$$M(\theta_1, \ldots, \theta_m) \cong \underbrace{S \oplus \cdots \oplus S}_{m} \quad \Longleftrightarrow \quad \left\langle \prod_i \theta_i \right\rangle = \langle 1 \rangle.$$

We now restrict S to be of the form $S = (0 \to S_- \to \Lambda \to S_+ \to 0)$; that is, $S \cong \Lambda$. With this restriction we obtain two statements which are familiar from Swan's original context but are perhaps surprising in this degree of generality:

Theorem 15.25 *Let* $M = M(\theta_1, \ldots, \theta_m)$ *be completely decomposable and projective of rank* m; *then for some projective Swan module* P *of rank* 1, M *is a direct sum*

$$M \cong P \oplus \Lambda^{m-1}.$$

And also:

Theorem 15.26 *If* $M(\theta_1, \ldots, \theta_m)$ *is stably free then it is free.*

Theorem 15.27 (Decomposition Theorem) *If* Θ *is generalized Euclidean then, relative to* S, *every generalized Swan module is completely decomposable.*

Proof Let $\delta : \mathcal{D}_m(\Theta) \to E_m(\Theta)\backslash E_m(\Theta)$, $\natural : \mathcal{D}_m(\Theta) \to GL_m(T_-)\backslash M_m(\Theta)/GL_m(T_+)$ denote the canonical mappings. It is straightforward to see that the following diagram commutes where ν_1, ν_2 are as in (15.12).

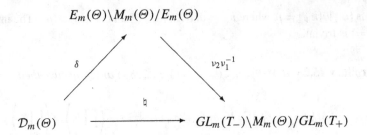

It follows directly from the generalized Euclidean hypothesis that δ is surjective whilst $v_2 v_1^{-1}$ is surjective as in the proof of Corollary 15.13. Thus $\natural$ is surjective.

$\square$

Chapter 16
Parametrizing $\Omega_1(\mathbf{Z}) : G = C_\infty \times \Phi$

Both first and second minimality criteria fail in a case of particular interest, namely when G is a direct product $G = F \times \Phi$ in which F is a free group and Φ is finite. As we observed in Proposition 13.2, when Φ is the trivial group, the conclusion also fails. Nevertheless, using a rather more intricate argument, we are still able to show that the conclusion is sustained when the finite factor Φ is nontrivial; that is we shall show:

Third minimality criterion: $\mathcal{I}$ lies at the minimal level of $\Omega_1(\mathbf{Z})$ when G is a direct product $F_m \times \Phi$ where F_m is a free group of rank $m \geq 1$ and Φ is finite and nontrivial.

The results of this section first appeared in [58]. The proof requires a knowledge of all the syzygies $\Omega_r(\mathbf{Z})$ over $\mathbf{Z}[F_n \times C_m]$ so we begin by giving a complete resolution of $\mathbf{Z}$ in this case.

16.1 The Syzygies of $F_m \times C_n$

Let F_m denote the free group of rank $m \geq 1$ given in the obvious presentation

$$F_m = \langle t_1, \ldots, t_m | \emptyset \rangle.$$

Then the algebraic Cayley complex gives a complete resolution of $\mathbf{Z}$ over $\mathbf{Z}[F_m]$ as follows:

$$0 \to \mathbf{Z}[F_m]^m \xrightarrow{T} \mathbf{Z}[F_m] \xrightarrow{\epsilon} \mathbf{Z} \to 0,$$

where

$$T = (t_1 - 1, \ldots, t_m - 1).$$

For any group Φ, we may identify $\mathbf{Z}[F_m \times \Phi] = \mathbf{Z}[F_m] \otimes \mathbf{Z}[\Phi]$ where tensor product is taken over $\mathbf{Z}$. Given a complete resolution

$$\mathbf{A} = (\cdots \to A_{n+1} \xrightarrow{\partial_{n+1}} A_n \xrightarrow{\partial_n} A_{n-1} \xrightarrow{\partial_{n-1}} \cdots \to A_1 \xrightarrow{\partial_1} A_0 \xrightarrow{\epsilon} \mathbf{Z} \to 0)$$

F.E.A. Johnson, *Syzygies and Homotopy Theory*, Algebra and Applications 17, DOI 10.1007/978-1-4471-2294-4_16, © Springer-Verlag London Limited 2012

for $\mathbf{Z}$ over $\mathbf{Z}[\Phi]$. We proceed to construct a complete resolution $\mathcal{C}$ for $\mathbf{Z}$ over $\mathbf{Z}[F_m \times \Phi]$:

Put $R(m) = \underbrace{R \oplus \cdots \oplus R}_{m}$ where $R = \mathbf{Z}[F_m]$. Put $\mathcal{C}_0 = R \otimes A_0$ and for $n \geq 1$ write

$$C_n^+ = R(m) \otimes A_{n-1}, \qquad C_n^- = R \otimes A_n.$$

When $n = 1$ we put $\Delta_1 = (T \otimes 1, 1 \otimes \partial_1)$. For any $n \geq 2$ and any signs σ, τ we define $\mathbf{Z}[F_m \times \Phi]$-linear maps $(\Delta_n)_{\sigma\tau} : C_n^\tau \to C_{n-1}^\sigma$ as follows:

$$(\Delta_n)_{++} = -(1 \otimes \partial_{n-1}); \qquad (\Delta_n)_{+-} = 0;$$

$$(\Delta_n)_{-+} = T \otimes 1; \qquad (\Delta_n)_{--} = 1 \otimes \partial_n$$

and put

$$\Delta_n = \begin{pmatrix} (\Delta_n)_{++} & (\Delta_n)_{+-} \\ (\Delta_n)_{-+} & (\Delta_n)_{--} \end{pmatrix} = \begin{pmatrix} -(1 \otimes \partial_{n-1}) & 0 \\ T \otimes 1 & 1 \otimes \partial_n \end{pmatrix}.$$

We obtain homomorphisms $\Delta_n : \mathcal{C}_n \to \mathcal{C}_{n-1}$ over $\mathbf{Z}[F_m \times \Phi]$ where $\mathcal{C}_n = C_n^+ \oplus C_n^-$:

Theorem 16.1 $\mathcal{C} = (\cdots \to \mathcal{C}_{n+1} \xrightarrow{\Delta_{n+1}} \mathcal{C}_n \xrightarrow{\Delta_n} \mathcal{C}_{n-1} \xrightarrow{\Delta_{n-1}} \cdots \xrightarrow{\Delta_2} \mathcal{C}_1 \xrightarrow{\Delta_1} \mathcal{C}_0 \xrightarrow{\epsilon} \mathbf{Z} \to 0)$ *is a complete resolution for* $\mathbf{Z}$ *over* $\mathbf{Z}[F_m \times \Phi]$.

We now specialise to the case where Φ is the cyclic group of order n; $\Phi = C_n = \langle y | y^n = 1 \rangle$. Take the usual periodic resolution of $\mathbf{Z}$ over $\mathbf{Z}[C_n]$

$$\cdots \xrightarrow{\Sigma} \mathbf{Z}[C_n] \xrightarrow{y-1} \mathbf{Z}[C_n] \xrightarrow{\Sigma} \cdots \xrightarrow{y-1} \mathbf{Z}[C_n] \xrightarrow{\Sigma} \mathbf{Z}[C_n] \xrightarrow{y-1} \mathbf{Z}[C_n] \xrightarrow{\epsilon} \mathbf{Z} \to 0,$$

where $\Sigma = \sum_{r=1}^n y^r$. The tensor product resolution of Theorem 16.1 then assumes the form:

$$\mathcal{C} = (\cdots \to \Lambda^2 \xrightarrow{\Delta_{2k}} \Lambda^2 \xrightarrow{\Delta_{2k-1}} \cdots \xrightarrow{\Delta_3} \Lambda^2 \xrightarrow{\Delta_2} \Lambda^2 \xrightarrow{\Delta_1} \Lambda \xrightarrow{\epsilon} \mathbf{Z} \to 0),$$

where $\Delta_1 = (T \otimes 1, 1 \otimes (y-1))$ whilst for $k \geq 1$

$$\Delta_{2k} = \begin{pmatrix} -1 \otimes (y-1) & 0 \\ T \otimes 1 & 1 \otimes \Sigma \end{pmatrix}; \qquad \Delta_{2k+1} = \begin{pmatrix} -1 \otimes \Sigma & 0 \\ T \otimes 1 & 1 \otimes (y-1) \end{pmatrix}.$$

Evidently this resolution is periodic in dimensions ≥ 2. In particular for all $k \geq 1$ we have:

$$\mathrm{Im}(\Delta_{2k}) \cong \mathrm{Im}(\Delta_2), \tag{16.2}$$

$$\mathrm{Im}(\Delta_{2k+1}) \cong \mathrm{Im}(\Delta_3). \tag{16.3}$$

In the odd case we can improve on this. We first make an elementary observation: suppose X, M_1, M_2 are modules over a ring Λ and that $h = \binom{h_1}{h_2} : X \to M_1 \oplus M_2$ is a Λ-homomorphism. Let $\pi : M_1 \oplus M_2 \to M_2$ be the projection; then with this notation:

Proposition 16.4 *The sequence* $0 \to \mathrm{Im}(h_{1|\mathrm{Ker}(h_2)}) \to \mathrm{Im}(h) \overset{\pi}{\to} \mathrm{Im}(h_2) \to 0$ *is exact.*

We now obtain

Theorem 16.5 $\mathrm{Im}(\Delta_{2k+1}) \cong \mathrm{Im}(\Delta_1) = \mathcal{I}$ *for all* $k \geq 1$.

Proof Observe that $\Delta_{2k+1} = \left(\begin{smallmatrix} g \\ \Delta_1 \end{smallmatrix} \right)$ where $g = (-1 \otimes \Sigma, 0)$. We may now apply Proposition 16.4 to get an exact sequence

$$0 \to \mathrm{Im}(g_{|\mathrm{Ker}(\Delta_1)}) \to \mathrm{Im}(\Delta_{2k+1}) \overset{\pi}{\to} \mathrm{Im}(\Delta_1) \to 0.$$

Observe that $\mathrm{Im}(\Delta_1) = \mathcal{I}$. Moreover, one calculates easily that $g \circ \Delta_2 \equiv 0$; that is, $g_{|\mathrm{Im}(\Delta_2)} = 0$. However, $\mathrm{Im}(\Delta_2) = \mathrm{Ker}(\Delta_1)$ by exactness of $\mathcal{C}$ so that the above exact sequence reduces to an isomorphism $\mathrm{Im}(\Delta_{2k+1}) \cong \mathrm{Im}(\Delta_1) = \mathcal{I}$ as claimed. $\square$

The conclusions of (16.2) and Theorem 16.5 are worthy of emphasis; when $G = F_m \times C_n$ the syzygies $\Omega_k(\mathbf{Z})$ $(k > 0)$ are *completely periodic of period two*: that is;

$$\Omega_k(\mathbf{Z}) = \begin{cases} [\mathcal{I}] & k \text{ odd,} \\ \mathrm{Im}(\Delta_2) & k \text{ even.} \end{cases} \tag{16.6}$$

This syzygetic periodicity should be contrasted with the cohomological behaviour. Whilst it is true (a special case of Farrell-Tate cohomology [29]) that the cohomology of $G = F_m \times C_n$ is periodic in dimensions ≥ 2, this cohomological periodicity breaks down in dimension 1; for example, $\mathrm{Ext}^1_\Lambda(\mathbf{Z}, \mathbf{Z}[G]) \neq 0$ whilst $\mathrm{Ext}^3_\Lambda(\mathbf{Z}, \mathbf{Z}[G]) = 0$.

16.2 Two Calculations

Given a ring R and a finite group Φ we consider R as a bimodule over the group ring $\Lambda = R[\Phi]$ where Φ acts trivially.

Proposition 16.7 $\mathrm{End}_{\mathcal{D}\mathrm{er}(\Lambda)}(R) \cong R/|\Phi|$.

Proof Any Λ-homomorphism $\beta : \Lambda \to R$ is a multiple $\beta = b\epsilon$ where $b \in R$ and ϵ is the R-augmentation $R[\Phi] \to R$. Any Λ-homomorphism $\gamma : R \to \Lambda$ is a multiple $\gamma = c\epsilon^*$ where $c \in \Lambda$ and $\epsilon^* : R \to \Lambda$ is the R-dual of ϵ; that is $\epsilon^*(1) = \sum_{\phi \in \Phi} \hat{\phi}$ where $\{\hat{\phi}\}_{\phi \in \Phi}$ is the canonical R-basis of $\Lambda = R[\Phi]$. Observe that $\epsilon^*(1)$ lies in the centre of Λ and that $\epsilon\epsilon^*(1) = |\Phi|$. Suppose that $\alpha = \beta\gamma$ is a factorization of α through Λ^m where

$$\gamma = \begin{pmatrix} c_1 \epsilon^* \\ \vdots \\ c_m \epsilon^* \end{pmatrix} : R \to \Lambda^m \quad \text{and} \quad \beta = (b_1 \epsilon, \ldots, b_m \epsilon) : \Lambda^m \to R.$$

Then α is completely determined by $\alpha(1) = \sum_i b_i \epsilon \epsilon^*(1) c_i = (\sum_i b_i c_i)|\Phi|$. Conversely, if $\alpha = \lambda|\Phi|$ for some $\lambda \in \Lambda$ then α factors through Λ since $\alpha = \lambda \epsilon \circ \epsilon^*$; with the above notation

$$\alpha : R \to R \quad \text{factors through } \Lambda^m \quad \Longleftrightarrow \quad \alpha = \lambda|\Phi| \quad \text{for some } \lambda \in \Lambda.$$

The result now follows as $\alpha \in \text{End}_\Lambda(R)$ factorizes through a projective module if and only if it factorizes through some Λ^m. □

We now specialize to the case where R is the integral group ring $R = \mathbf{Z}[F_m]$ where F_m is free group of rank $m \geq 1$ and where $\Phi = C_n$ so that $\Lambda = R[C_n] = \mathbf{Z}[F_m \times C_n]$. We denote by $\mathcal{I}$ the integral augmentation ideal of $\mathbf{Z}[F_m \times C_n]$. From the exact sequence $0 \to \mathcal{I} \to \Lambda \to \mathbf{Z} \to 0$ we get, by dimension shifting, that

Proposition 16.8 $\text{Ext}_\Lambda^{k+1}(\mathbf{Z}, N) \cong \text{Ext}_\Lambda^k(\mathcal{I}, N)$ *for any Λ-module N.*

Proposition 16.9 $\text{Ext}_\Lambda^{k+1}(\mathbf{Z}, \mathbf{Z}) \cong \text{Ext}_\Lambda^{k+1}(\mathcal{I}, \mathcal{I})$ *for $k \geq 1$.*

Proof Clearly $\text{Ext}_R^k(\mathbf{Z}, R) = 0$ for $k \geq 2$ since F_m has cohomological dimension one. Moreover, as F_m is a subgroup of finite index in $G = F_m \times \Phi$ it follows by the Eckmann-Shapiro Lemma that $\text{Ext}_\Lambda^k(\mathbf{Z}, \Lambda) = 0$ for $k \geq 2$. By dimension shifting as in Proposition 16.8, we see that $\text{Ext}_\Lambda^k(\mathcal{I}, \Lambda) = 0$ for $k \geq 1$. Hence the exact sequence

$$\text{Ext}^k(\mathcal{I}, \Lambda) \to \text{Ext}^k(\mathcal{I}, \mathbf{Z}) \to \text{Ext}^{k+1}(\mathcal{I}, \mathcal{I}) \to \text{Ext}^{k+1}(\mathcal{I}, \Lambda)$$

reduces to an isomorphism $\text{Ext}_\Lambda^k(\mathcal{I}, \mathbf{Z}) \cong \text{Ext}_\Lambda^{k+1}(\mathcal{I}, \mathcal{I})$. However, again by dimension shifting, $\text{Ext}_\Lambda^{k+1}(\mathbf{Z}, \mathbf{Z}) \cong \text{Ext}_\Lambda^k(\mathcal{I}, \mathbf{Z})$ so that $\text{Ext}_\Lambda^{k+1}(\mathbf{Z}, \mathbf{Z}) \cong \text{Ext}_\Lambda^{k+1}(\mathcal{I}, \mathcal{I})$ for $k \geq 1$. □

Proposition 16.10 $\text{Ext}_\Lambda^3(\mathbf{Z}, \mathcal{I}) \cong \mathbf{Z}/n$.

Proof The Künneth Theorem applied to $G = F_m \times C_n$ shows that $\text{Ext}_\Lambda^2(\mathbf{Z}, \mathbf{Z}) \cong \mathbf{Z}/n$; thus $\text{Ext}_\Lambda^2(\mathcal{I}, \mathcal{I}) \cong \mathbf{Z}/n$ by Proposition 16.9; now apply dimension shifting as in Proposition 16.8. □

By (16.6) $\mathcal{I}$ is a representative of $\Omega_3(\mathbf{Z})$ over $\Lambda = \mathbf{Z}[F_m \times C_n]$. As $\text{Ext}_\Lambda^3(\mathbf{Z}, \Lambda) = 0$ the corepresentation formula Theorem 5.37 gives an isomorphism $\text{Hom}_{\mathcal{D}\text{er}}(\mathcal{I}, N) \cong \text{Ext}_\Lambda^3(\mathbf{Z}, N)$ for any Λ-module N; on taking $N = \mathcal{I}$ we obtain:

Corollary 16.11 $\text{End}_{\mathcal{D}\text{er}}(\mathcal{I}) \cong \mathbf{Z}/n$.

16.3 The Third Minimality Criterion

Let G be a direct product of groups $G = \Psi \times \Phi$ and make the abbreviations

$$\Lambda = \mathbf{Z}[G]; \qquad R = \mathbf{Z}[\Psi]; \qquad \mathcal{I} = I_{\mathbf{Z}}(G).$$

With the identifications $\Lambda = \mathbf{Z}[\Psi \times \Phi] \cong \mathbf{Z}[\Psi] \otimes_{\mathbf{Z}} \mathbf{Z}[\Phi] \cong R[\Phi]$ we may write $\epsilon = \epsilon_{\mathbf{Z}, \Psi \times \Phi} = \epsilon_{\mathbf{Z}, \Psi} \epsilon_{R, \Phi}$; we obtain a commutative diagram of Λ-homomorphisms in which the rows and the right hand column are exact.

$$
\begin{array}{ccccccccc}
 & & & & 0 & & & & \\
 & & & & \downarrow & & & & \\
0 & \to & I_R(\Phi) & \to & \mathcal{I} & \longrightarrow & \mathrm{Ker}(\epsilon_{\mathbf{Z},\psi}) & \to & 0 \\
 & & \| & & \cap & & \cap & & \\
0 & \to & I_R(\Phi) & \xrightarrow{\epsilon_{R,\Phi}} & \Lambda & & R & \to & 0 \\
 & & & & & & \downarrow \epsilon_{\mathbf{Z},\psi} & & \\
 & & & & & & \mathbf{Z} & & \\
 & & & & & & \downarrow & & \\
 & & & & & & 0 & &
\end{array}
$$

In particular Λ is an extension of the form:

$$0 \to I_R(\Phi) \to \Lambda \to R \to 0. \tag{16.12}$$

Specializing to the case where $\Psi = F_m = \langle x_1, \ldots, x_m \rangle$ is the free group of rank m we obtain a complete resolution $(0 \to R^m \xrightarrow{X} R \xrightarrow{\epsilon_{\mathbf{Z},F_m}} \mathbf{Z} \to 0)$ for $\mathbf{Z}$ over R where $X = (x_1 - 1, \ldots, x_m - 1)$. Then $\mathrm{Ker}(\epsilon_{\mathbf{Z},F_m}) \cong R^m$ so that

$\mathcal{I}$ is an extension of the form $0 \to I_R(\Phi) \to \mathcal{I} \to R^m \to 0.$ $\qquad$ (16.13)

Now specialize further to the case where Φ is a nontrivial finite group and put $n = |\Phi| > 1$.

Proposition 16.14 *If* $\mathcal{L} \in [\mathcal{I}]$ *then* $\mathcal{L} \neq 0$.

Proof Otherwise one would have $\mathcal{I} \oplus \Lambda^r \cong \Lambda^s$ for some $r, s \geq 1$. That is, $\mathcal{I}$ is stably free and so G has cohomological dimension 1. This is a contradiction, since $G = F_m \times \Phi$ has infinite cohomological dimension. $\qquad \square$

We note the following:

Proposition 16.15 $\mathrm{Hom}_\Lambda(I_R(\Phi), R) = 0$.

Proof By (14.1) it suffices to show that $\mathrm{Ext}^1_{R[\Phi]}(R, R) = 0$. However, by (14.4), as R is free over $\mathbf{Z}$, $\mathrm{Ext}^1_{R[G]}(R, R) \cong (\Phi^{ab}/\mathrm{Torsion}) \otimes_{\mathbf{Z}} R = 0$ since Φ is finite. $\qquad \square$

Now suppose that $\mathcal{L} \in [\mathcal{I}]$, so that $\mathcal{L} \oplus \Lambda^a \cong \mathcal{I} \oplus \Lambda^b$ for some $a, b \geq 0$. We shall establish a sequence of increasingly better estimates for the relative sizes of $\mathcal{I}$ and $\mathcal{L}$:

Proposition 16.16 $\quad a \leq b + m.$

Proof From the exact sequence

$$0 \to \operatorname{Hom}_\Lambda(R^m, R) \to \operatorname{Hom}_\Lambda(\mathcal{I}, R) \to \operatorname{Hom}_\Lambda(I_R(\Phi), R)$$

and Proposition 16.15 we see that $\operatorname{Hom}_\Lambda(\mathcal{I}, R) \cong R^m$. It follows that $\operatorname{Hom}_\Lambda(\mathcal{I} \oplus \Lambda^b, R) \cong R^{b+m}$; since $\mathcal{L} \oplus \Lambda^a \cong \mathcal{I} \oplus \Lambda^b$ then $\operatorname{Hom}_\Lambda(\mathcal{L} \oplus \Lambda^a, R) \cong \operatorname{Hom}_\Lambda(\mathcal{L}, R) \oplus R^a \cong R^{b+m}$. Thus $\operatorname{Hom}_\Lambda(\mathcal{L}, R)$ is a projective R-module. By the Bass-Sheshadri Theorem [1, 2] $\operatorname{Hom}_\Lambda(\mathcal{L}, R)$ is free and so $\operatorname{Hom}_\Lambda(\mathcal{L}, R) \cong R^{b+m-a}$ since R has the invariant basis property. Hence $a \leq b + m$. □

Next we show:

Proposition 16.17 $a < b + m$ and $\operatorname{Hom}_\Lambda(\mathcal{L}, R) \cong R^{b+m-a} \neq 0$.

Proof Choose an isomorphism $h : \mathcal{L} \oplus \Lambda^a \to \mathcal{I} \oplus \Lambda^b$. Since $\operatorname{Hom}_\Lambda(\mathcal{I} \oplus \Lambda^b, R) \cong R^{b+m}$ there exists a surjective homomorphism $p : \mathcal{I} \oplus \Lambda^b \to R^{b+m}$. We know from Proposition 16.16 that $a \leq b+m$, so suppose that $a = b+m$. Then $\operatorname{Hom}_\Lambda(\mathcal{L}, R) = 0$ so that the restriction $p \circ h_{|\mathcal{L}} : \mathcal{L} \to R$ is zero. Likewise, we may choose a surjective homomorphism $q : \Lambda^a \to R^a$ in which $\operatorname{Ker}(q) \cong I_R(\Phi)^a$. Abbreviating $I_R(\Phi)$ to I_R then in the following diagram

$$\begin{array}{ccccccccc} 0 \to & \mathcal{L} \oplus I_R^a & \xrightarrow{j} & \mathcal{L} \oplus \Lambda^a & \xrightarrow{(0,q)} & R^a & \to 0 \\ & & & \downarrow h & & & \\ 0 \to & I_R \oplus I_R^b & \xrightarrow{i} & \mathcal{I} \oplus \Lambda^b & \xrightarrow{p} & R^{b+m} & \to 0 \end{array}$$

$p \circ h$ vanishes on $\mathcal{L} \oplus I_R^a$. Thus there exist unique homomorphisms h_- and h_+ making the following diagram commute:

$$\begin{array}{ccccccccc} 0 \to & \mathcal{L} \oplus I_R^a & \xrightarrow{j} & \mathcal{L} \oplus \Lambda^a & \xrightarrow{(0,q)} & R^a & \to 0 \\ & \downarrow h_- & & \downarrow h & & \downarrow h_+ & \\ 0 \to & I_R \oplus I_R^b & \xrightarrow{i} & \mathcal{I} \oplus \Lambda^b & \xrightarrow{p} & R^{b+m} & \to 0 \end{array}$$

As h is bijective and the rows are exact $h_+ : R^a \to R^{b+m}$ is surjective and, by hypothesis, $a = b + m$. Now $R = \mathbf{Z}[F_m]$, being an integral group ring, is weakly finite [75]. Thus h_+ is an isomorphism. It follows from the Five Lemma (extending the rows to the left by zeroes) that $h_- : \mathcal{L} \oplus I_R^a \to I_R^{b+1}$ is also an isomorphism. Now I_R is free of rank $n - 1$ over R where $n = |\Phi| > 1$. As $\mathcal{L} \oplus I_R^a \cong I_R^{b+1}$ it follows that $\mathcal{L}$ is stably free and hence (by the theorem of Bass-Sheshadri [1, 2]) free over R. In particular

$$\operatorname{rk}_R(\mathcal{L}) = (n-1)(b+m-a) < (n-1)(b+m-a) = 0.$$

This contradicts Proposition 16.14. Hence $a < b + m$ and $\operatorname{Hom}_\Lambda(\mathcal{L}, R) \cong R^{b+m-a} \neq 0$. □

Proposition 16.18 If $\mathcal{L} \oplus \Lambda^a \cong \mathcal{I} \oplus \Lambda^b$ then $a \leq b + 1$.

Proof Since $\operatorname{Hom}_\Lambda(\mathcal{L}, R) \cong R^{b+m-a}$ choose $\pi : \mathcal{L} \to R^{b+m-a}$ to be a surjective Λ-homomorphism and put $\mathcal{L}_0 = \operatorname{Ker}(\pi)$. Let $g : \mathcal{L} \oplus \Lambda^a \to \mathcal{I} \oplus \Lambda^b$ be the inverse of the isomorphism h considered above, and consider the following diagram with exact rows:

$$
\begin{array}{ccccccc}
0 \to & I_R \oplus I_R^b & \xrightarrow{i} & \mathcal{I} \oplus \Lambda^b & \xrightarrow{p} & R^{b+m} & \to 0 \\
& & & \downarrow g & & & \\
0 \to & \mathcal{L}_0 \oplus I_R^a & \xrightarrow{j} & \mathcal{L} \oplus \Lambda^a & \xrightarrow{(\pi, \mathrm{Id})} & R^{b+m-a} \oplus R^a & \to 0
\end{array}
$$

Making the obvious identification of $R^{b+m-a} \oplus R^a$ with R^{b+m}, we note that $(\pi, \mathrm{Id}) \circ g$ vanishes on $I_R \oplus I_R^b$ since $\operatorname{Hom}(I_R, R) = 0$ so that, again using the fact that R is weakly finite, g induces an isomorphism of exact sequences

$$
\begin{array}{ccccccc}
0 \to & I_R \oplus I_R^b & \xrightarrow{i} & \mathcal{I} \oplus \Lambda^b & \xrightarrow{p} & R^{b+m} & \to 0 \\
& \downarrow g_- & & \downarrow g & & \downarrow g_+ & \\
0 \to & \mathcal{L}_0 \oplus I_R^a & \xrightarrow{j} & \mathcal{L} \oplus \Lambda^a & \xrightarrow{(\pi, q)} & R^{b+m} & \to 0
\end{array}
$$

Thus $\mathcal{L}_0 \oplus I_R^a \cong I_R^{b+1}$. Computing R-ranks we obtain

$$
\operatorname{rk}(\mathcal{L}_0) + (n-1)a = (n-1)(b+1)
$$

so that $\operatorname{rk}(\mathcal{L}_0) = (n-1)(b+1-a)$. Hence $0 \leq b+1-a$ and so $a \leq b+1$. $\square$

We first consider the special case where $\Phi \cong C_n$.

Proposition 16.19 $\mathcal{I}$ *is minimal in* $\Omega_1(\mathbf{Z})$ *when* $G \cong F_m \times C_n$.

Proof Suppose that $\mathcal{L} \in [\mathcal{I}]$ and that $\mathcal{L} \oplus \Lambda^a \cong \mathcal{I} \oplus \Lambda^b$; then $a \leq b+1$ by Proposition 16.18. Suppose that $a = b+1$. Then $b + m - a = m - 1$, so that, as in Proposition 16.17, there exists a surjection $\pi : \mathcal{L} \to R^{m-1}$ with $\operatorname{Ker}(\pi) = \mathcal{L}_0$. As in the proof of Proposition 16.18, $\operatorname{rk}_R(\mathcal{L}_0) = (n-1)(b+1-a) = 0$; thus $\mathcal{L}_0 = 0$ so that the surjection $\pi : \mathcal{L} \to R^{m-1}$ is an isomorphism of Λ-modules. Thus

$$
\operatorname{End}_{\mathcal{D}er}(\mathcal{L}) \cong M_{m-1}(\operatorname{End}_{\mathcal{D}er}(R)).
$$

Now as $\mathcal{L}$ is stably equivalent to $\mathcal{I}$ then, by Corollary 16.11, $\operatorname{End}_{\mathcal{D}er}(\mathcal{L}) \cong \operatorname{End}_{\mathcal{D}er}(\mathcal{I}) \cong \mathbf{Z}/n$. In particular, $\operatorname{End}_{\mathcal{D}er}(\mathcal{L})$ is finite. However, by Proposition 16.7, $\operatorname{End}_{\mathcal{D}er}(R) \cong R/n$ which is an infinite ring so that $M_{m-1}(\operatorname{End}_{\mathcal{D}er}(R))$ is also infinite. From this contradiction we conclude that $a \leq b$ and that $\mathcal{I}$ is minimal in $\Omega_1(\mathbf{Z})$. $\square$

Before proceeding to the general case we make a general observation. Suppose G is a group and let $i : H \subset G$ be the inclusion of a subgroup H with finite index $k \geq 2$. Let

$$
\mathcal{I} = \operatorname{Ker}(\epsilon_G : \mathbf{Z}[G] \to \mathbf{Z}); \qquad \mathcal{I}_0 = \operatorname{Ker}(\epsilon_H : \mathbf{Z}[H] \to \mathbf{Z})
$$

be the respective integral augmentation ideals and let $\natural : i^*(\mathcal{I}) \to i^*(\mathcal{I})/\mathcal{I}_0$ be the canonical mapping. If $\{x_0, x_1, \ldots, x_{k-1}\}$ is a complete set of coset representatives for G/H with $x_0 = 1$ then $i^*(\mathcal{I})/\mathcal{I}_0$ is free of rank $k - 1$ over $\mathbf{Z}[H]$ on the basis $\{\natural(x_r - 1)\}_{1 \leq r \leq k-1}$. It follows immediately that:

Proposition 16.20 $i^*(\mathcal{I}) \cong \mathcal{I}_0 \oplus \mathbf{Z}[H]^{k-1}$.

We can now establish the final minimality criterion in the general case:

(16.21) Third Minimality Criterion $\mathcal{I}$ lies at the minimal level of $\Omega_1(\mathbf{Z})$ when G is the direct product $F_m \times \Phi$ where Φ is finite and nontrivial and $m \geq 1$.

Proof Put $\Lambda = \mathbf{Z}[F_m \times \Phi]$ where $G = F_m \times \Phi$ and Φ is a nontrivial finite group. As usual let $\mathcal{I} = \mathrm{Ker}(\epsilon : \mathbf{Z}[F_m \times \Phi] \to \mathbf{Z})$ denote the integral augmentation ideal. We shall prove that if $\mathcal{L} \oplus \Lambda^a \cong \mathcal{I} \oplus \Lambda^b$ then $a \leq b$. We may write $\Lambda = R[\Phi]$ where $R = \mathbf{Z}[F_m]$. By the special case already established we may suppose that Φ is not cyclic. Take $C_n \subset \Phi$ to be a nontrivial cyclic subgroup and put $H = F_m \times C_n$ and $k = |G/H| = |\Phi|/n$. Put $\Lambda_0 = R[C_n]$ and let $\mathcal{I}_0 = \mathrm{Ker}(\epsilon : \mathbf{Z}[F_m \times C_n] \to \mathbf{Z})$ be the integral augmentation ideal of $F_m \times C_n$. From the hypothesis $\mathcal{L} \oplus \Lambda^a \cong \mathcal{I} \oplus \Lambda^b$ it follows that

$$i^*(\mathcal{L}) \oplus i^*(\Lambda)^a \cong i^*(\mathcal{I}) \oplus i^*(\Lambda)^b.$$

However, $i^*(\Lambda) \cong \Lambda_0^k$ and by Proposition 16.20, $i^*(\mathcal{I}) \cong \mathcal{I}_0 \oplus \Lambda_0^{k-1}$. Thus $i^*(\mathcal{L}) \oplus \Lambda_0^{ka} \cong \mathcal{I}_0 \oplus \Lambda_0^{kb+k-1}$. Now, by Proposition 16.18, $ka \leq kb + (k - 1)$ and so $a \leq b$. $\square$

16.4 Decomposition in a Special Case

In this section we consider generalized Swan modules defined relative to a quasi augmentation sequence in which $S = \Lambda$ is the ambient ring; that is

$$\mathcal{S} = (0 \to S_- \to \Lambda \xrightarrow{\eta} S_+ \to 0).$$

We denote by Θ the characteristic ring. When Θ is generalized Euclidean we saw in Sect. 15.4 that all generalized Swan modules decompose as a sum of Swan modules of rank 1. Unfortunately this hypothesis is too restrictive for our intended application. For the remainder of this section we shall impose the following hypothesis $\mathcal{E}$ (= Existence) which will ensure the existence of a decomposition of suitable generalized Swan modules:

$\mathcal{E}$: The characteristic ring Θ is *commutative* and there exists a surjective ring homomorphism $\varphi : \Theta \to \Theta_1 \times \cdots \times \Theta_n$ such that each Θ_r is *generalized Euclidean* and such that φ has the strong lifting property for units.

We denote by $\varphi_r : \Theta \to \Theta_r$ the composition of φ with the projection

$$\Theta_1 \times \cdots \times \Theta_n \to \Theta_r.$$

In consequence of Proposition 2.35 we note that:

$$\Theta_1 \times \cdots \times \Theta_n \text{ is generalized Euclidean.} \tag{16.22}$$

A fortiori, $\Theta_1 \times \cdots \times \Theta_n$ is weakly Euclidean, so that, as φ has the strong lifting property for units then by Proposition 2.43:

$$\Theta \text{ is weakly Euclidean.} \tag{16.23}$$

We observed, (15.14), that a generalized Swan module J defines an ideal $\langle \det(J) \rangle \triangleleft \Theta$ by the rule $\langle \det(J) \rangle = (\det(X))$ where $X \in M_m(\Theta)$ is any matrix representing J.

Theorem 16.24 *Let $\mathbf{b} \in \Theta$ and let J be a generalized Swan module of height k over S such that $\langle \det(J) \rangle = (\mathbf{b})$; suppose that $\varphi(\mathbf{b})$ is not a zero divisor and for each r that $\varphi_r(\mathbf{b})$ is indecomposable in Θ_r; then for some $\gamma \in \Theta^*$*

$$J \cong M(\gamma \mathbf{b}) \oplus \Lambda^k.$$

Proof Let J be classified by a matrix $X \in M_{k+1}(\Theta)$. We may identify the matrix $\varphi(X)$ with the sequence $(\varphi_1(X), \ldots, \varphi_n(X))$ where $\varphi_r(X) \in M_{k+1}(\Theta_r)$. Observe that $\varphi_r(X)$ has a Smith Normal Form as, by hypothesis, Θ_r is generalized Euclidean. As $\langle \det(J) \rangle = (\mathbf{b})$ then $(\det(X_r)) = (\varphi_r(\mathbf{b}))$ and as $\varphi_r(\mathbf{b})$ is indecomposable in Θ_r then

$$\varphi_r(X) \sim \begin{pmatrix} \gamma_r \varphi_r(\mathbf{b}) & & & 0 \\ & 1 & & \\ & & \ddots & \\ 0 & & & 1 \end{pmatrix}$$

for some $\gamma_r \in \Theta_r^*$. By Theorem 2.32 $\varphi(X)$ has a Smith Normal Form of very restricted type

$$\varphi(X) \sim \begin{pmatrix} \check{\gamma} \varphi(\mathbf{b}) & & & 0 \\ & 1 & & \\ & & \ddots & \\ 0 & & & 1 \end{pmatrix},$$

where $\check{\gamma} = (\gamma_1, \ldots, \gamma_n) \in (\Theta_1 \times \cdots \times \Theta_n)^*$. Hence by Lemma 2.42 and (2.38) X also has a Smith Normal Form of very restricted type

$$X \sim \begin{pmatrix} \delta & & & 0 \\ & 1 & & \\ & & \ddots & \\ 0 & & & 1 \end{pmatrix}.$$

Moreover $\delta = \det(X)$ as Θ is commutative and likewise $\check{\gamma}\varphi\mathbf{b} = \det(\varphi(X))$. However $\langle\det(J)\rangle = (\mathbf{b})$ so that $\det(X) = \gamma\mathbf{b}$ for some $\gamma \in \Theta$. Hence

$$J \cong M(\gamma\mathbf{b}) \oplus \Lambda^k.$$

It only remains to show that $\gamma \in \Theta^*$. However $\varphi(\gamma)\varphi(\mathbf{b}) = \check{\gamma}\varphi(\mathbf{b})$ and by hypothesis, $\varphi(\mathbf{b})$ is not a zero divisor. Thus $\varphi(\gamma) = \check{\gamma} \in (\Theta_1 \times \cdots \times \Theta_n)^*$. As φ has the strong lifting property for units it follows that $\gamma \in \Theta^*$ as required.   □

16.5 Eliminating Ambiguity in the Description

With the hypotheses of Theorem 16.24 we see that $\mathbf{b}$ is not a zero divisor in Θ. It is natural to consider the following:

Question Let $\mathbf{b}$ be a non-zero divisor in Θ and let $\gamma \in \Theta^*$ be a unit. When is it true that $M(\gamma\mathbf{b}) \oplus \Lambda^k \cong M(\mathbf{b}) \oplus \Lambda^k$?

As in Sect. 16.4, in this section we consider generalized Swan modules defined relative to a quasi augmentation sequence in which $S = \Lambda$ is the ambient ring; that is

$$\mathcal{S} = (0 \to S_- \to \Lambda \xrightarrow{\eta} S_+ \to 0).$$

However, the arguments of this section require us only to impose the following hypothesis:

$\mathcal{WE}$: The characteristic ring Θ is commutative and weakly Euclidean.

By (16.23) the property $\mathcal{WE}$ follows from the hypothesis $\mathcal{E}$ considered in Sect. 16.4. Here we are concerned specifically with the relations which exist between $M(\gamma)$, $M(\gamma\mathbf{b})$, $M(1)(= \Lambda)$ and $M(\mathbf{b})$ when $\gamma \in \Theta^*$. We recall from (15.16) that the condition $\gamma \in \Theta^*$ is precisely equivalent to $M(\gamma)$ being projective. If $\gamma \in \Theta^*$ and $m \geq 0$ we say that γ is $(m+1)$-*liftable* when, for some $\alpha_\sigma \in GL_{m+1}(T_\sigma)$

$$\gamma = \det[\varphi_+(\alpha_+)\varphi_-(\alpha_-)],$$

where $T_+ = \mathrm{End}_\Lambda(S_+)$, $T_- = \mathrm{End}_\Lambda(S_-)$ and $\varphi_\sigma : GL_{m+1}(T_\sigma) \to GL_{m+1}(\Theta)$ is the canonical representation.

Proposition 16.25 *Suppose that $\gamma \in \Theta^*$ and that $\mathbf{b} \in \Theta$ is not a zero divisor; then* $M(\gamma\mathbf{b}) \oplus \Lambda^m \cong M(\mathbf{b}) \oplus \Lambda^m \iff \gamma$ *is $(m+1)$-liftable.*

Proof ($\Longrightarrow$) For arbitrary $\mathbf{c} \in \Theta$ the module $M(\mathbf{c}) \oplus \Lambda^m$ is parametrized by the matrix

$$\Delta(\mathbf{c}) = \begin{pmatrix} \mathbf{c} & 0 \\ 0 & I_m \end{pmatrix}$$

given in block $1 \times m$ form. The condition that $M(\mathbf{c}) \oplus \Lambda^m \cong M(\mathbf{d}) \oplus \Lambda^m$ is precisely that there should exist $\alpha_\sigma \in GL_{m+1}(T_\sigma)$ such that

$$\Delta(\mathbf{c}) = \varphi_+(\alpha_+)\Delta(\mathbf{d})\varphi_-(\alpha_-).$$

However, Θ is weakly Euclidean so that we may write

$$\varphi_+(\alpha_+) = E_+\Delta(\det[\varphi_+(\alpha_+)]); \qquad \varphi_-(\alpha_-) = \Delta(\det[\varphi_-(\alpha_-)])E_-$$

for some E_+, $E_- \in E_{m+1}(\Theta)$. Now suppose that $\gamma \in \Theta^*$, $\mathbf{b} \in \Theta$ is not a zero divisor and $M(\gamma\mathbf{b}) \oplus \Lambda^m \cong M(\mathbf{b}) \oplus \Lambda^m$; then

$$\Delta(\gamma\mathbf{b})) = E_+\Delta(\det[\varphi_+(\alpha_+)])\Delta(\mathbf{b})\Delta(\det[\varphi_-(\alpha_-])E_-.$$

Taking determinants we see that

$$\gamma\mathbf{b} = \det[\varphi_+(\alpha_+)\varphi_-(\alpha_-)]\mathbf{b}.$$

However, as $\mathbf{b}$ is not a zero divisor in Θ then

$$\gamma = \det[\varphi_+(\alpha_+)(\varphi_-(\alpha_-)]$$

and γ is $(m + 1)$-liftable. This proves $(\Longrightarrow)$.

To prove $(\Longleftarrow)$, suppose that $\gamma \in \Theta^*$ is $(m + 1)$-liftable so that for some $\alpha_\sigma \in GL_{m+1}(T_\sigma)$

$$\gamma = \det[\varphi_+(\alpha_+)(\varphi_-(\alpha_-)].$$

Hence

$$\Delta(\gamma\mathbf{b}) = \Delta(\det[\varphi_+(\alpha_+))\Delta(\mathbf{b})\Delta(\det[\varphi_-(\alpha_-)). \qquad (*)$$

As above write $\varphi_+(\alpha_+) = E_+\Delta(\det[\varphi_+(\alpha_+)]); \varphi_-(\alpha_-) = \Delta(\det[\varphi_-(\alpha_-)])E_-$ for some E_+, $E_- \in E_{m+1}(\Theta)$. Equivalently

$$(E_+)^{-1}\varphi_+(\alpha_+) = \Delta(\det[\varphi_+(\alpha_+)]); \qquad \Delta(\det[\varphi_-(\alpha_-)]) = \varphi_-(\alpha_-)E_-^{-1}.$$

Substituting in $(*)$ gives

$$\Delta(\gamma\mathbf{b}) = (E_+)^{-1}\varphi_+(\alpha_+)\Delta(\mathbf{b})\varphi_-(\alpha_-)E_-^{-1}. \qquad (**)$$

Now the canonical maps $T_\sigma \to \Theta$ are surjective so that there exist $\beta_\sigma \in E_{m+1}(T_\sigma)$ such that $\varphi_\sigma(\beta_\sigma) = E_\sigma$. Thus

$$\Delta(\gamma\mathbf{b}) = \varphi_+(\beta_+^{-1}\alpha_+)\Delta(\mathbf{b})\varphi_-(\alpha_-\beta_-^{-1}) \qquad (***)$$

from which it follows that $M(\gamma\mathbf{b}) \oplus \Lambda^m \cong M(\mathbf{b}) \oplus \Lambda^m$. This proves $(\Longleftarrow)$ and completes the proof. $\qquad\square$

As a special case, 1 is not a zero divisor in Θ and $M(1) = \Lambda$; hence:

$$M(\gamma) \oplus \Lambda^m \cong \Lambda^{m+1} \quad \Longleftrightarrow \quad \gamma \in \Theta^* \text{ is } (m+1)\text{-liftable.} \qquad (16.26)$$

Comparing Proposition 16.25 and (16.26) we see:

$$M(\gamma \mathbf{b}) \oplus \Lambda^m \cong M(\mathbf{b}) \oplus \Lambda^m \quad \Longleftrightarrow \quad M(\gamma) \oplus \Lambda^m \cong \Lambda^{m+1}. \qquad (16.27)$$

We note the special case of (16.27) when $m = 0$:

$$M(\gamma \mathbf{b}) \cong M(\mathbf{b}) \quad \Longleftrightarrow \quad M(\gamma) \cong \Lambda. \qquad (16.28)$$

We note that in (16.27) and (16.28) it is not necessary to specify that $\gamma \in \Theta^*$. In each case this is forced, either by the projectivity criterion (15.16) or (recall that Θ is commutative) by taking determinants and using the fact that $\mathbf{b}$ is not a zero divisor.

16.6   Complete Description of the First Syzygy (Tame Case)

Let $G = C_\infty \times \Phi$ where Φ is a nontrivial finite group. In this section, subject to certain restictions on Φ, we complete the description of the first syzygy $\Omega_1(\mathbf{Z})$ over $\mathbf{Z}[G]$. We adopt the following notation

$$R = \mathbf{Z}[C_\infty] = \mathbf{Z}[t, t^{-1}],$$

$$\Lambda = R[\Phi] = \mathbf{Z}[C_\infty \times \Phi],$$

$$\Sigma = \sum_{\varphi \in \Phi} \widehat{\varphi} \in R[\Phi],$$

$$\mathcal{R} = (0 \to I_R \xrightarrow{j} R[\Phi] \xrightarrow{\epsilon_R} R \to 0).$$

Evidently the augmentation sequence $\mathcal{R}$ is a quasi-augmentation. Observe that Σ is central in $\Lambda = R[\Phi]$ and spans a two-sided ideal (Σ) so that $R[\Phi]/(\Sigma)$ is naturally a ring. We shall further assume the following conditions (I) and (II); these constitute the 'tame case'.

 (I) $R[\Phi]/(\Sigma)$ has the *SFC* property;
(II) Every stably free Λ-module of rank 1 is free.

We observe that the characteristic ring Θ $(= \Theta(\mathcal{R}))$ is commutative and takes the form

$$\Theta = R/|\Phi| \cong (\mathbf{Z}/|\Phi|)[t, t^{-1}].$$

Writing $|\Phi|$ as a product of powers of distinct primes $|\Phi| = p_1^{e_1} \cdots p_n^{e_n}$ we see that $\mathbf{Z}/|\Phi|$ is a product of local rings $\mathbf{Z}/|\Phi| \cong \mathbf{Z}/p_1^{e_1} \times \cdots \times \mathbf{Z}/p_n^{e_n}$; we make the identification

$$\Theta \cong (\mathbf{Z}/p_1^{e_1})[t, t^{-1}] \times \cdots \times (\mathbf{Z}/p_n^{e_n})[t, t^{-1}].$$

The ring homomorphism $\nu_r : \mathbf{Z}/p_r^{e_r}[t, t^{-1}] \to \mathbf{F}_{p_r}[t, t^{-1}]$, obtained by factoring out the radical of $\mathbf{Z}/p_r^{e_r}$, is surjective and it follows from Proposition 2.50 that each ν_r has the strong lifting property for units. Hence the surjective homomorphism $\varphi = \nu_1 \times \cdots \times \nu_n : \Theta \to \Theta_1 \times \cdots \times \Theta_n$ also has the strong lifting property for units. As $\Theta_r = \mathbf{F}_{p_r}[t, t^{-1}]$ is a Euclidean domain then $\Theta_1 \times \cdots \times \Theta_n$ is generalized Euclidean and hence:

The augmentation sequence $\mathcal{R}$ satisfies the hypothesis $\mathcal{E}$ of Sect. 16.4.   (16.29)

Proposition 16.30 $\mathrm{Hom}_\Lambda(\mathcal{I}, R) \cong R$.

Proof Applying $\mathrm{Hom}_\Lambda(-, R)$ to the exact sequence $(0 \to I_R \to \mathcal{I} \overset{\eta}{\to} R \to 0)$ gives an exact sequence $0 \to \mathrm{Hom}_\Lambda(R, R) \to \mathrm{Hom}_\Lambda(\mathcal{I}, R) \to \mathrm{Hom}_\Lambda(I_R, R)$. However $\mathrm{Hom}_\Lambda(I_R, R) = 0$ so that $\mathrm{Hom}_\Lambda(\mathcal{I}, R) \cong \mathrm{Hom}_\Lambda(R, R) \cong R$. $\square$

Proposition 16.31 *If* $J \in \Omega_1(\mathbf{Z})$ *has height* k *then* $\mathrm{Hom}_\Lambda(J, R) \cong R^{k+1}$.

Proof To say that J is at height k means that $J \oplus \Lambda^n \cong \mathcal{I} \oplus \Lambda^{n+k}$. Thus

$$\mathrm{Hom}_\Lambda(J, R) \oplus \mathrm{Hom}_\Lambda(\Lambda, R)^n \cong \mathrm{Hom}_\Lambda(\mathcal{I}, R) \oplus \mathrm{Hom}_\Lambda(\Lambda, R)^{n+k}.$$

Evidently $\mathrm{Hom}_\Lambda(\Lambda, R) \cong R$ so that, by Proposition 16.30, $\mathrm{Hom}_\Lambda(J, R) \oplus R^n \cong R^{n+k+1}$. Thus $\mathrm{Hom}_\Lambda(J, R)$ is a stably free R-module of rank $k + 1$. As previously observed, R has the SFC property hence $\mathrm{Hom}_\Lambda(J, R) \cong R^{k+1}$ as claimed. $\square$

In what follows we employ duality, converting the left Λ-module $M^* = \mathrm{Hom}_\Lambda(M, \Lambda)$ to a right Λ-module via the involution on $\Lambda = R[\Phi]$ obtained from $\varphi \to \varphi^{-1}$.

I_R^* has a natural ring structure; in fact $I_R^* \cong \Lambda/(\Sigma)$.   (16.32)

There is a natural transformation given by $\natural : M \to M^{**}$; $\natural(x)(\alpha) = \alpha(x)$. In general $\natural_M$ need not be an isomorphism; it is however in a number of important cases:

$$\natural : \Lambda \to \Lambda^{**} \text{ is an isomorphism;} \qquad (16.33)$$

$$\natural : R \to R^{**} \text{ is an isomorphism.} \qquad (16.34)$$

On extending the exact sequence $\mathcal{R}$ one place to the left by zeroes and applying $\natural$ we obtain a commutative diagram:

$$
\begin{array}{ccccccccc}
0 & \to & 0 & \to & I_R & \to & \Lambda & \overset{\epsilon_R}{\to} & R \\
& & \downarrow & & \downarrow & & \downarrow \natural & & \downarrow \natural & & \downarrow \natural \\
0 & \to & 0 & \to & I_R^{**} & \to & \Lambda^{**} & \overset{\epsilon_R}{\to} & R^{**}
\end{array}
$$

Applying the Five Lemma in conjunction with (16.33) and (16.34) we see also that:

$$\natural : I_R \to I_R^{**} \text{ is an isomorphism.} \qquad (16.35)$$

We now assume hypothesis (I) of the preamble to prove:

Proposition 16.36 *If K is a Λ-module satisfying $K \oplus I_R^a \cong I_R^b$ then $K \cong I_R^{b-a}$.*

Proof Choosing a Λ-isomorphism $h : K \oplus I_R^a \to I_R^b$ we obtain a commutative diagram

$$
\begin{array}{ccc}
K \oplus I_R^a & \xrightarrow{\left(\begin{smallmatrix} \natural_K & 0 \\ 0 & \natural^a \end{smallmatrix}\right)} & K^{**} \oplus (I_R^{**})^a \\
\downarrow h & & \downarrow h^{**} \\
I_R^b & \xrightarrow{\natural^b} & (I_R^{**})^b
\end{array}
$$

in which h, h^{**}, $\natural^a$ and $\natural^b$ are isomorphisms. Thus $\natural_K : K \to K^{**}$ is an isomorphism. Also the dual mapping $h^* : (I_R^*)^b \to K^* \oplus (I_R^*)^a$ is an isomorphism; thus K^* is a stably free module over the ring I_R^*. By hypothesis (I) $K^* \cong (I_R^*)^{b-a}$ and so $K^{**} \cong (I_R^{**})^{b-a} \cong I_R^{b-a}$. However, $K \cong K^{**}$ so the result follows. $\square$

Corollary 16.37 *Any module $J \in \Omega_1(\mathbf{Z})$ is a generalized Swan module over $\mathcal{R}$. Moreover*

$$\mathrm{rk}_{\mathcal{R}}(J) = \mathrm{height}(J) + 1.$$

Proof The condition that J has height $= k$ in $\Omega_1(\mathbf{Z})$ means that there is a Λ-isomorphism $\upsilon : \mathcal{I} \oplus \Lambda^{m+k} \xrightarrow{\cong} J \oplus \Lambda^m$. Now $\mathrm{Hom}_\Lambda(J, R) \cong R^{k+1}$ by Proposition 16.31. Choosing an R-basis $\{\eta_1, \ldots, \eta_{k+1}\}$ for $\mathrm{Hom}_\Lambda(J, R)$ gives an exact sequence $0 \to K \to J \xrightarrow{\eta} R^{k+1} \to 0$ where

$$
\eta(x) = \begin{pmatrix} \eta_1(x) \\ \eta_2(x) \\ \vdots \\ \eta_{k+1}(x) \end{pmatrix} \in R^{k+1} \quad \text{and} \quad K = \mathrm{Ker}(\eta).
$$

It will suffice to show that $K \cong I_R^{k+1}$. Let μ denote the surjective homomorphism obtained by regarding $\mathcal{I}$ as a Swan module over $\mathcal{R}$ thus; $0 \to I_R \to \mathcal{I} \xrightarrow{\mu} R \to 0$. Consider the following diagram:

$$
\begin{array}{ccccc}
0 \to I_R \oplus I_R^{m+k} & \to & \mathcal{I} \oplus \Lambda^{m+k} & \xrightarrow{\widehat{\mu}} & R \oplus R^{m+k} \to 0 \\
& & \downarrow \upsilon & & \\
0 \to K \oplus I_R^m & \to & J \oplus \Lambda^m & \xrightarrow{\widehat{\eta}} & R^{k+1} \oplus R^m \to 0
\end{array}
$$

where $\widehat{\mu} = \left(\begin{smallmatrix} \mu & 0 \\ 0 & \epsilon_R^{m+k} \end{smallmatrix}\right)$ and $\widehat{\eta} = \left(\begin{smallmatrix} \eta & 0 \\ 0 & \epsilon_R^m \end{smallmatrix}\right)$. As $\mathrm{Hom}_\Lambda(I_R, R) = 0$ we see that $\widehat{\eta} \circ \upsilon$ restricts to zero on $I_R \oplus I_R^{m+k}$; thus υ induces Λ-homomorphisms υ_-, υ_+ making the following commute:

$$0 \to I_R \oplus I_R^{m+k} \to \mathcal{I} \oplus \Lambda^{m+k} \xrightarrow{\widehat{\mu}} R \oplus R^{m+k} \to 0$$
$$\downarrow \upsilon_- \qquad\qquad \downarrow \upsilon \qquad\qquad \downarrow \upsilon_+$$
$$0 \to K \oplus I_R^m \to J \oplus \Lambda^m \xrightarrow{\widehat{\eta}} R^{k+1} \oplus R^m \to 0$$

As υ is an isomorphism υ_+ is necessarily is surjective. However, R^{m+k+1} has the strong Hopfian property so that υ_+ is an isomorphism. By extending the sequence one place to the left by zeroes one may apply the Five Lemma to show that $\upsilon_- : I_R^{m+k+1} \to K \oplus I_R^m$ is an isomorphism. By Proposition 16.36, $K \cong I_R^{k+1}$ and this completes the proof. $\qquad\qquad\qquad\qquad\qquad\qquad\qquad\qquad\qquad\qquad\qquad\quad\square$

We next show that the modules in $\Omega_1(\mathbf{Z})$ at levels $k \geq 1$ are entirely determined by those at level 0.

Theorem 16.38 *Let $G = C_\infty \times \Phi$ where Φ is a nontrivial finite group and suppose that hypothesis (I) is satisfied; if $J \in \Omega_1(\mathbf{Z})$ is a module at height k then there exists a unit $\gamma \in \Theta^*$ such that*

$$J \cong M(\gamma(t-1)) \oplus \Lambda^k.$$

Proof Let $J \in \Omega_1(\mathbf{Z})$ be a module at height k. We have seen in Corollary 16.37 that J is a generalized Swan module of rank $k + 1$ over $\mathcal{R}$. Moreover, as $J \in \Omega_1(\mathbf{Z})$ then $J \sim \mathcal{I}$. Hence $\langle \det(J) \rangle = \langle \det(\mathcal{I}) \rangle = (t-1)$. However $t-1$ is indecomposable in $\Theta_r \cong \mathbf{F}_{p_r}[t, t^{-1}]$ and $t-1$ is not a zero divisor in the product $\Theta_1 \times \cdots \times \Theta_n \cong \mathbf{F}_{p_1}[t, t^{-1}] \times \cdots \times \mathbf{F}_{p_n}[t, t^{-1}]$. The result now follows from Theorem 16.24. $\qquad\qquad\qquad\qquad\qquad\qquad\qquad\qquad\qquad\qquad\qquad\qquad\qquad\quad\square$

As $\mathcal{I} = M(t-1)$ and $(t-1)$ is not a zero divisor in Θ it follows from (16.26) that:

$$M(\gamma(t-1)) \oplus \Lambda^m \cong \mathcal{I} \oplus \Lambda^m \quad \Longleftrightarrow \quad \gamma \in \Theta^* \text{ is } (m+1)\text{-liftable.} \qquad (16.39)$$

This is a specific instance of (16.26) combined with (16.27). Moreover, we have the following specific instances of (16.27) and (16.28):

$$M(\gamma(t-1)) \oplus \Lambda^m \cong \mathcal{I} \oplus \Lambda^m \quad \Longleftrightarrow \quad M(\gamma) \oplus \Lambda^m \cong \Lambda^{m+1}. \qquad (16.40)$$

We note the special case of (16.40) when $m = 0$:

$$M(\gamma(t-1)) \cong \mathcal{I} \quad \Longleftrightarrow \quad M(\gamma) \cong \Lambda. \qquad (16.41)$$

Finally we adopt the hypothesis (II) to arrive at:

Theorem 16.42 *Let $G = C_\infty \times \Phi$ where Φ is a nontrivial finite group and suppose that conditions (I) and (II) are satisfied; then every $J \in \Omega_1(\mathbf{Z})$ is uniquely of the* *form*

$$J \cong \mathcal{I} \oplus \Lambda^k$$

for some $k \geq 0$; in particular, $\Omega_1(\mathbf{Z}) = [\mathcal{I}]$ is straight.

Proof Let $J \in \Omega_1(\mathbf{Z})$ have height$(J) = k$; then by Theorem 16.38, for some $\gamma \in \Theta^*$, $J \cong M(\gamma(t-1)) \oplus \Lambda^k$. As J is stably equivalent to $\mathcal{I}$ then for some $m \geq 1$, $J \oplus \Lambda^m \cong \mathcal{I} \oplus \Lambda^{m+k}$. It follows that

$$M(\gamma(t-1)) \oplus \Lambda^{m+k} \cong \mathcal{I} \oplus \Lambda^{m+k}.$$

By (16.27) $M(\gamma) \oplus \Lambda^{m+k} \cong \Lambda^{m+k+1}$; that is, $M(\gamma)$ is a stably free Λ-module of rank 1. By hypothesis (II), $M(\gamma) \cong \Lambda$; hence $M(\gamma(t-1)) \cong \mathcal{I}$ by (16.41). Hence $J \cong \mathcal{I} \oplus \Lambda^k$ which is the desired conclusion. $\qquad\square$

Chapter 17
Conclusion

In this chapter we draw our results together. We first present the solution of the $\mathcal{R}(2)$-problem for $C_\infty \times C_m$. This was first achieved by Edwards in his thesis [25, 26]. The account given here simplifies Edwards' argument at a number of points. We then present some duality results for higher syzygies. We conclude by giving a survey of the current status of the $\mathcal{R}(2)$–$\mathcal{D}(2)$ problem.

17.1 The $\mathcal{R}(2)$ Problem for $C_\infty \times C_m$

Recall from the Introduction the statement of the 2-dimensional realization problem:

$\mathcal{R}(2)$: Let G be a finitely presented group. Is every algebraic 2-complex over $\mathbf{Z}[G]$ geometrically realizable?

Similarly we say that $J \in \Omega_3(\mathbf{Z})$ is geometrically realizable when there exists a finite connected 2-complex X with $\pi_1(X) = G$ such that $\pi_2(X) \cong J$; this leads to an analogous problem:

π_2 Realization Problem: Let G be a finitely presented group. Is every $J \in \Omega_3(\mathbf{Z})$ geometrically realizable?

We saw in Proposition 8.18 that if $\text{Ext}^3(\mathbf{Z}, \mathbf{Z}[G]) = 0$ then $\pi_2 : \mathbf{Alg}_2(\mathbf{Z}) \to \Omega_3(\mathbf{Z})$ is surjective. In this case, an affirmative answer to the $\mathcal{R}(2)$ problem thus gives an affirmative answer to the π_2 realization problem or, in the contrapositive, a negative answer to the π_2 realization problem gives a negative answer to the $\mathcal{R}(2)$ problem. Moreover, in some cases the two problems are equivalent. We shall say the finitely presented group G is of type $FT(3)$ when the trivial module $\mathbf{Z}$ is of type $FT(3)$ over $\mathbf{Z}[G]$:

Theorem 17.1 *Let G be an infinite finitely presented group of type $FT(3)$ such that $\text{Ext}^3(\mathbf{Z}, \mathbf{Z}[G]) = 0$; then the $\mathcal{R}(2)$ problem for G is equivalent to the π_2 realization problem for G.*

F.E.A. Johnson, *Syzygies and Homotopy Theory*, Algebra and Applications 17, DOI 10.1007/978-1-4471-2294-4_17, © Springer-Verlag London Limited 2012

Proof By the above remarks it suffices to show that if each $J \in \Omega_3(\mathbf{Z})$ is realizable then each algebraic 2-complex over $\mathbf{Z}[G]$ is realized up to homotopy equivalence. Thus suppose that $A_* = (0 \to J \to A_2 \xrightarrow{\partial_2} A_1 \xrightarrow{\partial_1} A_0 \to \mathbf{Z} \to 0)$ is an algebraic 2-complex over $\mathbf{Z}[G]$. By Theorem 8.22 the set of homotopy classes of algebraic 2-complexes B_* which realize J as 'algebraic π_2' in 1–1 correspondence with $\mathrm{Im}(\nu_J)\backslash\mathrm{Ker}(S_J)$ where

$$S_J : \mathrm{Aut}_{\mathcal{D}\mathrm{er}}(J) \to \widetilde{K}_0(\mathbf{Z}[G])$$

is the Swan mapping and $\nu_J : \mathrm{Aut}_\Lambda(J) \to \mathrm{Aut}_{\mathcal{D}\mathrm{er}}(J)$ is the canonical mapping. When G is infinite, $\mathrm{End}_{\mathcal{D}\mathrm{er}}(J) \cong \mathbf{Z}$ so that $\mathrm{Aut}_{\mathcal{D}\mathrm{er}}(J) = \{\pm 1\}$ and, in fact, $\mathrm{Im}(\nu_J)\backslash\mathrm{Ker}(S_J)$ consists of a single point. Thus A_* represents the unique homotopy type which realizes J as algebraic π_2. By hypothesis, J is realized geometrically as $J = \pi_2(X)$ where X is a finite 2-complex with $\pi_1(X) = G$, so that $C_*(\widetilde{X})$ also realizes J as algebraic π_2. By uniqueness of A_*, $C_*(\widetilde{X}) \simeq A_*$ and so A_* is realized geometrically. $\qquad\square$

We now specialise to the case $G = C_\infty \times C_m$. By Theorem 10.20, $\Lambda = \mathbf{Z}[C_\infty \times C_m]$ has the SFC property. Write $C_m = \langle y \mid y^m \rangle$ and let Σ denote the sum of elements in $C_m = \langle y \mid y^m \rangle$, $\Sigma = \sum_{a=0}^{m-1} y^a$. Put $R = \mathbf{Z}[C_\infty]$ and $S = \mathbf{Z}[C_m]/(\Sigma)$; then S is a cyclotomic ring,

$$S = \mathbf{Z}[y]/q_m(y) \quad \text{where } q_m(y) = y^{m-1} + \cdots + y + 1.$$

By Corollary 10.19, $R[C_m]/(\Sigma) \cong S[C_\infty]$ has the SFC property; it follows that:

$C_\infty \times C_m$ satisfies the tameness conditions (I), (II) of Sect. 16.6. $\qquad$ (17.2)

By (16.21) the augmentation ideal $\mathcal{I}$ is a minimal element of $\Omega_1(\mathbf{Z})$. Moreover, we saw, (16.6), that in this case $\Omega_3(\mathbf{Z}) = \Omega_1(\mathbf{Z})$ so that $\mathcal{I}$ is also a minimal element of $\Omega_3(\mathbf{Z})$; then:

Theorem 17.3 $\mathcal{I}$ *is geometrically realizable when* $G = C_\infty \times C_m$.

Proof Write $C_\infty = \langle t \mid \emptyset \rangle$, $C_m = \langle y \mid y^m = 1 \rangle$ and put $\Lambda = \mathbf{Z}[C_\infty \times C_m]$. Then $C_\infty \times C_m$ has the (balanced) presentation

$$\mathcal{G} = \langle t, y \mid ty = yt, y^m = 1 \rangle$$

whose algebraic Cayley complex takes the form

$$C_*(\mathcal{G}) = (0 \to \mathrm{Ker}(\Delta_2) \to \Lambda^2 \xrightarrow{\Delta_2} \Lambda^2 \xrightarrow{\Delta_1} \Lambda \xrightarrow{\epsilon} \mathbf{Z} \to 0),$$

where

$$\Delta_1 = ((t-1)\otimes 1, 1\otimes(y-1)); \qquad \Delta_2 = \begin{pmatrix} -1\otimes(y-1) & 0 \\ (t-1)\otimes 1 & 1\otimes\Sigma \end{pmatrix}.$$

However we saw in Sect. 16.1 that there is a complete resolution for $\mathbf{Z}$ over Λ which begins:

$$\cdots \to \Lambda^2 \xrightarrow{\Delta_3} \Lambda^2 \xrightarrow{\Delta_2} \Lambda^2 \xrightarrow{\Delta_1} \Lambda \xrightarrow{\epsilon} \mathbf{Z} \to 0$$

with Δ_1, Δ_2 as above and with

$$\Delta_3 = \begin{pmatrix} -1 \otimes \Sigma & 0 \\ (t-1) \otimes 1 & 1 \otimes (y-1) \end{pmatrix}.$$

By exactness we may re-write $C_*(\mathcal{G}) = (0 \to \mathrm{Im}(\Delta_3) \to \Lambda^2 \xrightarrow{\Delta_2} \Lambda^2 \xrightarrow{\Delta_1} \Lambda \xrightarrow{\epsilon} \mathbf{Z} \to 0)$. However, by Theorem 16.5, $\mathrm{Im}(\Delta_3) \cong \mathrm{Im}(\Delta_1)$. Again, by exactness, $\mathrm{Im}(\Delta_1) = \mathrm{Ker}(\epsilon) = \mathcal{I}$ so that $C_*(\mathcal{G})$ finally takes the form

$$C_*(\mathcal{G}) = (0 \to \mathcal{I} \to \Lambda^2 \xrightarrow{\Delta_2} \Lambda^2 \xrightarrow{\Delta_1} \Lambda \xrightarrow{\epsilon} \mathbf{Z} \to 0).$$

Now $C_*(\mathcal{G}) = C_*(X_\mathcal{G})$ where $X_\mathcal{G}$ is the geometric Cayley complex of $\mathcal{G}$ so that $\mathcal{I}$ is geometrically realized.   □

We now arrive at the theorem of Edwards [25, 26]:

Theorem 17.4 (T.M. Edwards) *The $\mathcal{R}(2)$ problem for $G = C_\infty \times C_m$ has an affirmative solution for any $m \geq 2$.*

Proof Put $\Lambda = \mathbf{Z}[G]$ and $R = \mathbf{Z}[C_\infty]$; then $\Lambda = i_*(R)$ where i is the standard inclusion $C_\infty \to C_\infty \times C_m$. By the Eckmann-Shapiro Lemma, $\mathrm{Ext}^3_\Lambda(\mathbf{Z}, \Lambda) \cong \mathrm{Ext}^3_R(\mathbf{Z}, R)$. However $\mathrm{Ext}^k_R(\mathbf{Z}, R) = 0$ for $k \neq 1$. Hence $\mathrm{Ext}^3_\Lambda(\mathbf{Z}, \Lambda) = 0$. By Theorem 17.1, it suffices to show that each $J \in \Omega_3(\mathbf{Z})$ is geometrically realizable. By Theorem 17.3 $\mathcal{I}$ is geometrically realised as $\pi_2(\mathcal{G}(0))$ where

$$\mathcal{G}(0) = \langle t, y \mid ty = yt, y^m \rangle.$$

Take $W(1), \ldots, W(k)$ to be trivial relators (for example, $W(r) = (tyt^{-1}y^{-1})^r$) and put

$$\mathcal{G}(k) = \langle t, y \mid ty = yt, y^m, W(1), \ldots, W(k) \rangle.$$

One sees easily that $\pi_2(\mathcal{G}(k)) \cong \mathcal{I} \oplus \Lambda^k$ and so each $\mathcal{I} \oplus \Lambda^k$ is geometrically realisable.

Now in this case, from (16.6), $\Omega_3(\mathbf{Z}) = \Omega_1(\mathbf{Z})$. Also we showed in Theorem 16.42 that $\Omega_1(\mathbf{Z})$ is straight and that $\mathcal{I}$ is the unique module at the minimal level. Thus each $J \in \Omega_3(\mathbf{Z})$ has the form $J \cong \mathcal{I} \oplus \Lambda^k$ and so is geometrically realizable.   □

Edwards' original proof deals directly with $\Omega_3(\mathbf{Z})$ and does not make use of the identity $\Omega_3(\mathbf{Z}) = \Omega_1(\mathbf{Z})$. Also, the theorem is technically easier to prove when m is square free; the Swan module arguments required for a detailed description of

$\Omega_3(\mathbf{Z})$ are then less complicated. In particular, the characteristic ring Θ is generalized Euclidean, all generalized Swan modules are completely decomposable and the elaboration of Sect. 16.4 becomes unnecessary.

17.2 A Duality Theorem for Syzygies

Let G be a finitely presented group with integral group ring $\Lambda = \mathbf{Z}[G]$ and such that $\mathbf{Z}$ has a finitely generated free resolution over Λ of the form

$$0 \to F_n \to F_{n-1} \to \cdots \to F_0 \to \mathbf{Z} \to 0. \tag{17.5}$$

We say that G is a Poincaré Duality group of dimension n ($=$ PD^n-group) when in addition

$$\mathrm{Ext}^r_\Lambda(\mathbf{Z}, \Lambda) = \begin{cases} \mathbf{Z} & r = n, \\ 0 & r \neq n. \end{cases}$$

The fundamental group of a closed aspherical n-manifold is a PD^n-group [60]. The following is a straightforward consequence of iterating the 'de-stabilization lemma' Proposition 5.17.

Proposition 17.6 *Let G be a PD^n-group; then for $2 \leq k \leq n-1$, $J \in \Omega_k(\mathbf{Z})$ if and only if there exists an exact sequence of the form* (17.5) *in which*

$$J \cong \mathrm{Ker}(\partial_{k-1} : F_{k-1} \to F_{k-2}).$$

Theorem 17.7 *Let G be a PD^n-group; then for each k, $2 \leq k \leq n-1$, the duality map $J \mapsto J^* = \mathrm{Hom}_\Lambda(J, \Lambda)$ defines a mapping of trees*

$$\delta_k : \Omega_k(\mathbf{Z}) \to \Omega_{n+1-k}(\mathbf{Z}).$$

Proof Let $(0 \to F_n \xrightarrow{\partial_n} F_{n-1} \xrightarrow{\partial_{n-1}} \cdots \xrightarrow{\partial_2} F_1 \xrightarrow{\partial_1} F_0 \xrightarrow{\epsilon} \mathbf{Z} \to 0)$ be an exact sequence as in Proposition 17.9 and consider its canonical decomposition into short exact sequences

$$0 \to J_k \xrightarrow{j_k} F_{k-1} \xrightarrow{\pi_{k-1}} J_{k-1} \to 0 \tag{F(k)}$$

where $J_0 = \mathbf{Z}$, $\pi_0 = \epsilon$, $J_1 = \mathrm{Ker}(\epsilon)$, $J_n = F_n$ and $j_n = \partial_n$, and where $J_k = \mathrm{Ker}(\partial_{k-1})$ for $2 \leq k \leq n-1$. By (1.16), $J_k \in \Omega_k(\mathbf{Z})$ for $2 \leq k \leq n-1$. Also

$$\mathrm{Ext}^r(\mathbf{Z}, \Lambda) = 0 \quad \text{for } r \neq n. \tag{$*$}$$

since G is a duality group of dimension n. Dualizing the exact sequence **F(1)**, and using the fact that $\mathrm{Hom}_\Lambda(\mathbf{Z}, \Lambda) = 0$ and $\mathrm{Ext}^1(\mathbf{Z}, \Lambda) = 0$, we obtain an isomorphism

$$0 \to F_0 \xrightarrow{j_1^*} J_0^* \to 0. \tag{F(1)*}$$

Since $\operatorname{Ext}^1(J_{k-1}, \Lambda) = \operatorname{Ext}^k(\mathbf{Z}, \Lambda) = 0$, dualization of $\mathbf{F}(k)$ for $2 \le k \le n-1$, gives an exact sequence

$$0 \to J_{k-1}^* \overset{\pi_{k-1}^*}{\to} F_{k-1}^* \overset{j_k^*}{\to} J_k^* \to 0. \qquad\qquad \mathrm{F(k)}^*$$

Finally, on putting $\Delta = \operatorname{Ext}^n(\mathbf{Z}, \Lambda) = \operatorname{Ext}^1(J_{n-1}, \Lambda)$ then dualization of $\mathbf{F}(n)$

$$0 \to J_{n-1}^* \overset{\pi_{n-1}^*}{\to} F_{n-1}^* \overset{\partial_n^*}{\to} F_n^* \to \Delta \to 0. \qquad\qquad \mathrm{F(n)}^*$$

Splicing the sequences $\mathrm{F(1)}^*, \ldots, \mathrm{F(n)}^*$ together gives an exact sequence

$$0 \to F_0^* \overset{\partial_1^*}{\to} F_1^* \overset{\partial_2^*}{\to} \cdots \overset{\partial_{n-1}^*}{\to} F_{n-1}^* \overset{\partial_n^*}{\to} F_n^* \to \Delta \to 0, \qquad\qquad \mathrm{F}^*$$

whose the canonical decomposition into short exact sequences consists precisely of the sequences $\mathrm{F(1)}^*, \ldots, \mathrm{F(n)}^*$. In particular, for $2 \le k \le n-1$

$$J_k^* = \operatorname{Ker}(\partial_{k+1}^* : F_k^* \to F_{k+1}^*) \in \Omega_{n+1-k}(\Delta)$$

and $J \mapsto J^*$ defines a mapping $\delta_k : \Omega_k(\mathbf{Z}) \to \Omega_{n+1-k}(\Delta)$. However, $\Delta = \operatorname{Ext}^n(\mathbf{Z}, \Lambda)$ from which the stated result follows. $\qquad\square$

Repeating the argument we obtain:

Theorem 17.8 *Let G be PD^n-group; then for $2 \le k \le n-1$ the duality map $J \mapsto J^* = \operatorname{Hom}_\Lambda(J, \Lambda)$ defines an isomorphism of trees $\delta = \delta_k : \Omega_k(\mathbf{Z}) \to \Omega_{n+1-k}(\mathbf{Z})$; moreover, $\delta \circ \delta = \operatorname{Id}$.*

Proof Fix k such that $2 \le k \le n-1$; as in Theorem 17.7, there is a finite free resolution

$$0 \to E_n \overset{\partial_n}{\to} E_{n-1} \overset{\partial_{n-1}}{\to} \cdots \overset{\partial_2}{\to} E_1 \overset{\partial_1}{\to} E_0 \overset{\epsilon}{\to} \mathbf{Z} \to 0 \qquad\qquad \mathrm{E}$$

in which $J = \operatorname{Ker}(\partial_k : E_k \to E_{k-1})$. Dualization, as in Theorem 17.7, gives a finite free resolution

$$0 \to E_0^* \overset{\partial_1^*}{\to} E_1^* \overset{\partial_2^*}{\to} \cdots \overset{\partial_{n-1}^*}{\to} E_{n-1}^* \overset{\partial_n^*}{\to} E_n^* \to \Delta \to 0 \qquad\qquad \mathrm{E}^*$$

in which $J^* = \operatorname{Ker}(\partial_{k+1}^* : E_k^* \to E_{k+1}^*)$. However, $\Delta = \mathbf{Z}$, so we may repeat the argument to obtain a finite free resolution

$$0 \to E_n^{**} \overset{\partial_n^{**}}{\to} E_{n-1}^{**} \overset{\partial_{n-1}^{**}}{\to} \cdots \overset{\partial_2^{**}}{\to} E_1^{**} \overset{\partial_1^{**}}{\to} E_0^{**} \overset{\epsilon}{\to} \mathbf{Z} \to 0 \qquad\qquad \mathrm{E}^{**}$$

in which $\operatorname{Ker}(\partial_k^{**} : E_k^{**} \to E_{k-1}^{**}) = J^{**}$. Now for homomorphisms between free modules, dualization is self inverse. In particular, $\partial_k^{**} : E_k^{**} \to E_{k-1}^{**}$ is equivalent to

$\partial_k : E_k \to E_{k-1}$, and so $J^{**} = \text{Ker}(\partial_k^{**}) \cong \text{Ker}(\partial_k) = J$. Thus $\delta \circ \delta = \text{Id} : \Omega_k(\mathbf{Z}) \to \Omega_k(\mathbf{Z})$. This completes the proof.   □

A generalization of Theorem 17.8 to more general duality groups (cf. [9]) may be found in the thesis of Humphreys [44]. We stress that the dimension shift in Theorem 17.8 is correctly stated; Ω_k is dual to Ω_{n+1-k} rather than to Ω_{n-k}. Also, in Theorem 17.8, δ_k preserves heights in the sense that $\mathbf{h}(\delta_k(J)) = \mathbf{h}(J)$ for all $J \in \Omega_k(\mathbf{Z})$. In the case where G is a PD^4 group Theorem 17.8 reduces to an iso-morphism of trees

$$\delta : \Omega_3(\mathbf{Z}) \xrightarrow{\approx} \Omega_2(\mathbf{Z}).$$

In fact, the result proved is slightly more precise, namely δ gives an identifi-cation of $\Omega_3(\mathbf{Z})$, the tree of 'algebraic π_2', with $\Omega_2(\mathbf{Z})^*$, the tree formed by 'dual relation modules'. Likewise, for PD^5 groups duality gives a self-isomorphism $\delta : \Omega_3(\mathbf{Z}) \xrightarrow{\approx} \Omega_3(\mathbf{Z})$ and an isomorphism $\delta : \Omega_4(\mathbf{Z}) \xrightarrow{\approx} \Omega_2(\mathbf{Z})$. However, for $n \geq 6$, duality provides no obvious simplification for low dimensional homotopy theory.

Finally, when G is a PD^3 group then $\Omega_3(\mathbf{Z}) = \mathcal{SF}_+$ and duality gives a self-isomorphism of trees

$$\delta : \Omega_2(\mathbf{Z}) \xrightarrow{\approx} \Omega_2(\mathbf{Z}).$$

This result can be regarded in two ways: *either* as a self isomorphism as indicated *or* as an isomorphism of the stable module $\Omega_2(\mathbf{Z})$ with its dual $\Omega_2(\mathbf{Z})^*$. The first in-terpretation seems related to the duality theory for 3-manifolds observed by Hempel [40] and Turaev [96]. The second, with a somewhat different justification, extends to a larger class of groups as we now see.

17.3 A Duality Theorem for Relation Modules

In the traditional language of combinatorial group theory by a *relation module*[1] of a finitely presented group G we mean any module J which occurs in an exact sequence over $\mathbf{Z}[G]$ where F_0, F_1 are finitely generated and free thus:

$$0 \to J \to F_1 \to F_0 \to \mathbf{Z} \to 0.$$

Again from the 'de-stabilization lemma' Proposition 5.17 it follows easily that:

Proposition 17.9 *Let Λ be a ring, let n be an integer ≥ 2, and let M be a finitely generated Λ-module such that $\text{Ext}^r(M, \Lambda) = 0$ for $1 \leq r \leq n$, and such that $\Omega_r(M)$ is finitely generated for $1 \leq r \leq n-1$; then for each $J \in \Omega_n(M)$ there exists an exact sequence of the form $0 \to J \to F_{n-1} \to \cdots \to F_0 \to M \to 0$ where $F_0, \ldots, F_{n-1}$ are finitely generated and free over Λ.*

---

[1]Sometimes, in view of [30], called a *Fox ideal*.

Theorem 17.10 *Let G be an infinite finitely presented group such that*

$$\text{Ext}^1(\mathbf{Z}, \Lambda) = \text{Ext}^2(\mathbf{Z}, \Lambda) = 0;$$

then, for each $K \in \Omega_2(\mathbf{Z})$, $K^{**} \cong K$; moreover $K^* \not\cong K$ if $\text{Ext}^3(\mathbf{Z}, \Lambda) \not\cong \mathbf{Z}$.

Proof Let $K \in \Omega_2(\mathbf{Z})$; by Proposition 17.9, there exists an exact sequence

$$0 \to K \to F_1 \xrightarrow{\partial_1} F_0 \to \mathbf{Z} \to 0 \tag{0}$$

where F_1, F_0 are finitely generated and free over Λ; (0) decomposes into a pair of short exact sequences $(0 \to K \xrightarrow{k} F_1 \xrightarrow{\pi} L \to 0)$, $(0 \to L \xrightarrow{\lambda} F_0 \xrightarrow{\eta} \mathbf{Z} \to 0)$. Since G is infinite and $\text{Ext}^1(L), \Lambda) = \text{Ext}^2(\mathbf{Z}, \Lambda) = 0$ these dualize to give respectively an exact sequence

$$0 \to L^* \xrightarrow{\pi^*} F_1^* \xrightarrow{k^*} K^* \to 0 \tag{I}$$

and an isomorphism

$$0 \to F_0^* \xrightarrow{\lambda^*} L^* \to 0. \tag{II}$$

Transforming (I) using the isomorphism λ^* gives an exact sequence

$$0 \to F_0^* \xrightarrow{\partial_1^*} F_1^* \xrightarrow{k^*} K^* \to 0, \tag{0*}$$

where, of course, $\partial_1^* = (\lambda \circ \pi)^* = \pi^* \circ \lambda^*$. Dualizing $(0)^*$ gives an exact sequence

$$0 \to K^{**} \to F_1^{**} \xrightarrow{\partial_1^{**}} F_0^{**} \to \text{Ext}^1(K^*, \Lambda) \to 0. \tag{0**}$$

For homomorphisms between free modules, dualization is self-inverse so that $\partial_1^{**} : F_1^{**} \to F_0^{**}$ is equivalent to $\partial_1 : F_1 \to F_0$. In particular, we have $K^{**} = \text{Ker}(\partial_1^{**}) \cong \text{Ker}(\partial_1) = K$, proving (i). In addition, $\text{Coker}(\partial_1^{**}) \cong \mathbf{Z}$, so that $\text{Ext}^1(K^*, \Lambda) \cong \mathbf{Z}$. If $\text{Ext}^3(\mathbf{Z}, \Lambda) \not\cong \mathbf{Z}$, then $\text{Ext}^1(K, \Lambda) \not\cong \mathbf{Z}$, and so $K^* \not\cong K$, completing the proof. $\square$

17.4 The Current State of the $\mathcal{R}(2)$–$\mathcal{D}(2)$ Problem

In the wake of Perelman's solution of the Poincaré conjecture the $\mathcal{R}(2)$–$\mathcal{D}(2)$ problem is perhaps the most significant and recalcitrant problem left in low-dimensional topology. In its $\mathcal{R}(2)$ formulation a successful affirmative solution requires a number of steps to be carried out; having prescribed a fundamental group G one needs to be able to give explicit descriptions of a number of things; firstly the stable module $\Omega_3(\mathbf{Z})$; secondly $\text{Alg}_2(\mathbf{Z}[G])$ and the fibres of $\pi_2 : \text{Alg}_2(\mathbf{Z}[G]) \to \Omega_3(\mathbf{Z})$; finally, for each algebraic 2-complex A_* over $\mathbf{Z}[G]$ one needs to be able to construct a finite presentation $\mathcal{G}$ of G such that $C_*(\mathcal{G}) \simeq A_*$.

By contrast, a successful negative solution would require a choice of fundamental group G and a description of a specific algebraic 2-complex $A_* \in \mathrm{Alg}_2(\mathbf{Z}[G])$ together with a proof that no finite presentation of G realizes A_*. However, as at present no one has any idea what the details of such a proof might involve, until some progress in this direction is forthcoming one's best hope is to try to solve the problem affirmatively along the lines already indicated. To review progress to date we first consider the case of finite fundamental groups.

The first successful solution of any $\mathcal{R}(2)$ problem was that obtained for finite cyclic groups by Cockroft and Swan in [14]. Thereafter, the problem was taken up in a series of papers by Dyer and Sieradski (cf. [24]) and subsequently the (sadly unpublished) work of Browning [12] achieved an affirmative solution of the $\mathcal{R}(2)$ problem for finite abelian groups. A major difficulty in Browning's approach was that it subjected the real representation theory of the finite group G to the severe technical restriction of satisfying the Eichler condition, whereby in the Wedderburn decomposition

$$\mathbf{R}[G] \cong \prod_{i=1}^{m} M_{d_i}(\mathcal{D}_i)$$

the case $d_i = 1$ $\mathcal{D}_i = \mathbf{H}$ is not allowed. This restriction precludes the possibility of considering the generalized quaternion groups and thereby many interesting examples are excluded.

In [52] the present author circumvented Browning's approach and, for groups of free period 4, gave an explicit faithful parametrization of $\mathrm{Alg}_2(\mathbf{Z}[G])$ by the tree $\mathcal{SF}_+$ of nonzero stably free modules. This led to the existence, in [50], of an exotic 2-complex over the group ring $\mathbf{Z}[Q(32)]$. More generally this parametrization, coupled with the calculations of Swan [94], implies the existence of exotic 2-complexes over the group rings $\mathbf{Z}[Q(4m)]$ of the quaternion groups $Q(4m)$ for $m \geq 6$ (cf. [52], Chaps. 9 and 10). These exotic complexes are *not obviously* realized by finite presentations and provide the first really serious candidates for a negative solution of $\mathcal{R}(2)$–$\mathcal{D}(2)$. The calculations of Beyl and Waller in the case of $Q(28)$ would seem to indicate that any realizing presentation would have to be very complicated [8].

Since [52] was published there has been some small progress in the finite case. Recent work [82] by the author's student Jonathan Remez has solved the $\mathcal{R}(2)$ problem for $G(21)$, the non-abelian group of order 21, in such a way as seems likely to generalize to other metacyclic groups. Remez' result relies heavily on cohomological periodicity; $G(21)$ is the smallest group of minimal cohomological period 6. To date the only published solution of $\mathcal{R}(2)$–$\mathcal{D}(2)$ for a finite non-abelian non-periodic group seems to be Mannan's solution [69, 70] for the dihedral group of order 8. However, as this book was being typeset (October 2011) the author's student Seamus O'Shea successfully generalized Mannan's approach to answer the question affirmatively for some other non-periodic dihedral groups, including D_{12}, D_{20}, D_{28} and D_{68}.

In all cases where the $\mathcal{R}(2)$ problem has been solved the first and main difficulty to be overcome is to obtain an accurate description of the minimal level in $\Omega_3(\mathbf{Z})$. However, the difficulties when G is finite are as nothing compared to those when G is infinite. In [52] the author gave an affirmative solution for free groups. It is approximately the same level of difficulty (cf. [25]) to give an affirmative solution for the free abelian groups of ranks 2 and 3. However, beyond these examples the only other infinite groups for which the $\mathcal{R}(2)$ problem has been successfully solved are the groups $C_\infty \times C_m$ for which the solution was given in Sect. 17.1.

For infinite groups our present ignorance of the extent to which non-cancellation occurs in $\Omega_3(\mathbf{Z})$ is almost total. As a first approximation we have concentrated on documenting this phenomon in $\Omega_1(\mathbf{Z})$ where there is a direct relation with non-cancellation in the stably class of 0; that is, with the existence of non-trivial stably free modules.

The groups considered in detail here are of the form $F_n \times \Phi$ where F_n is the free group of rank n and Φ is finite. Stably free modules over other infinite groups were previously studied by Martin Dunwoody both individually [22] and in association with his student Paul Berridge [7], and by Harlander and Jensen [37]. The groups studied by Dunwoody were the trefoil group $G = \langle x, y \mid x^3 = y^2 \rangle$ and the quotient $G/\mathcal{D}^2(G)$ by the second commutator subgroup $\mathcal{D}^2(G)$; Harlander and Jensen additionally studied the Baumslag-Solitar group $\langle x, y \mid xy^2x^{-1} = y^3 \rangle$.

In relation to the $\mathcal{R}(2)$–$\mathcal{D}(2)$ problem, however, the results of Dunwoody-Berridge-Harlander-Jensen are deceptive; the stably free modules obtained are constructed using exotic presentations of the groups involved. Nor is it clear that in any of these cases the $\mathcal{R}(2)$ problem has definitely been solved; only that some exotic homotopy groups are realized by some correspondingly exotic presentations.

By contrast, our construction of stably free modules over groups of the form $F_n \times \Phi$ is done purely module theoretically. These modules are *not obviously* associated with any group presentation.[2] The situation is perhaps most clearly illustrated for the groups $F_n \times C_m$ where $n \geq 2$ and some p^2 divides m. The kernel construction of Theorem 13.4 still gives a mapping of trees $\mathcal{SF}_+ \to \Omega_1(\mathbf{Z})$ and (as in this case $\Omega_1(\mathbf{Z}) = \Omega_3(\mathbf{Z})$ cf. Sect. 16.1) so also a mapping of trees $\kappa : \mathcal{SF}_+ \to \Omega_3(\mathbf{Z})$.

O'Shea's observation (cf. Sect. 10.4) shows that for these groups $\mathcal{SF}_+$ has infinite branching at the minimal level $\mathcal{SF}_1$. It seems likely that $\kappa(\mathcal{SF}_1)$ is also infinite. However, at the time of writing this is not known. The implications for the $\mathcal{R}(2)$–$\mathcal{D}(2)$ problem show up very clearly in this case; if it could be shown that $\kappa(\mathcal{SF}_1)$ is infinite then to solve the $\mathcal{R}(2)$ problem affirmatively for $F_n \times C_m$ one would be required to find an infinity of explicit finite presentations to realize algebraic 2-complexes which are obtained via a purely module theoretic construction.

---

[2] Added in proof, October 2011: A very recent paper of Harlander and Misseldine, *Homology, Homotopy and Applications* 13 (2011), 63–72, takes a similar approach. They consider algebraic 2-complexes defined purely algebraically over the fundamental group of the Klein bottle, using earlier results of V.A. Artamonov on the existence of stably free modules in this case.

Appendix A
A Proof of Dieudonné's Theorem

In this appendix we give a proof of the result of Dieudonné [21] that any division ring is fully determinantal. Throughout this section, without further mention, $\mathcal{D}$ will denote a division ring and we continue the notation of Chap. 2. The arguments, though still of course valid when $\mathcal{D}$ is commutative, are non-trivial only when $\mathcal{D}$ is non-commutative. First consider the set $\mathcal{T}_n^+$ of upper triangular matrices over $\mathcal{D}$

$$\mathcal{T}_n^+ = \{X \in M_n(\mathcal{D}) : X_{ji} = 0 \text{ for } 1 \le i < j \le n\}.$$

Also the set $\mathcal{T}_n^-$ of lower triangular matrices

$$\mathcal{T}_n^- = \{X \in M_n(\mathcal{D}) : X_{ij} = 0 \text{ for } 1 \le i < j \le n\}.$$

Both sets are multiplicative submonoids of $M_n(\mathcal{D})$. Moreover one has:

Proposition A.1 *Let* $X \in \mathcal{T}_n^s$ *where* $s = \{+, -\}$; *then*

(i) X *is invertible* $\iff X_{ii} \ne 0$ *for each* i;
(ii) *if* X *is invertible then* $X^{-1} \in \mathcal{T}_n^s$.

It is necessary to study the interaction of $\mathcal{T}_n^+$, $\mathcal{T}_n^-$ with the permutation matrices $P(\sigma)$. We first note that the two sets are conjugate under the action of a specific permutation matrix, namely the following product of disjoint 2-cycles

$$\theta = \prod_{r=1}^{[n/2]} (r, n-r).$$

Then $\theta^2 = \text{Id}$ and

Proposition A.2 $P(\theta)\mathcal{T}_n^- P(\theta) = \mathcal{T}_n^+$.

F.E.A. Johnson, *Syzygies and Homotopy Theory*, Algebra and Applications 17,
DOI 10.1007/978-1-4471-2294-4, © Springer-Verlag London Limited 2012

One also has:

Proposition A.3 *Let $U \in \mathcal{T}_n^+$ be invertible and suppose for some $\sigma, \tau \in \Sigma_n$ and some $L \in \mathcal{T}_n^-$ that $P(\sigma)UP(\tau) = L$; then $\sigma = \tau^{-1}$ and $L_{i,i} = U_{\tau(i),\tau(i)}$ for all i.*

Proof Put $\rho = \sigma \circ \tau$. If $\rho \neq \mathrm{Id}$ then there exists i such that $\rho(i) < i$ so that $L_{\rho(i),i} = 0$ as L is lower triangular. However a straightforward computation shows that

$$U_{\tau(i),\tau(i)} = L_{\rho(i),i} = 0$$

contradicting invertibility of U. Hence $\rho = \mathrm{Id}$, $\sigma = \tau^{-1}$ and $L_{\rho(i),i} = U_{\tau(i),\tau(i)}$ for all i.   □

We define $\mathcal{U}_n = \{X \in \mathcal{T}_n^+ : X_{i,i} = 1 \text{ for all } i\}$. It follows from Proposition A.1 that $\mathcal{U}_n$ is a subgroup of $GL_n(\mathcal{D})$. In fact, each $X \in \mathcal{T}_n^+$ is a product of elementary matrices $E(i, j; \mathcal{D})$ with $i < j$ so that, in fact, $\mathcal{U}_n$ is a subgroup of $E_n(\mathcal{D})$. A straightforward calculation shows that:

$$\text{If } X \in \mathcal{U}_n \text{ and } D \in \Delta_n(\mathcal{D}) \text{ then } D^{-1}XD \in \mathcal{U}_n. \tag{A.4}$$

Similarly we put $\mathcal{L}_n = \{X \in \mathcal{T}_n^- : X_{i,i} = 1 \text{ for all } i\}$. Then again $\mathcal{L}_n$ is a subgroup of $E_n(\mathcal{D})$ and:

$$\text{If } X \in \mathcal{L}_n \text{ and } D \in \Delta_n(\mathcal{D}) \text{ then } D^{-1}XD \in \mathcal{L}_n. \tag{A.5}$$

We note the following:

Lemma A.6 *Let $L' \in \mathcal{L}_{n-1}$, put $L = \begin{pmatrix} L' & 0 \\ 0 & 1 \end{pmatrix} \in \mathcal{L}_n$ and let σ be the cyclic permutation $\sigma = (m, m+1, \ldots, n-1, n)$ where $m < n$. Then $P(\sigma)LP(\sigma)^{-1} \in \mathcal{L}_n$.*

Proof Put $X = P(\sigma)LP(\sigma)^{-1}$. Then for all r, s $X_{r,s} = L_{\sigma^{-1}(r),\sigma^{-1}(s)}$. In particular, for all r, $X_{r,r} = L_{\sigma^{-1}(r),\sigma^{-1}(r)} = 1$ so that it suffices to show that if $r < s$ then $L_{\sigma^{-1}(r),\sigma^{-1}(s)} = 0$. Put $J = \{1, \ldots, n\} - \{m\}$; there are three cases to consider:

Case I $r, s \in J$ and $r < s$:
As σ^{-1} is increasing on J then $\sigma^{-1}(r) < \sigma^{-1}(s)$ and so $L_{\sigma^{-1}(r),\sigma^{-1}(s)} = 0$ as L is lower triangular.
Case II $1 \leq r < s = m$:
Then $\sigma^{-1}(r) = r < n = \sigma^{-1}(s)$ and so again $L_{\sigma^{-1}(r),\sigma^{-1}(s)} = 0$ as L is lower triangular.
Case III $r = m < s$:
As L is the stabilization $L = \begin{pmatrix} L' & 0 \\ 0 & 1 \end{pmatrix}$ then $L_{n,s-1} = 0$. However, in this case $\sigma^{-1}(r) = n$ and $\sigma^{-1}(s) = s - 1$ so that $L_{\sigma^{-1}(r),\sigma^{-1}(s)} = L_{n,s-1} = 0$. This completes the proof.   □

By a $\mathcal{UCL}$ decomposition of $X \in GL_n(\mathcal{D})$ we mean a product

$$X = UCL,$$

where $U \in \mathcal{U}_n$, $C \in \mathcal{C}_n(\mathcal{D})$ and $L \in \mathcal{L}_n$. There are well known existence and uniqueness theorems for $\mathcal{UCL}$ decompositions. We include proofs for completeness.

Theorem A.7 (Existence of $\mathcal{UCL}$ decompositions) *Let $\mathcal{D}$ be a division ring and let $X \in GL_n(\mathcal{D})$ where $n \geq 2$; then X has a $\mathcal{UCL}$ decomposition.*

Proof When $n = 2$ write $X = \begin{pmatrix} a & b \\ c & d \end{pmatrix}$. If $d \neq 0$ then take

$$X = \begin{pmatrix} 1 & bd^{-1} \\ 0 & 1 \end{pmatrix} \begin{pmatrix} a - bd^{-1}c & 0 \\ 0 & d \end{pmatrix} \begin{pmatrix} 1 & 0 \\ d^{-1}c & 1 \end{pmatrix}$$

whilst if $d = 0$ and $c \neq 0$ take

$$X = \begin{pmatrix} 1 & ac^{-1} \\ 0 & 1 \end{pmatrix} \begin{pmatrix} 0 & b \\ c & 0 \end{pmatrix} \begin{pmatrix} 1 & 0 \\ 0 & 1 \end{pmatrix}.$$

Both of these decompositions are $\mathcal{UCL}$ and as X is invertible one of these two possibilities holds.

Suppose $n \geq 3$ and that the existence of a $\mathcal{UCL}$ decomposition is established for $n - 1$. For $X \in GL_n(\mathcal{D})$ put

$$m = \max\{s : 1 \leq s \leq n \text{ and } X_{ns} \neq 0\},$$

$$d = X_{nm} \quad \text{so that } d \neq 0,$$

$$U_1 = \prod_{s=1}^{n} E(s, n : X_{sm}d^{-1}),$$

$$L_1 = \begin{cases} I_n & \text{if } m = 1, \\ \prod_{s=1}^{m-1} E(m, s; d^{-1}X_{n,s}) & \text{if } 2 \leq m. \end{cases}$$

Note that the definitions of U_1 and L_1 are independent of the order in which the products are taken and that $U_1 \in \mathcal{U}_n$ and $L_1 \in \mathcal{L}_n$. Put

$$X_1 = U_1^{-1}XL_1^{-1}.$$

Then X_1 is invertible and has a single nonzero entry in row n, namely

$$(X_1)_{n,m} = d \ (= X_{n,m})$$

and that this is also the unique nonzero entry in column m. We proceed via two special cases.

Case I $m < n$ and $d = 1$. Then define $X' = (X'_{r,s})_{1 \le r,s \le n-1} \in M_{n-1}(\mathcal{D})$ by

$$X'_{r,s} = \begin{cases} (X_1)_{r,s} & s \le m-1, \\ (X_1)_{r,s+1} & m \le s. \end{cases}$$

Then

$$\begin{pmatrix} X' & 0 \\ 0 & 1 \end{pmatrix} = X_1 P(\sigma),$$

where σ is the cyclic permutation $\sigma = (m, m+1, \ldots, n-1, n)$. As X' is evidently invertible we may write inductively

$$X' = U'C'L',$$

where $U' \in \mathcal{U}_{n-1}$, $C' \in \mathcal{C}_{n-1}(\mathcal{D})$ and $L' \in \mathcal{L}_{n-1}$. Now put

$$U_2 = \begin{pmatrix} U' & 0 \\ 0 & 1 \end{pmatrix}; \qquad C'_2 = \begin{pmatrix} C' & 0 \\ 0 & 1 \end{pmatrix}; \qquad L'_2 = \begin{pmatrix} L' & 0 \\ 0 & 1 \end{pmatrix}.$$

Then $X_1 P(\sigma) = U_2 C'_2 L'_2$ is a $\mathcal{UCL}$ decomposition. Put $C_2 = C'_2 P(\sigma)^{-1} \in \mathcal{C}_n(\mathcal{D})$ and $L_2 = P(\sigma) L'_2 P(\sigma)^{-1}$. Evidently $X_1 = U_2 C_2 L_2$ and this is a $\mathcal{UCL}$ decomposition as, by Lemma A.6, $L_2 \in \mathcal{L}_n$. However, $X_1 = U_1^{-1} X L_1^{-1}$ so that

$$X = UCL$$

is a $\mathcal{UCL}$ decomposition where $U = U_1 U_2$, $C = C_2$ and $L = L_2 L_1$. This completes the proof in Case I.

Case II $m = n$ and $d = 1$.

The details in Case II are similar to Case I but simpler as we do not need to involve the permutation σ. In this case we may write X_1 directly as a stabilization of some $X' \in GL_{n-1}(\mathcal{D})$ thus:

$$X_1 = \begin{pmatrix} X' & 0 \\ 0 & 1 \end{pmatrix}.$$

Taking a $\mathcal{UCL}$ decomposition $X' = U'C'L'$ we then write $X_1 = U_2 C_2 L_2$ where

$$U_2 = \begin{pmatrix} U' & 0 \\ 0 & 1 \end{pmatrix}; \qquad C_2 = \begin{pmatrix} C' & 0 \\ 0 & 1 \end{pmatrix}; \qquad L_2 = \begin{pmatrix} L' & 0 \\ 0 & 1 \end{pmatrix}.$$

Then again $X = UCL$ is a $\mathcal{UCL}$ decomposition where $U = U_1 U_2$, $C = C_2$ and $L = L_2 L_1$ and this completes the proof in Case II.

General Case Here we know only that $d \ne 0$. Then putting $Y = XD(n, d^{-1})$ we see that Y is in one or other of the special cases just considered and so has a $\mathcal{UCL}$ decomposition

$$Y = U\widehat{C}\widehat{L}.$$

Put $C = \widehat{C}D(n,d) \in \mathcal{C}_n(\mathcal{D})$ and put $L = D(n,d)^{-1}\widehat{L}D(n,d)$. Then $L \in \mathcal{L}_n$ by (A.5) and so

$$X = UCL$$

is a $\mathcal{UCL}$ decomposition. This completes the proof.   □

There is also the following weak uniqueness theorem for $\mathcal{UCL}$ decompositions.

Theorem A.8 *Let $\mathcal{D}$ be a division ring and let $X = U_1C_1L_1$ and $X = U_2C_2L_2$ be $\mathcal{UCL}$ decompositions of $X \in GL_n(\mathcal{D})$; then $C_1 = C_2$.*

Proof Write $C_1 = P(\tau)\Delta$ and $C_2 = P(\sigma)\Gamma$ for some σ, $\tau \in \Sigma_n$ and Γ, $\Delta \in \Delta_n(\mathcal{D})$. Put

$$U = U_2^{-1}U_1 \in \mathcal{U}_n; \qquad L = L_2L_1^{-1} \in \mathcal{L}_n; \qquad W = \Gamma L\Delta^{-1}.$$

Then W is invertible lower triangular and $P(\sigma^{-1})UP(\tau) = W$. By Proposition A.3 $\sigma^{-1} = \tau^{-1}$ so that $\sigma = \tau$. It now suffices to show that $\Gamma = \Delta$. Now again by Proposition A.3 for each i

$$W_{i,i} = U_{\tau(i),\tau(i)} = 1$$

so that $W \in \mathcal{L}_n$. However, $W = (\Gamma L\Gamma^{-1})(\Gamma\Delta^{-1})$ and $(\Gamma L\Gamma^{-1}) \in \mathcal{L}_n$ by (A.5). Thus $\Gamma\Delta^{-1} \in \mathcal{L}_n$. However $\Gamma\Delta^{-1}$ is diagonal. Thus $\Gamma\Delta^{-1} = I_n$ and $\Gamma = \Delta$ as required.   □

Given a division ring $\mathcal{D}$ we define a *function* $d_n : GL_n(\mathcal{D}) \to (\mathcal{D}^*)^{ab}$ thus: if $X \in GL_n(\mathcal{D})$ write X as a $\mathcal{UCL}$-decomposition $X = UCL$ and put

$$d_n(X) = \mathrm{prot}_n(C).$$

By Theorem A.8 this expression for $d_n(X)$ is independent of the particular $\mathcal{UCL}$-decomposition chosen. We proceed to show that $\{d_n\}_{2 \leq n}$ is a determinant for $\mathcal{D}$. The first requirement is to show that each d_n is a homomorphism. It follows immediately from the definition that:

Proposition A.9 *If $X \in GL_n(\mathcal{D})$, $U \in \mathcal{U}_n$ and $L \in \mathcal{L}_n$ then $d_n(UXL) = d_n(X)$.*

In particular, when restricted to $\mathcal{C}_n(\mathcal{D})$, d_n coincides with prot_n so that;

$$d_n(\Delta(\delta_1, \ldots, \delta_n)P(\tau)) = [\mathrm{sign}(\tau)][\delta_1] \cdots [\delta_n]. \tag{A.10}$$

We also have:

Proposition A.11 *If $X \in GL_n(\mathcal{D})$ and $\Delta \in \Delta_n(\mathcal{D})$ then*

$$d_n(\Delta X) = d_n(\Delta)d_n(X) = d_n(X\Delta).$$

Proof Write X as a $\mathcal{UCL}$ decomposition $X = UCL$. Then $\Delta X = (\Delta U \Delta^{-1})(\Delta C)L$ is a $\mathcal{UCL}$ decomposition as $\Delta U \Delta^{-1} \in \mathcal{U}_n$ and $\Delta C \in \mathcal{C}_n(\mathcal{D})$. Thus

$$d_n(\Delta X) = \mathrm{prot}_n(\Delta C) = \mathrm{prot}_n(\Delta)\,\mathrm{prot}_n(C) = d_n(\Delta)d_n(X).$$

Similarly $X\Delta = U(C\Delta)(\Delta^{-1}L\Delta)$ where $\Delta^{-1}L\Delta \in \mathcal{L}_n$ and $C\Delta \in \mathcal{C}_n(\mathcal{D})$; thus

$$d_n(X\Delta) = \mathrm{prot}_n(C\Delta) = \mathrm{prot}_n(C)\,\mathrm{prot}_n(\Delta) = d_n(X)d_n(\Delta). \qquad \square$$

The calculation below is the most delicate part of the proof that d_n is a homomorphism:

Lemma A.12 *Let* $1 \le i < j \le n$; *then for any* $\sigma \in \Sigma_n$

$$d_n(P(i,j)E(i,j;\mathcal{D})P(\sigma)) = [-1][\mathrm{sign}(\sigma)].$$

Proof If $\mathcal{D} = 0$ then $E(i,j;\mathcal{D}) = I_n$ and the result follows from (A.10) on taking $\tau = (i,j)\sigma$. Suppose that $\mathcal{D} \ne 0$. By (2.14) and (2.15) we may write $P(i,j)E(i,j;\mathcal{D})P(\sigma)$ in two distinct forms namely

$P(i,j)E(i,j;\mathcal{D})P(\sigma)$

$$= \begin{cases} E(i,j;\mathcal{D}^{-1})D(i,-\mathcal{D}^{-1})D(j,\mathcal{D})P(\sigma)E(\sigma^{-1}(j),\sigma^{-1}(i);\mathcal{D}^{-1}) & \text{(A)} \\ \text{or} \\ P(i,j)P(\sigma)E(\sigma^{-1}(i),\sigma^{-1}(j);\mathcal{D}) & \text{(B)} \end{cases}$$

Both forms are always correct but their usefulness varies depending upon the interaction of σ with (i,j). There are two cases:

Case I $\sigma^{-1}(i) < \sigma^{-1}(j)$.
The (A) form is a $\mathcal{UCL}$ decomposition as $E(i,j;\mathcal{D}^{-1}) \in \mathcal{U}_n$, $D(i,-\mathcal{D}^{-1}) \times D(j,\mathcal{D})P(\sigma) \in \mathcal{C}_n(\mathcal{D})$ and $E(\sigma^{-1}(j),\sigma^{-1}(i);\mathcal{D}^{-1}) \in \mathcal{L}_n$. In this case

$$d_n(P(i,j)E(i,j;\mathcal{D})P(\sigma)) = \mathrm{prot}_n(D(i,-\mathcal{D}^{-1})D(j,\mathcal{D})P(\sigma))$$
$$= \mathrm{prot}_n(D(i,-\mathcal{D}^{-1})D(j,\mathcal{D}))\,\mathrm{prot}_n(P(\sigma))$$
$$= [-1][\mathrm{sign}(\sigma)].$$

Case II $\sigma^{-1}(i) > \sigma^{-1}(j)$.
The (B) form is a $\mathcal{UCL}$ decomposition as $P(i,j)P(\sigma) \in \mathcal{C}_n(\mathcal{D})$ and $E(\sigma^{-1}(i), \sigma^{-1}(j);\mathcal{D}) \in \mathcal{L}_n$. Moreover

$$d_n(P(i,j)E(i,j;\mathcal{D})P(\sigma)) = \mathrm{prot}_n(P(i,j)P(\sigma))$$
$$= [-1][\mathrm{sign}(\sigma)].$$

The result is the same in each case. $\square$

We are now in a position to prove:

Theorem A.13 *Let $X \in GL_n(\mathcal{D})$; then for any $\tau \in \Sigma_n$*

$$d_n(P(\tau)X) = [\text{sign}(\tau)]d_n(X).$$

Proof We first take $\tau = (i, j)$ where $i < j$. Write a $\mathcal{UCL}$ decomposition X in the form

$$X = U(P(\sigma)\Delta)L,$$

where $U \in \mathcal{U}_n$ $L \in \mathcal{L}_n$ and $\Delta \in \Delta_n(\mathcal{D})$ so that

$$d_n(X) = [\text{sign}(\sigma)]d_n(\Delta).$$

Put $\mathcal{D} = U_{i,j}$ and define $U' = UE(i, j; -\mathcal{D})$; $U'' = P(i, j)U'P(i, j)$. Then $U' \in \mathcal{L}_n$ and $U'_{i,j} = 0$. A straightforward calculation now shows that $U'' \in \mathcal{U}_n$. However,

$$P(i, j)X = U''P(i, j)E(i, j, \mathcal{D})P(\sigma)\Delta L$$

so that applying (A.10), Proposition A.11 and Lemma A.12 in succession we see that

$$d_n(P(i, j)X) = d_n(P(i, j)E(i, j, \mathcal{D})P(\sigma))d_n(\Delta)$$
$$= [-1][\text{sign}(\sigma)]d_n(\Delta)$$
$$= [-1]d_n(X).$$

The result in general follows by induction on writing τ as a product of transpositions. $\square$

Corollary A.14 *Let $X \in GL_n(\mathcal{D})$; then for any $L \in \mathcal{L}_n$*

$$d_n(LX) = d_n(X).$$

Proof Let θ be the product of 2-cycles $\theta = \prod_{r=1}^{[n/2]}(r, n - r)$ and put $U = \theta L\theta$. Then $L = \theta U\theta$ and as in Proposition A.1 $U \in \mathcal{U}_n$. Moreover applying Proposition A.9 and Theorem A.13 we see that

$$d_n(LX) = d_n(\theta U\theta X)$$
$$= [\text{sign}(\theta)]d_n(U\theta X)$$
$$= [\text{sign}(\theta)]d_n(\theta X)$$
$$= [\text{sign}(\theta)]^2 d_n(X)$$
$$= d_n(X). \qquad \square$$

Corollary A.15 *For any division ring* $\mathcal{D}$, $d_n : GL_n(\mathcal{D}) \to (\mathcal{D}^*)^{ab}$ *is a homomorphism.*

Proof Let $Y, Z \in GL_n(\mathcal{D})$ and write Y as a $\mathcal{UCL}$ decomposition thus

$$Y = U(P(\sigma)\Delta)L$$

so that $d_n(Y) = [\text{sign}(\sigma)]d_n(\Delta)$. Now applying (A.?), (A.?), (A.?) and Corollary A.14 in succession we obtain

$$
\begin{aligned}
d_n(YZ) &= d_n(U P(\sigma)\Delta L Z) \\
        &= d_n(P(\sigma)\Delta L Z) \\
        &= [\text{sign}(\sigma)]d_n(\Delta L Z) \\
        &= [\text{sign}(\sigma)]d_n(\Delta)d_n(L Z) \\
        &= [\text{sign}(\sigma)]d_n(\Delta)d_n(Z).
\end{aligned}
$$

Thus $d_n(YZ) = d_n(Y)d_n(Z)$ as required.   □

When $\mathcal{D}$ is a division ring the homomorphism $\text{prot}_n : \mathcal{C}_n(\mathcal{D}) \to (\mathcal{D}^*)^{ab}$ extends to a homomorphism $d_n : GL_n(\mathcal{D}) \to (\mathcal{D}^*)^{ab}$. This is the theorem of Dieudonné [21]. It is straightforward to check that $\{d_n\}_{2 \leq n}$ is compatible with stabilization in that the diagram below commutes for each k, n with $k \geq 1$.

$$
\begin{array}{ccc}
GL_{n+k}(\mathcal{D}) & \overset{\det_{n+k}}{\to} & (\mathcal{D}^*)^{ab} \\
\uparrow {\scriptstyle s_{n,k}} & & \uparrow {\scriptstyle \text{Id}} \\
GL_n(\mathcal{D}) & \overset{\det_n}{\to} & (\mathcal{D}^*)^{ab}
\end{array}
$$

Appendix B
Change of Ring

B.1 Extension and Restriction of Scalars

Let $\varphi : R \to S$ be a ring homomorphism; if $M \in \mathcal{M}od_R$ we put $\varphi_*(M) = M \otimes_\varphi S$ where, in addition to bi-additivity, the tensor symbol for $M \otimes_\varphi S$ satisfies the identity

$$mr \otimes s = m \otimes \varphi(r)s.$$

Then $M \otimes_\varphi S$ acquires the structure of a module over S under the action $(m \otimes 1) \cdot s = m \otimes s$. We obtain a functor $\varphi_* : \mathcal{M}od_R \to \mathcal{M}od_S$ ('extension of scalars') by[1]

$$\varphi_*(M) = M \otimes_\varphi S,$$

where the action on an R-homomorphism $h : M_1 \to M_2$ is given by

$$\varphi_*(h)(m \otimes s) = h(m) \otimes s.$$

There is also a functor $\varphi^* : \mathcal{M}od_S \to \mathcal{M}od_R$ ('restriction of scalars') obtained by allowing R to act on the S-module N by means of $n \cdot r = n\varphi(r)$ where $n \in N$ and $r \in R$. If $M \in \mathcal{M}od_R$, $N \in \mathcal{M}od_S$ there is a natural mapping

$$\begin{array}{ccc}
\mathrm{Hom}_R(M, \varphi^*(N)) & \xrightarrow{\ \nu\ } & \mathrm{Hom}_S(\varphi_*(M), N), \\
\nu(f)(m \otimes s) & = & f(m) \cdot s.
\end{array}$$

It is straightforward to check that ν is additive. Suppose that $\nu(f) = 0$; then in particular, for all $m \in M$ $f(m) = \nu(f)(m \otimes 1) = 0$ and so $f = 0$. Hence ν is injective. Finally, suppose that $\alpha : \varphi_*(M) \to N$ is a homomorphism of S-modules. We obtain a homomorphism of R-modules $\tilde{\alpha} : M \to \varphi^*(N)$ by means of $\tilde{\alpha}(m) = \alpha(m \otimes 1)$ so that $\nu(\tilde{\alpha}) = \alpha$. Thus ν is also surjective and we obtain the following *adjointness isomorphism*:

$$\nu : \mathrm{Hom}_R(M, \varphi^*(N)) \xrightarrow{\ \simeq\ } \mathrm{Hom}_S(\varphi_*(M), N). \qquad \text{(B.1)}$$

---

[1] φ_* and φ^* are also called 'base change' and 'co-base change' respectively.

F.E.A. Johnson, *Syzygies and Homotopy Theory*, Algebra and Applications 17, DOI 10.1007/978-1-4471-2294-4, © Springer-Verlag London Limited 2012

The extension of scalars functor φ_* enjoys the following properties:

φ_* is additive; that is $\varphi_*(\bigoplus_{i \in I} M_i) \cong \bigoplus_{i \in I} \varphi_*(M_i)$ for an arbitrary collection

$(M_i)_{i \in I}$ of R-modules.   (B.2)

$\varphi_*(R) = S$.   (B.3)

If $F_X(R)$ denotes the free R-module on the set X, then it follows from (B.2), (B.3) that $\varphi_*(F_X(R)) \cong F_X(S)$. Moreover, if P is a projective R-module then P is isomorphic to a direct summand of some $F_X(R)$. By additivity, $\varphi_*(P)$ is isomorphic to a direct summand of $F_X(S)$; that is:

φ_* preserves both free modules and projective modules.   (B.4)

Moreover φ_* preserves the 'relative size' of modules in that:

$\varphi_*(M)$ is finitely generated (resp. countably generated) over S

if M is finitely generated (resp. countably generated) over R.   (B.5)

In general, φ_* is not exact; that is, it does not take short exact sequences to short exact sequences. In the contexts we encounter, however, S is free as a module over R and then φ_* is exact; more generally:

φ_* is exact if S is flat as a module over R.   (B.6)

Turning to the restriction of scalars functor φ^* we have:

φ^* is additive;   (B.7)

φ^* is exact.   (B.8)

In general, however, φ^* enjoys fewer nice properties than φ_* and, without extra hypotheses, φ^* fails to preserve either free modules or projectives. If, however, S is free over R, say $S \cong F_X(R)$, then $\varphi^*(F_Y(S)) \cong F_{X \times Y}(R)$. Moreover, by additivity, if P is isomorphic to a direct summand of $F_Y(R)$ then $\varphi^*(P)$ is isomorphic to a direct summand of $F_{X \times Y}(R)$:

φ^* preserves both free modules and projective modules if S is free over R. (B.9)

B.2 Adjointness in Cohomology

Here we assume given a ring homomorphism $\varphi : R \to S$ in which S is flat over R. Given a free resolution

$$(F_* \to M) = (\cdots \to F_{n+1} \to F_n \to \cdots \to F_1 \to F_0 \to M \to 0)$$

of M over R then, as φ_* is exact,

$$\varphi_*(F_* \to M)$$

$$= (\cdots \to \varphi_*(F_{n+1}) \to \varphi_*(F_n) \to \cdots \to \varphi_*(F_1) \to \varphi_*(F_0) \to \varphi_*(M) \to 0)$$

is also exact and by (B.4), $\varphi_*(F_*)$ is free over S. Hence the cohomology groups $H^n(\varphi_*(M), N)$ may be computed in the form

$$H^n(\varphi_*(M), N) \cong \frac{\mathrm{Ker}(\mathrm{Hom}_S(\varphi_*(F_n), N) \overset{\partial^*_{n+1}}{\to} \mathrm{Hom}_S(\varphi_*(F_{n+1}), N))}{\mathrm{Im}(\mathrm{Hom}_S((\varphi_*(F_{n-1}), N) \overset{\partial^*_n}{\to} \mathrm{Hom}_S((\varphi_*(F_n), N))}.$$

However, the adjoint isomorphism $\mathrm{Hom}_S(\varphi_*(F_n), N) \cong \mathrm{Hom}_S(F_n, \varphi^*(N))$ induces corresponding isomorphisms

$$\mathrm{Ker}(\mathrm{Hom}_S(\varphi_*(F_n), N) \to \mathrm{Hom}_S(\varphi_*(F_{n+1}), N))$$

$$\cong \mathrm{Ker}(\mathrm{Hom}_S(F_n, \varphi^*(N)) \to \mathrm{Hom}_S(F_{n+1}, \varphi^*(N)));$$

$$\mathrm{Im}(\mathrm{Hom}_S(\varphi_*(F_{n-1}), N) \to \mathrm{Hom}_S(\varphi_*(F_n), N))$$

$$\cong \mathrm{Im}(\mathrm{Hom}_S(F_{n-1}, \varphi^*(N)) \to \mathrm{Hom}_S(F_n, \varphi^*(N)))$$

which descend, by naturality, to isomorphisms of quotients to give the following cohomology adjointness isomorphism for any R-module M and any S-module N:

$$H^n(\varphi_*(M), N) \cong H^n(M, \varphi^*(N)). \tag{B.10}$$

From the cohomological interpretation of Ext^1 it follows that

$$\mathrm{Ext}^1_S(\varphi_*(M), F) \cong \mathrm{Ext}^1_R(M, \varphi^*(F)).$$

Under the hypothesis that S is free over R, if F is a free S-module then $\varphi^*(F)$ is free over S and it follows that $\mathrm{Ext}^1_S(\varphi_*(M), S) = 0 \iff \mathrm{Ext}^1_R(M, R) = 0$. Hence, with the proviso that S is free over R it follows that

$$M \text{ is coprojective over } R \iff \varphi_*(M) \text{ is coprojective over } S. \tag{B.11}$$

B.3 Adjointness in the Derived Module Category

Recall (Sect. 5.1) that for R-modules M_1, M_2 we define

$$\langle M_1, M_2 \rangle = \{ f \in \mathrm{Hom}_R(M_1, M_2) : f \approx 0 \},$$

where '$f \approx 0$' means that f factors through a projective. If $M \in \mathcal{M}od_R$, $N \in \mathcal{M}od_S$ then, by (B.1), there is a bijective mapping $v : \mathrm{Hom}_R(M, \varphi^*(N)) \to \mathrm{Hom}_S(\varphi_*(M), N)$. Moreover, as φ_* preserves projective modules it follows that

$\nu(\langle M, \varphi^*(N)\rangle) \subset \langle \varphi_*(M), N\rangle$. Clearly ν is already injective on $\langle M, \varphi^*(N)\rangle$. Now suppose that the S-homomorphism $f : \varphi_*(M) \to N$ factors through the S-projective Q. If $\tilde{f} : M \to \varphi^*(N)$ is the R-homomorphism such that $\nu(\tilde{f}) = f$ then $\tilde{f}$ factors through $\varphi^*(Q)$. If S is free over R then $\varphi^*(Q)$ is R-projective so that $\nu : \langle M, \varphi^*(N)\rangle \to \langle \varphi_*(M), N\rangle$ is surjective. Thus if S is free over R then ν restricts to a bijective mapping $\nu : \langle M, \varphi^*(N)\rangle \to \langle \varphi_*(M), N\rangle$ and so induces an isomorphism of abelian groups:

$$\nu : \operatorname{Hom}_{\mathcal{D}er(R)}(M, \varphi^*(N)) \xrightarrow{\simeq} \operatorname{Hom}_{\mathcal{D}er(S)}(\varphi_*(M), N). \qquad (B.12)$$

B.4 Preservation of Syzygies and Generalized Syzygies by φ_*, φ^*

We now assume the blanket condition that S is free over R. Then φ_* is exact and preserves projective modules. In particular, if

$$0 \to K \xrightarrow{i} P \xrightarrow{p} N \to 0$$

is an exact sequence of R modules in which P is projective then $0 \to \varphi_*(K) \xrightarrow{i} \varphi_*(P) \xrightarrow{p} \varphi_*(N) \to 0$ is an exact sequence of S modules in which $\varphi_*(P)$ is projective. Thus we see

$$\varphi_*(D_1(N)) \cong D_1(\varphi_*(N)). \qquad (B.13)$$

Likewise if $0 \to J \xrightarrow{i} F \xrightarrow{p} M \to 0$ is an exact sequence of R modules in which F is finitely generated and free then $0 \to \varphi_*(J) \xrightarrow{i} \varphi_*(F) \xrightarrow{p} \varphi_*(M) \to 0$ is an exact sequence of S modules in which $\varphi_*(F)$ is finitely generated and free and so $\varphi_*(\Omega_1(M)) \cong \Omega_1(\varphi_*(M))$. More generally, if $\Omega_n(M)$ is defined then so is $\Omega_n(\varphi_*(M))$ and

$$\varphi_*(\Omega_n(M)) \cong \Omega_n(\varphi_*(M)). \qquad (B.14)$$

Similarly, as S is free over R then φ^* preserves projectives so that:

$$\varphi^*(D_n(N)) \cong D_n(\varphi^*(M)). \qquad (B.15)$$

Finally, if in addition S has finite rank over R then φ^* preserves finite generation so that if $\Omega_n(N)$ is defined so also is $\Omega_n(\varphi^*(N))$ and

$$\varphi^*(\Omega_n(N)) \cong \Omega_n(\varphi^*(N)). \qquad (B.16)$$

B.5 Co-adjointness and the Eckmann-Shapiro Lemma

The inclusion $H \hookrightarrow G$ of a subgroup H in a group G induces a ring homomorphism denoted $i : \mathbf{Z}[H] \hookrightarrow \mathbf{Z}[G]$ and so an 'extension of scalars' functor $i_* : \mathcal{M}od_{\mathbf{Z}[H]} \to$

$\mathcal{M}\text{od}_{\mathbf{Z}[G]}$ and a 'restriction of scalars' functor $i^* : \mathcal{M}\text{od}_{\mathbf{Z}[G]} \to \mathcal{M}\text{od}_{\mathbf{Z}[H]}$. Evidently $\mathbf{Z}[G]$ is free of rank $|G/H|$ over $\mathbf{Z}[H]$. Moreover, the adjointness formula of (B.1) translates into isomorphisms

$$\nu : \text{Hom}_{\mathbf{Z}[H]}(M, i^*(N)) \to \text{Hom}_{\mathbf{Z}[G]}(i_*(M), N). \tag{B.17}$$

In the context of group theory the adjointness theorem is traditionally called 'Frobenius reciprocity'. In view of (B.9) there are corresponding isomorphisms in cohomology:

$$\nu : \text{Ext}^n_{\mathbf{Z}[H]}(M, i^*(N)) \to \text{Ext}^n_{\mathbf{Z}[G]}(i_*(M), N). \tag{B.18}$$

However, in the special case where H has finite index in G there is the unusual circumstance that in addition to the canonical adjointness isomorphism there is a second co-adjointness isomorphism

$$\nu : \text{Hom}_{\mathbf{Z}[G]}(N, i_*(M)) \to \text{Hom}_{\mathbf{Z}[H]}(i^*(N), M);$$

that is, i_* and i^* are simultaneously mutual left and right adjoints without being mutually inverse. To see this, note that for any right module $\mathbf{Z}[H]$-module M, the $\mathbf{Z}[H]$-dual $\text{Hom}_{\mathbf{Z}[H]}(M, \mathbf{Z}[H])$ admits a natural left $\mathbf{Z}[H]$-module structure, given by

$$(h \cdot \varphi)(m) = h \cdot \varphi(m).$$

However, in the special case where $M = \mathbf{Z}[G]$, $\text{Hom}_{\mathbf{Z}[H]}(M, \mathbf{Z}[H])$ is naturally a $(\mathbf{Z}[H] - \mathbf{Z}[G])$ bimodule, with natural right $\mathbf{Z}[G]$-module structure given by

$$(\varphi \cdot g)(x) = \varphi(g \cdot x).$$

Evidently $\mathbf{Z}[G]$ also has a natural $(\mathbf{Z}[H] - \mathbf{Z}[G])$ bimodule structure, given by translation on either side. In fact, we have:

Proposition B.19 *There is an isomorphism of* $(\mathbf{Z}[H] - \mathbf{Z}[G])$ *bimodules*

$$\mathbf{Z}[G] \xrightarrow{\simeq} \text{Hom}_{\mathbf{Z}[H]}(\mathbf{Z}[G], \mathbf{Z}[H]).$$

Proof Let $\rho = \{\rho_1, \ldots, \rho_n\}$ be a complete set of representatives for the quotient set $H \backslash G$; that is:

$$G = \bigcup_{i=1}^n H\rho_i \quad \text{where } H\rho_i \cap H\rho_j = \emptyset \text{ if } i \neq j.$$

Put $\lambda_i = \rho_i^{-1}$. Then $\lambda = \{\lambda_1, \ldots, \lambda_n\}$ is a complete set of representatives for G/H. As a left $\mathbf{Z}[H]$-module, $\mathbf{Z}[G]$ is free on $\{\rho_1, \ldots, \rho_n\}$. Moreover, as a left $\mathbf{Z}[H]$-module, $\text{Hom}_{\mathbf{Z}[H]}(\mathbf{Z}[G], \mathbf{Z}[H])$ is free on $\{\hat{\lambda}_1, \ldots, \hat{\lambda}_n\}$, where $\hat{\lambda}_i : \mathbf{Z}[G] \to$

$\mathbf{Z}[H]$ be the right $\mathbf{Z}[H]$-homomorphism given by

$$\hat{\lambda}_i \left(\sum_{j=1}^n \lambda_j h_j \right) = h_i.$$

It follows that there is an isomorphism of left $\mathbf{Z}[H]$-modules $\nu : \mathbf{Z}[G] \to$ $\mathrm{Hom}_{\mathbf{Z}[H]}(\mathbf{Z}[G], \mathbf{Z}[H])$ given by $\nu(\rho_i) = \hat{\lambda}_i$, and straightforward computation shows that ν is equivariant with respect to the right G action. □

Proposition B.20 *When H has finite index in G there are natural isomorphisms*

$$\mathrm{Hom}_{\mathbf{Z}[H]}((i^*N), M) \cong \mathrm{Hom}_{\mathbf{Z}[G]}(N, i_*(M)).$$

Proof We make use of a different model for i_*. Put $\mathcal{E}(M) = \mathrm{Hom}_{\mathbf{Z}[H]}(\mathbf{Z}[G], M)$ where the right $\mathbf{Z}[G]$ structure on $\mathcal{E}(M)$ is given by

$$(\alpha \cdot g)(x) = \alpha(xg^{-1}).$$

There is a preliminary natural equivalence $\nu_1 : \mathcal{E}(M) \to M \otimes_{\mathbf{Z}[H]} \mathrm{Hom}_{\mathbf{Z}[H]}(\mathbf{Z}[G], \mathbf{Z}[H])$. By Proposition B.19 above, there is now an equivalence

$$\nu_2 : M \otimes_{\mathbf{Z}[H]} \mathrm{Hom}_{\mathbf{Z}[H]}(\mathbf{Z}[G], \mathbf{Z}[H]) \to M \otimes_{\mathbf{Z}[H]} \mathbf{Z}[G] = i_*(M)$$

which is natural both in M and with respect to the right $\mathbf{Z}[G]$ action. Then

$$\nu = \nu_2 \circ \nu_1 : \mathcal{E}(M) \to i_*(M)$$

is a natural equivalence. The homomorphism $\psi : \mathrm{Hom}_{\mathbf{Z}[G]}(N, \mathcal{E}(M)) \to$ $\mathrm{Hom}_{\mathbf{Z}[H]}(i^*(N), M)$ given by $[\psi(\alpha)](n) = \alpha(n)(1)$ is an isomorphism. Composing with the induced map from ν^{-1} gives the co-adjoint isomorphism

$$\psi\nu^{-1} : \mathrm{Hom}_{\mathbf{Z}[G]}(N, i_*(M)) \to \mathrm{Hom}_{\mathbf{Z}[H]}(i^*(N), M).$$ □

Still assuming that H has finite index in G then again from (B.9) there are corresponding 'Eckmann-Shapiro' isomorphisms in cohomology:

$$\nu : \mathrm{Ext}^n_{\mathbf{Z}[G]}(N, i_*(M)) \to \mathrm{Ext}^n_{\mathbf{Z}[H]}(i^*(N), M). \tag{B.21}$$

Appendix C
Group Rings with Trivial Units

In what follows, R will denote a (possibly noncommutative) ring with unit group R^*. When G is a group, the group ring $R[G]$ is described formally as the set of functions $\alpha : G \to R$ whose support $\mathrm{supp}(\alpha)$ is finite; addition and multiplication in $R[G]$ are then given by

$$(\alpha + \beta)(g) = \alpha(g) + \beta(g),$$

$$(\alpha\beta)(g) = \sum_{h \in G} \alpha(gh^{-1})\beta(h).$$

With each $g \in G$ one associates an element $\widehat{g} \in R[G]$ by

$$\widehat{g}(h) = \begin{cases} 1 & h = g, \\ 0 & h \neq g \end{cases}$$

and every element $\alpha \in R[G]$ is written uniquely as a sum $\alpha = \sum_{g \in \mathrm{supp}(\alpha)} \alpha(g)\widehat{g}$. When $a \in R^*$ the element $a\widehat{g}$ is a unit in $R[G]$. Units of this form are said to be *trivial*; otherwise expressed, if $\alpha \in R[G]^*$ then:

$$\alpha \text{ is trivial} \quad \Longleftrightarrow \quad |\mathrm{supp}(\alpha)| = 1. \tag{C.1}$$

We say that $R[G]$ *has only trivial units* when every unit $\alpha \in R[G]^*$ is trivial. Let A, B be nonempty subsets of G and $g \in AB = \{ab : a \in A, b \in B\}$; we say that g is *uniquely represented in AB* when given $a, a' \in A$, $b, b' \in B$ then

$$g = ab = a'b' \quad \Longrightarrow \quad a = a' \quad \text{and} \quad b = b'.$$

We put $\mathcal{U}(A, B) = \{g \in G : g \text{ is uniquely represented in } AB\}$. The group G is said to have the *two unique products* property when, for any finite subsets A, B of G:

$$\mathcal{TUP}: \quad 2 \leq |A| \quad \text{and} \quad 2 \leq |B| \quad \Longrightarrow \quad 2 \leq |\mathcal{U}(A, B)|.$$

F.E.A. Johnson, *Syzygies and Homotopy Theory*, Algebra and Applications 17,
DOI 10.1007/978-1-4471-2294-4, © Springer-Verlag London Limited 2012

We note that for any $x, y \in G$ $x\mathcal{U}(A, B)y = \mathcal{U}(xA, By)$. The following is useful in simplifying calculations:

$$\text{For any } x, y \in G \quad |\mathcal{U}(xA, By)| = |\mathcal{U}(A, B)|. \tag{C.2}$$

The following [80] gives a convenient sufficient condition for $R[G]$ to have trivial units.

Theorem C.3 *Let R be a (possibly noncommutative) integral domain; if G satisfies the $\mathcal{TUP}$ condition then $R[G]$ has only trivial units.*

Proof Let $\alpha \in R[G]^*$; put $\beta = \alpha^{-1}$ and write $\alpha = \sum_{r=1}^{m} a_r \widehat{g_r}$; $\beta = \sum_{s=1}^{n} b_s \widehat{h_s}$ where $\text{supp}(\alpha) = \{g_1, \ldots, g_m\}$, $\text{supp}(\beta) = \{h_1, \ldots, h_n\}$ so that $a_r, b_s \neq 0$ for each r, s. Now suppose that α nontrivial so that $2 \leq m$. If $n = 1$ write $\beta = b\widehat{h}$ so that $\alpha = b^{-1}\widehat{h}^{-1}$ is trivial, contradiction. Thus $2 \leq n$; by the $\mathcal{TUP}$ condition $2 \leq |\mathcal{U}(\text{supp}(\alpha), \text{supp}(\beta))|$.

Without loss of generality we may suppose that $\text{supp}(\alpha), \text{supp}(\beta)$ are indexed so that $g_1 h_1$ and $g_m h_n$ are uniquely represented as products in $\text{supp}(\alpha)\,\text{supp}(\beta)$; then in the expression

$$\alpha\beta = \sum_{r=1}^{m}\sum_{s=1}^{n} a_r b_s \widehat{g_r h_s}$$

the coefficients of $g_1 h_1$ and $g_m h_n$ are respectively $a_1 b_1$ and $a_m b_n$. As A is an integral domain these are both nonzero. Thus $2 \leq |\text{supp}(\alpha\beta)|$. In particular, $\alpha\beta \neq 1$, which is a contradiction. Thus α is a trivial unit. $\qquad\square$

One sees easily that no nontrivial finite group can be $\mathcal{TUP}$; it follows that:

$$\text{Every } \mathcal{TUP} \text{ group is torsion free.} \tag{C.4}$$

The $\mathcal{TUP}$ notion originates in the thesis of Higman but has undergone some refinements since (cf. [80]). In particular, the following, though not explicitly stated in this way is essentially due to Higman [42]:

Theorem C.5 *Suppose that for every nontrivial finitely generated subgroup H of a group G there exists a $\mathcal{TUP}$ group Γ_H and a nontrivial homomorphism $\varphi_H : H \to \Gamma_H$; then G is $\mathcal{TUP}$.*

Proof Let A, B be finite subsets of G such that $2 \leq \min\{|A|, |B|\}$. We show by induction on $|A| + |B|$ that $2 \leq |\mathcal{U}(A, B)|$. For the induction base take $|A| = |B| = 2$. By left translating A by some element $x \in A$, right translating B by some element $y \in B$ and appealing to (C.2) we may assume $A = \{1, a\}$, $B = \{1, b\}$, then $AB = \{1, a, b, ab\}$. If $a \neq b$ then $\{a, b\} \subset \mathcal{U}(A, B)$ and so $2 \leq |\mathcal{U}(A, B)|$. If $a = b$ then $AB\{1, a, a^2\}$. As $a \neq 1$ then also $a \neq a^2$ so the only way to obtain $\mathcal{U}(A, B) < 2$ is to have $a^2 = 1$. But then $AB \cong C_2$ is a finitely generated subgroup of G which admits no nontrivial homomorphism to any torsion free group, contradicting the hypothesis on G. Hence $a^2 \neq 1$ and $2 \leq |\mathcal{U}(A, B)|$.

For the induction step suppose that A, B are finite subsets of G with $2 \leq \min\{|A|, |B|\}$ and assume, given finite subsets A', B' of G, that

$$2 \leq |\mathcal{U}(A', B')| \text{ provided that } 2 \leq \min\{|A'|, |B'|\} \text{ and } |A'| + |B'| < |A| + |B|. \quad (*)$$

We must show that $2 \leq |\mathcal{U}(A, B)|$. After suitable left and right translation to A, B respectively we may suppose that $1 \in A \cap B$. Let H be the subgroup of G generated by $A \cup B$ and let $\varphi : H \to \Gamma$ be a nontrivial homomorphism to a $\mathcal{TUP}$ group Γ. We first claim that:

> There exist $a_1, a_2 \in A$ and $b_1, b_2 \in B$ such that $\varphi(a_1)\varphi(b_1)$ and
>
> $\varphi(a_2)\varphi(b_2)$ are uniquely represented in $\varphi(A)\varphi(B)$. $\quad (**)$

If it is the case that $2 \leq \min\{|\varphi(A)|, |\varphi(B)|\}$ then $(**)$ follows from the $\mathcal{TUP}$ property for Γ. When $|\varphi(B)| = 1$ then, by nontriviality of φ, $|\varphi(A)| \neq 1$. Thus choosing $a_1, a_2 \in A$ so that $\varphi(a_1)\varphi(1)$, $\varphi(a_2)\varphi(1)$ are uniquely represented in $\varphi(A)\varphi(B)$ verifying $(**)$ in this case. Similarly, when $|\varphi(A)| = 1$ then $|\varphi(B)| \neq 1$ and we may again verify $(**)$.

Now put $A' = \varphi^{-1}(\varphi\{a_1, a_2\}) \cap A$ and $B' = \varphi^{-1}(\varphi\{b_1, b_2\}) \cap B$. If $\varphi_{|A'}$ and $\varphi_{|B'}$ are both injective then $a_1 b_1$ and $a_2 b_2$ are uniquely represented in AB. If $\varphi_{|A'}$ is not injective choose i such that $|\varphi^{-1}\varphi(a_i)| \geq 2$. There are two cases;

$$\text{either} \quad (i) \quad |\varphi^{-1}\varphi(b_i)| = 1 \quad \text{or} \quad (ii) \quad |\varphi^{-1}\varphi(b_i)| \geq 2.$$

If (i) then choosing $a_i' \in A'$ such that $\varphi(a_i') = \varphi(a_i)$ and $a_i' \neq a_i$ it is easy to see that $a_i b_i$ and $a_i' b_i$ are uniquely represented in AB. If (ii) then put $K = \{k \in \text{Ker}(\varphi) : a_i k \subset A\}$ and $L = \{l \in \text{Ker}(\varphi) : lb_i \subset B\}$. Then K, L are subsets of G with $2 \leq \min\{|K|, |L|\}$ and $|K| + |L| < |A| + |B|$. By induction, choose $k_1, k_2 \in K$, $l_1, l_2 \in L$ such that $k_1 l_1$, $k_2 l_2$ are uniquely represented in KL. Then $a_i k_1 l_1 b_i$ and $a_i k_2 l_2 b_i$ are uniquely represented in AB, so that in every case we have shown that $2 \leq |\mathcal{U}(A, B)|$. $\qquad\square$

The most obvious example is:

$$\text{The infinite cyclic group } C_\infty \text{ is } \mathcal{TUP}. \quad (C.6)$$

The group G is *locally indicible*[1] when every nontrivial finitely generated subgroup $H \subset G$ admits a surjective homomorphism $\varphi : H \to C_\infty$; this amounts to taking Γ_H uniformly to be C_∞ in the hypotheses of Theorem C.5. Thus as a consequence of Theorem C.5 we obtain the following, which was explicitly proved by Higman in [42]:

$$\text{Every locally indicible group has the } \mathcal{TUP} \text{ property.} \quad (C.7)$$

---

[1] Higman's original terminology [42] is 'indicible throughout'.

In particular, free groups are locally indicible so that:

$$\text{Free groups have the } \mathcal{TUP} \text{ property.} \tag{C.8}$$

One notes the following general properties:

$$\text{A subgroup of a } \mathcal{TUP} \text{ group is } \mathcal{TUP}. \tag{C.9}$$

$$\text{The class of } \mathcal{TUP} \text{ groups is closed under extension.} \tag{C.10}$$

$$\text{A locally } \mathcal{TUP} \text{ group is } \mathcal{TUP}. \tag{C.11}$$

$$\text{A right ordered group is } \mathcal{TUP}. \tag{C.12}$$

Higman also showed:

$$\text{The class of locally indicible groups is closed under both extension}$$

$$\text{and free product.} \tag{C.13}$$

It follows from this that a great many torsion free groups familiar from low dimensional topology are $\mathcal{TUP}$; for example the fundamental groups of surfaces of genus ≥ 1.

Appendix D
The Infinite Kernel Property

Throughout we work in the category of unitary associative rings which are augmented by means of a (necessarily surjective) ring homomorphism $\epsilon : \Lambda \to \mathbf{Z}$. Morphisms in this category are commutative triangles of ring homomorphisms

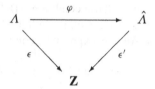

When M is a Λ-module, we write $M \otimes_\Lambda \mathbf{Z} = M \otimes_\epsilon \mathbf{Z}$. If P is a countably generated projective Λ-module, then, for any ring homomorphism $\varphi : \Lambda \to \hat{\Lambda}$, $P \otimes_\varphi \hat{\Lambda}$ is projective and countably generated over $\hat{\Lambda}$. Over $\mathbf{Z}$, every projective module is free of uniquely determined rank. Thus if P is a countably generated projective Λ-module then $P \otimes_\Lambda \mathbf{Z} \cong \mathbf{Z}^\alpha$ for some uniquely determined value of α ($\alpha = 1, 2, \ldots, \infty$), and we define the *rank*, $\mathrm{rk}(P)$ of P by means of $\mathrm{rk}(P) = \alpha = \mathrm{rk}_{\mathbf{Z}}(P \otimes_\Lambda \mathbf{Z})$. We say that Λ has property $\mathcal{K}(\infty)$ when there exists an exact sequence

$$0 \to P \to \Lambda^b \to \Lambda^a,$$

where a, b are positive integers and P is a projective Λ-module of infinite rank.

Proposition D.1 *Let $\Lambda \subset \hat{\Lambda}$ be an extension of augmented $\mathbf{Z}$-algebras, and suppose that $\hat{\Lambda}$ is free as a (left) Λ-module; if Λ has property $\mathcal{K}(\infty)$ then so also does $\hat{\Lambda}$.*

Proof Suppose that Λ has property $\mathcal{K}(\infty)$; that is, there exists an exact sequence of Λ-modules $0 \to P \xrightarrow{i} \Lambda^b \xrightarrow{\varphi} \Lambda^a$ where a, b are positive integers and P is Λ-projective of infinite rank. Since $\hat{\Lambda}$ is free as a left Λ-module, the functor $- \otimes_\Lambda \hat{\Lambda}$ is exact. Since $\Lambda^\alpha \otimes_\Lambda \hat{\Lambda} \cong \hat{\Lambda}^\alpha$ we obtain an exact sequence

$$0 \to P \otimes_\Lambda \hat{\Lambda} \xrightarrow{i} \hat{\Lambda}^b \xrightarrow{\varphi} \hat{\Lambda}^a.$$

F.E.A. Johnson, *Syzygies and Homotopy Theory*, Algebra and Applications 17, DOI 10.1007/978-1-4471-2294-4, © Springer-Verlag London Limited 2012

Moreover $P \otimes_\Lambda \hat{\Lambda}$ is $\hat{\Lambda}$-projective, and $(P \otimes_\Lambda \hat{\Lambda}) \otimes_{\hat{\Lambda}} \mathbf{Z} \cong P \otimes_\Lambda \mathbf{Z}$ so that $P \otimes_\Lambda \hat{\Lambda}$ also has infinite rank. Hence $\hat{\Lambda}$ has property $\mathcal{K}(\infty)$.   □

We recall the following from Sect. 6.3:

Theorem D.2 *The following conditions are equivalent for any ring* Λ:

(i) *if* M *is a finitely presented* Λ-*module and* $\Omega \in \Omega_1(M)$ *then* Ω *is also finitely presented*;

(ii) *if* M *is a finitely presented* Λ-*module and then* $\Omega_n(M)$ *is defined and finitely generated for all* $n \geq 2$;

(iii) *if* M *is a finitely generated* Λ-*module such that* $\Omega_1(M)$ *is finitely generated then* $\Omega_n(M)$ *is defined and finitely generated for all* $n \geq 2$;

(iv) *in any exact sequence of* Λ-*modules* $0 \to \Omega \to \Lambda^b \to \Lambda^a \to M \to 0$, *where* a, b *are positive integers, the module* Ω *is finitely generated*;

(v) *in any exact sequence of* Λ-*modules* $0 \to \Omega \to \Lambda^b \to \Lambda^a$, *where* a, b *are positive integers, the module* Ω *is finitely generated*.

A ring Λ which satisfies any of these conditions (i)–(v) is said to be *coherent*. Otherwise we shall say that Λ is *incoherent*. Let $\mathcal{M}od_{fp}(\Lambda)$ denote the category of finitely presented Λ-modules. We may express the condition alternatively thus:

$$\Lambda \text{ is coherent} \quad \Leftrightarrow \quad \text{the category } \mathcal{M}od_{fp}(\Lambda) \text{ is abelian.} \qquad (\text{D.3})$$

Clearly we have:

Proposition D.4 *If* Λ *has property* $\mathcal{K}(\infty)$ *then* Λ *is incoherent.*

Writing $cd(M)$ for the cohomological dimension of the Λ-module M we have:

Proposition D.5 *Let* M *be a finitely generated* Λ-*module such that, for some* $m \geq 2$,

(i) $\Omega_1(M), \ldots, \Omega_{m-1}(M)$ *are defined and finitely generated*;

(ii) $\Omega_m(M)$ *is infinitely generated*;

(iii) $cd(M) \leq m$;

then Λ *has property* $\mathcal{K}(\infty)$.

Proof By (i), there exists an exact sequence

$$0 \to \Omega \to \Lambda^{e_{m-1}} \to \cdots \to \Lambda^{e_1} \to \Lambda^{e_0} \to M \to 0, \qquad (*)$$

where $e_0, \ldots, e_{m-1}$ are positive integers. By (iii), there is an exact sequence

$$0 \to P_m \to P_{m-1} \to \cdots \to P_1 \to P_0 \to M \to 0. \qquad (**)$$

Comparing $(*)$ and $(**)$ by means of Swan's generalization of Schanuel's Lemma [91], we see that

$$\Omega \oplus Q \cong P_m \oplus Q'$$

for some projective modules Q, Q'. In particular, Ω, being a direct summand of the projective module $P_m \oplus Q'$, is necessarily projective. However, by (ii), Ω is not finitely generated. The sequence $0 \to \Omega \to \Lambda^{e_m-1} \to \Lambda^{e_m-2}$ shows that Λ has property $\mathcal{K}(\infty)$.   □

These considerations apply when Λ is the integral group ring $\Lambda = \mathbf{Z}[G]$ of a group G. We say that a group G has property $\mathcal{K}(\infty)$ when the integral group ring $\mathbf{Z}[G]$ has property $\mathcal{K}(\infty)$; likewise, we say that G is incoherent when $\mathbf{Z}[G]$ is incoherent. If H is a subgroup of G, then the induced ring extension $\mathbf{Z}[H] \subset \mathbf{Z}[G]$ is a morphism of augmented $\mathbf{Z}$-algebras. Furthermore, as a left $\mathbf{Z}[H]$-module, $\mathbf{Z}[G]$ is free on the basis G/H. From Proposition D.1 we see that:

Proposition D.6 *Let H be a subgroup of a group G; if H has property $\mathcal{K}(\infty)$ then so also does G.*

For subgroups of finite index, the relation is one of equivalence:

Proposition D.7 *Let H be a subgroup of finite index in a group G; then*

$$H \text{ has property } \mathcal{K}(\infty) \quad \Longleftrightarrow \quad G \text{ has property } \mathcal{K}(\infty).$$

Proof By Proposition D.6, it suffices to show $(\Longleftarrow)$; thus suppose that there is an exact sequence of $\mathbf{Z}[G]$-modules

$$0 \to P \to \mathbf{Z}[G]^b \to \mathbf{Z}[G]^a,$$

where P is a projective $\mathbf{Z}[G]$-module of infinite rank. Let $i : H \hookrightarrow G$ be the inclusion so that i^* is the functor which restricts scalars from $\mathbf{Z}[G]$ to $\mathbf{Z}[H]$. Then $i^*(\mathbf{Z}[G]) \cong \mathbf{Z}[H]^d$; applying i^* to the above gives an exact sequence $0 \to Q \to \mathbf{Z}[H]^{db} \to \mathbf{Z}[H]^{da}$ where $Q = i^*(P)$. However,

$$
\begin{aligned}
Q \otimes_{\mathbf{Z}[H]} \mathbf{Z} &\cong (P \otimes_{\mathbf{Z}[G]} \mathbf{Z}[G]) \otimes_{\mathbf{Z}[H]} \mathbf{Z} \\
&\cong P \otimes_{\mathbf{Z}[G]} (\mathbf{Z}[G] \otimes_{\mathbf{Z}[H]} \mathbf{Z}) \\
&\cong P \otimes_{\mathbf{Z}[G]} \mathbf{Z}^d \\
&\cong (\mathbf{Z}^{\infty})^d \\
&\cong \mathbf{Z}^{\infty}.
\end{aligned}
$$

Thus $\mathbf{Z}[H]$ also has property $\mathcal{K}(\infty)$.   □

In consequence, possession of property $\mathcal{K}(\infty)$ is an invariant of commensurability class.

If G, H are commensurable then

$$H \text{ has property } \mathcal{K}(\infty) \quad \Leftrightarrow \quad G \text{ has property } \mathcal{K}(\infty). \tag{D.8}$$

As usual we denote by $\mathbf{Z}$ the trivial $\mathbf{Z}[H]$-module having $\mathbf{Z}$ as underlying abelian group. Suppose that the group G is finitely generated by the set $\{x_1, \ldots, x_g\}$. Then we have an exact sequence of $\mathbf{Z}[G]$-modules $\mathbf{Z}[G]^g \overset{\partial}{\to} \mathbf{Z}[G] \overset{\epsilon}{\to} \mathbf{Z} \to 0$ where ϵ is the augmentation map, and ∂ is the $\mathbf{Z}[G]$-homomorphism defined by the $1 \times g$ matrix $\partial = (x_1 - 1, \ldots, x_g - 1)$. The augmentation ideal $\mathcal{I} = \mathrm{Ker}(\epsilon)$, being isomorphic to $\mathrm{Im}(\partial)$, is finitely generated. As $\Omega_1^G(\mathbf{Z})$ is represented by $\mathcal{I}$ we have:

$$\text{If } G \text{ is finitely generated then } \Omega_1^G(\mathbf{Z}) \text{ is finitely generated.} \tag{D.9}$$

Recall that the cohomological dimension $\mathrm{cd}(H)$ of the group H is the same as the cohomological dimension $\mathrm{cd}(\mathbf{Z})$ of the trivial $\mathbf{Z}[H]$-module $\mathbf{Z}$. If $\mathrm{cd}(H) \leq 2$, then any representative of $\Omega_2(\mathbf{Z})$ is projective; from Proposition D.5 we obtain immediately:

Proposition D.10 *Let H be a finitely generated group; if $\mathrm{cd}(H) \leq 2$ and $\Omega_2(\mathbf{Z})$ is infinitely generated, then $\mathbf{Z}[H]$ has property $\mathcal{K}(\infty)$.*

Corollary D.11 *Let H be a finitely generated group; if $\mathrm{cd}(H) \leq 2$ and $\Omega_2(\mathbf{Z})$ is infinitely generated, then H is incoherent.*

We proceed to produce a class of groups satisfying the hypotheses of Proposition D.10. Thus for $n \geq 2$ let F_n be the (nonabelian) free group of rank n, and let $F_1 = C_\infty$ be the infinite cyclic group. Choose $n \geq 2$, and let $\varphi : F_n \to C_\infty$ be a surjective homomorphism; we define $H(n, \varphi)$ to be the fibre product

$$H(n, \varphi) = F_n \underset{\varphi, \varphi}{\times} F_n = \{(x, y) \in F_n \times F_n : \varphi(x) = \varphi(y)\}.$$

Proposition D.12 *$H(n, \varphi)$ has property $\mathcal{K}(\infty)$.*

Proof It is easy to check that $H(n, \varphi)$ is both a normal subgroup and a subdirect product of $F_n \times F_n$. The finite generation of $H(n, \varphi)$ thus follows from [48] (1.21). The argument of Grunewald ([35], Proposition B) now shows that, over $H(n, \varphi)$, the derived module $\Omega_2(\mathbf{Z})$ is infinitely generated. Finally, since $\mathrm{cd}(F_n) = 1$ and $H(n, \varphi)$ is a subgroup of $F_n \times F_n$ then $\mathrm{cd}(H(n, \varphi)) \leq 2$.   □

From Propositions D.6, D.10 and D.12 we see that:

Corollary D.13 *Let G be a group which contains a subgroup isomorphic to $H(n, \varphi)$; then $\mathbf{Z}[G]$ has property $\mathcal{K}(\infty)$.*

In particular, $F_2 \times F_2$ contains a copy of every $H(2, \varphi)$ so that $F_2 \times F_2$ has property $\mathcal{K}(\infty)$. Furthermore, for $m, n \geq 2$, F_m is contained as a subgroup of index

$m - 1$ in F_2 so that as $F_m \times F_n$ is commensurable with $F_2 \times F_2$ then $F_m \times F_n$ also has property $\mathcal{K}(\infty)$. Thus from Propositions D.7 and D.10 we obtain:

Theorem D.14 *Let G be a group which contains a copy of $F_m \times F_n$ for some $m, n \geq 2$; then $\mathbf{Z}[G]$ fails to be coherent.*

One may observe that a group G contains a copy of $F_m \times F_n$ for some $m, n \geq 2$ precisely when G contains a copy of $F_\infty \times F_\infty$. Thus a group G which contains a copy of $F_\infty \times F_\infty$ also has property $\mathcal{K}(\infty)$, and again fails to be coherent.

It follows from these observations that many familiar infinite groups fail to be coherent; in particular, this is true of 'most' semisimple lattices. To see this, in general terms, note that by the Arithmeticity Theorem of Margulis [72], a typical lattice Γ in a general noncompact semisimple Lie group is arithmetic; that is, there is an algebraic group $\mathbf{G}$ defined and semisimple over $\mathbf{Q}$ such that Γ is commensurable with the group $\mathbf{G_Z}$ of points which stabilize an integer lattice under a faithful representation. It suffices to consider the case where $\mathbf{G}$ is $\mathbf{Q}$-simple of real rank ≥ 2. Then, except in low dimensional cases, $\mathbf{G}$ contains a proper semisimple algebraic subgroup $\mathbf{H} \times \mathbf{K}$. By a result of Tits [95], both $\mathbf{H_Z}$ and $\mathbf{K_Z}$ contain nonabelian free groups. Thus $\mathbf{G_Z}$ contains a copy of $F_m \times F_n$, and so has property $\mathcal{K}(\infty)$. Hence Γ, being commensurable with $\mathbf{G_Z}$, also has property $\mathcal{K}(\infty)$.

It is also true that 'most' poly-Surface groups fail to be coherent. For example, let

$$1 \to \Sigma_h \to G \to \Sigma_g \to 1$$

be a group extension where Σ_n denotes the fundamental group of a closed surface of genus $n \geq 2$. Using, for example, the arguments of [49], it is straightforward to see that if the operator homomorphism $c : \Sigma_g \to \mathrm{Out}(\Sigma_h)$ fails to be injective then G contains a subgroup of the form $F_\infty \times F_\infty$, and so fails to be coherent. However, our arguments do not settle those cases (cf. [36]) in which the operator homomorphism is injective.

References

1. Bass, H.: Projective modules over free groups are free. J. Algebra **1**, 367–373 (1964)
2. Bass, H.: Algebraic K-Theory. Benjamin, Elmsford (1968)
3. Bass, H., Murthy, M.P.: Grothendieck groups and Picard groups of abelian group rings. Ann. Math. **86**, 16–73 (1967)
4. Bass, H., Heller, A., Swan, R.G.: The Whitehead group of a polynomial extension. Publ. Math. IHÉS **22**, 61–79 (1964)
5. Baer, R.: Erweiterung von Gruppen und ihren Isomorphismen. Math. Z. **38**, 375–416 (1934)
6. Baer, R.: Abelian groups that are direct summands of every containing abelian group. Bull. Am. Math. Soc. **46**, 800–806 (1940)
7. Berridge, P.H., Dunwoody, M.: Non-free projective modules for torsion free groups. J. Lond. Math. Soc. (2) **19**, 433–436 (1979)
8. Beyl, F.R., Waller, N.: A stably free non-free module and its relevance to homotopy classification, case Q_{28}. Algebr. Geom. Topol. **5**, 899–910 (2005)
9. Bieri, R., Eckmann, B.: Groups with homological duality generalizing Poincaré Duality. Invent. Math. **20**, 103–124 (1973)
10. Birch, B.J.: Cyclotomic fields and Kummer extensions. In: Cassells, J.W.S., Fröhlich, A. (eds.) Algebraic Number Theory. Academic Press, San Diego (1967)
11. Bourbaki, N.: Commutative Algebra. Hermann/Addison Wesley, Paris/Reading (1972)
12. Browning, W.: Truncated projective resolutions over a finite group. ETH, April 1979 (unpublished notes)
13. Cartan, H., Eilenberg, S.: Homological Algebra. Princeton University Press, Princeton (1956)
14. Cockroft, W.H., Swan, R.G.: On the homotopy types of certain two-dimensional complexes. Proc. Lond. Math. Soc. (3) **11**, 194–202 (1961)
15. Cohn, P.M.: Some remarks on the invariant basis property. Topology **5**, 215–228 (1966)
16. Cohn, P.M.: On the structure of the GL_2 of a ring. Publ. Math. IHES **30**, 5–53 (1966)
17. Cohn, P.M.: Free Rings and Their Relations. LMS, 2nd edn. Academic Press, San Diego (1985)
18. Curtis, C.W., Reiner, I.: Methods of Representation Theory, vols. I & II. Wiley-Interscience, New York (1981/1987)
19. Dicks, W.: Free algebras over Bezout domains are Sylvester domains. J. Pure Appl. Algebra **27**, 15–28 (1983)
20. Dicks, W., Sontag, E.D.: Sylvester domains. J. Pure Appl. Algebra **13**, 243–275 (1978)
21. Dieudonné, J.: Les déterminants sur un corps noncommutatif. Bull. Soc. Math. Fr. **71**, 27–45 (1943)
22. Dunwoody, M.: Relation modules. Bull. Lond. Math. Soc. **4**, 151–155 (1972)
23. Dyer, M.N.: Homotopy trees with trivial classifying ring. Proc. Am. Math. Soc. **55**, 405–408 (1976)

F.E.A. Johnson, *Syzygies and Homotopy Theory*, Algebra and Applications 17,
DOI 10.1007/978-1-4471-2294-4, © Springer-Verlag London Limited 2012

24. Dyer, M.N., Sieradski, A.J.: Trees of homotopy types of two-dimensional CW complexes. Comment. Math. Helv. **48**, 31–44 (1973)

25. Edwards, T.M.: Algebraic 2-complexes over certain infinite abelian groups. Ph.D. Thesis, University College London (2006)

26. Edwards, T.M.: Generalised Swan modules and the D(2) problem. Algebr. Geom. Topol. **6**, 71–89 (2006)

27. Eichler, M.: Über die Idealklassenzahl total definiter Quaternionalgebren. Math. Z. **43**, 102–109 (1938)

28. Eisenbud, D.: The Geometry of Syzygies. Springer, Berlin (2005)

29. Farrell, F.T.: An extension of Tate cohomology to a class of infinite groups. J. Pure Appl. Algebra. **10** (1977)

30. Fox, R.H.: Free differential calculus V. Ann. Math. **71**, 408–422 (1960)

31. Fröhlich, A., Taylor, M.J.: Algebraic Theory of Numbers. Cambridge University Press, Cambridge (1991)

32. Gabel, M.R.: Stably free projectives over commutative rings. Ph.D. Thesis, Brandeis University (1972)

33. Gollek, S.: Computations in the derived module category. Ph.D. Thesis, University College London (2010)

34. Gruenberg, K.W.: Cohomological Topics in Group Theory. Lecture Notes in Mathematics, vol. 143. Springer, Berlin (1970)

35. Grunewald, F.J.: On some groups which cannot be finitely presented. J. Lond. Math. Soc. **17**, 427–436 (1978)

36. Gonzalez-Diez, G., Harvey, W.J.: Surface groups inside mapping class groups. Topology **38**, 57–69 (1999)

37. Harlander, J., Jensen, J.A.: Exotic relation modules and homotopy types for certain 1-relator groups. Algebr. Geom. Topol. **6**, 3001–3011 (2006)

38. Hasse, H.: Number Theory. Springer, Berlin (1962)

39. Heller, A.: Indecomposable modules and the loop space operation. Proc. Am. Math. Soc. **12**, 640–643 (1961)

40. Hempel, J.: Intersection calculus on surfaces with applications to 3-manifolds. Mem. Am. Math. Soc. (282) (1983)

41. Higgins, P.J.: Categories and Groupoids. Mathematical Studies. Van Nostrand/Reinhold, Princeton/New York (1971)

42. Higman, D.G.: The units of group rings. Proc. Lond. Math. Soc. (2) **46**, 231–248 (1940)

43. Hilbert, D.: Über die Theorie der Algebraischen Formen. Math. Ann. **36**, 473–534 (1890)

44. Humphreys, J.J.A.M.: Algebraic properties of semi-simple lattices and related groups. Ph.D. Thesis, University College London (2006)

45. Hurwitz, A.: Über die Zahlentheorie der Quaternionen. Collected Works (1896)

46. Jacobinski, H.: Genera and decompositions of lattices over orders. Acta Math. **121**, 1–29 (1968)

47. Jacobson, N.: Basic Algebra. Freeman, New York (1974)

48. Johnson, F.E.A.: On normal subgroups of direct products. Proc. Edinb. Math. Soc. **33**, 309–319 (1990)

49. Johnson, F.E.A.: Surface fibrations and automorphisms of non-abelian extensions. Q. J. Math. **44**, 199–214 (1993)

50. Johnson, F.E.A.: Minimal 2-complexes and the D(2)-problem. Proc. Am. Math. Soc. **132**, 579–586 (2003)

51. Johnson, F.E.A.: Stable modules and Wall's D(2) problem. I. Comment. Math. Helv. **78**, 18–44 (2003)

52. Johnson, F.E.A.: Stable Modules and the D(2)-Problem. LMS Lecture Notes in Mathematics, vol. 301. Cambridge University Press, Cambridge (2003)

53. Johnson, F.E.A.: Incoherence of integral groups rings. J. Algebra **286**, 26–34 (2005)

54. Johnson, F.E.A.: The stable class of the augmentation ideal. K-Theory **34**, 141–150 (2005)

55. Johnson, F.E.A.: Rigidity of hyperstable complexes. Arch. Math.. **90**, 123–132 (2008)

56. Johnson, F.E.A.: Homotopy classification and the generalized Swan homomorphism. J. K-Theory **4**, 491–536 (2009)
57. Johnson, F.E.A.: Stably free cancellation for group rings of cyclic and dihedral type. Q. J. Math. (2011). doi:10.1093/qmath/har006
58. Johnson, F.E.A.: Infinite branching in the first syzygy. J. Algebra **337**, 181–194 (2011)
59. Johnson, F.E.A.: Stably free modules over rings of Laurent polynomials. Arch. Math. **97**, 307–317 (2011)
60. Johnson, F.E.A., Wall, C.T.C.: On groups satisfying Poincaré Duality. Ann. Math. **96**, 592–598 (1972)
61. Kamali, P.: Stably free modules over infinite group algebras. Ph.D. Thesis, University College London (2010)
62. Kaplansky, I.: Projective modules. Ann. Math. **68**(2), 372–377 (1958)
63. Karoubi, M.: Localization des formes quadratiques. I. Ann. Sci. Éc. Norm. Super. (4) **7**, 359–404 (1974)
64. Klingenberg, W.: Die Struktur der linearen Gruppe über einem nichtkommutativen lokalen Ring. Arch. Math. **13**, 73–81 (1962)
65. Lam, T.Y.: Serre's Conjecture. Lecture Notes in Mathematics, vol. 635. Springer, Berlin (1978)
66. Lam, T.Y.: A First Course in Noncommutative Rings. Springer, Berlin (2001)
67. Lam, T.Y.: Serre's Problem on Projective Modules. Springer, Berlin (2006)
68. MacLane, S.: Homology. Springer, Berlin (1963)
69. Mannan, W.H.: Low dimensional algebraic complexes over integral group rings. Ph.D. Thesis, University College London (2007)
70. Mannan, W.H.: The D(2) property for D8. Algebr. Geom. Topol. **7**, 517–528 (2007)
71. Mannan, W.H.: Realizing algebraic 2-complexes by cell complexes. Math. Proc. Camb. Philos. Soc. (2009)
72. Margulis, G.A.: Discrete Subgroups of Semisimple Lie Groups. Ergebnisse der Mathematik. 3. Folge, Bd. 17. Springer, Berlin (1991)
73. Milnor, J.: Introduction to Algebraic K-Theory. Annals of Mathematics Studies, vol. 72. Princeton University Press, Princeton (1971)
74. Mitchell, B.: Theory of Categories. Academic Press, San Diego (1965)
75. Montgomery, M.S.: Left and right inverses in group algebras. Bull. Am. Meteorol. Soc. **75**, 539–540 (1969)
76. Ojanguran, M., Sridharan, R.: Cancellation of Azumaya algebras. J. Algebra **18**, 501–505 (1971)
77. O'Meara, O.T.: Introduction to Quadratic Forms. Springer, Berlin (1963)
78. O' Shea, S.: Term paper. University College London (2010)
79. Parimala, S., Sridharan, R.: Projective modules over polynomial rings over division rings. J. Math. Kyoto Univ. **15**, 129–148 (1975)
80. Passman, D.S.: The Algebraic Structure of Group Rings. Wiley, New York (1978)
81. Quillen, D.G.: Projective modules over polynomial rings. Invent. Math. **36**, 167–171 (1976)
82. Remez, J.J.: Term paper. University College London (2010)
83. Samuel, P.: Algebraic Theory of Numbers. Kershaw, London (1972)
84. Serre, J.P.: Cohomologie des groupes discrets. In: Prospects in Mathematics. Annals of Mathematics Studies, vol. 70, pp. 77–169. Princeton University Press, Princeton (1971)
85. Sheshadri, C.S.: Triviality of vector bundles over the affine space K^2. Proc. Natl. Acad. Sci. USA **44**, 456–458 (1958)
86. Smith, H.J.S.: On systems of linear indeterminate equations and congruences. Philos. Trans. **151**, 293–326 (1861)
87. Steinitz, E.: Rechteckige Systeme und Moduln in algebraischen Zahlkörpen. I. Math. Ann. **71**, 328–354 (1911)
88. Steinitz, E.: Rechteckige Systeme und Moduln in algebraischen Zahlkörpen. II. Math. Ann. **72**, 297–345 (1912)
89. Strouthos, I.: Stably free modules over group rings. Ph.D. Thesis, University College London (2010)

90. Suslin, A.A.: Projective modules over polynomial rings are free. Sov. Math. Dokl. **17**, 1160–1164 (1976)
91. Swan, R.G.: Periodic resolutions for finite groups. Ann. Math. **72**, 267–291 (1960)
92. Swan, R.G.: Projective modules over group rings and maximal orders. Ann. Math. **76**, 55–61 (1962)
93. Swan, R.G.: K-Theory of Finite Groups and Orders (notes by E.G. Evans). Lecture Notes in Mathematics, vol. 149. Springer, Berlin (1970)
94. Swan, R.G.: Projective modules over binary polyhedral groups. J. Reine Angew. Math. **342**, 66–172 (1983)
95. Tits, J.: Free subgroups in linear groups. J. Algebra **20**, 250–270 (1972)
96. Turaev, V.G.: Fundamental groups of manifolds and Poincaré complexes. Mat. Sb. **100**, 278–296 (1979) (in Russian)
97. Vignéras, M.F.: Arithmétique des algebres de quaternions. Lecture Notes in Mathematics, vol. 800. Springer, Berlin (1980)
98. Wall, C.T.C.: Finiteness conditions for CW Complexes. Ann. Math. **81**, 56–69 (1965)
99. Weber, M.: The Protestant Ethic and the Spirit of Capitalism. Harper Collins, New York (1930)
100. Whitehead, J.H.C.: Combinatorial homotopy II. Bull. Am. Math. Soc. **55**, 453–496 (1949)
101. Yoneda, N.: On the homology theory of modules. J. Fac. Sci. Tokyo, Sect. I **7**, 193–227 (1954)
102. Zassenhaus, H.: Neuer Beweis der Endlichkeit der Klassenzahl bei unimodularer Äquivalenz endkicher ganzzahliger Substitutionsgruppen. Hamb. Abh. **12**, 276–288 (1938)

Index

A
Algebraic n-complex, 151
Augmentation
    mapping, 228
    quasi, 227
    sequence, 228

B
Baer sum, 68
Bourbaki-Nakayama lemma, 170
Branching, xii, 7

C
Cancellation, xii, 10
Cayley complex, 239, 256
Co-augmentation, 152
Congruence, 63
Coprojective module, xvi, 96
Corepresentability
    cohomology, 113
    Ext^1, xvii, 100
Corner, 37
Cyclic algebra, 185, 207

D
$\mathcal{D}(2)$-problem, x, 261
Derived
    functor, xv, xvii
    module category, xiv, 90
    object, xvii
De-stabilization lemma, 97
Determinant
    Dieudonne, 265
    full, 34
    weak, 26
Duality, 4
    of relation modules, 260
    of syzygies, 258
    Poincaré, 258

E
Eckmann-Shapiro relations, 223, 278
Eichler condition, xiii, 262
$\overline{E}$-triviality, 47
$\overline{E_n}$-triviality, 48
Extension of scalars, 273

F
Farrell-Tate cohomology, 241
Fibre square, 43
$FT(n)$-condition, 121
Full
    determinant, 34
    module, 147

G
Generalized Swan module, 231
    classification, 234
    completely decomposable, 235
    rigidity, 231
Geometrically realizable, x, 162
Group
    dihedral, 185
    free, 23
    infinite cyclic, 23
    quaternionic, 176, 199
    restricted linear, 15

H
Heller operator $(= \Omega_n)$, xiv, 122
Homotopy, 152
Hopfian
    strongly, 227
Hyper-stability, xv, 12, 117

I
Injective module, xvi